PARADIGMATIC SHIFTS IN GEOGRAPHICAL THOUGHT

A Methodological Examination

By the same author:
—Geographical Thought: Classical to Contemporaty (Revised)

THE AUTHOR

Dr. Lalita Rana is the Associate Professor in the Department of Geography, Shivaji College, University of Delhi. Especially interested in the History and Philosophy of geography, she has been actively involved in the research and teaching of the subject for the past three decades. Written in simple language, her previous three books, namely, *Geographical thought: A Systematic Record of Evolution* (2008); *Ancient Tradition Geography* (2014); and *Geographical Thought: Classical to Contemporary* (2014) have been widely accepted and well read by the students and researchers alike.

Paradigmatic Shifts in Geographical Thought

A Methodological Examination

Lalita Rana

CONCEPT PUBLISHING COMPANY PVT. LTD.
NEW DELHI-110059

ISBN-13 : 978-93-88937-26-9 (HB)
ISBN-13 : 978-93-88937-27-6 (PB)

First Published 2020

Published and Printed by:

Concept Publishing Company Pvt. Ltd.
A/15-16, Commercial Block, Mohan Garden,
New Delhi-110059 (INDIA)
T: +91 11 25351460, +91 11 25351794
E: publishing@conceptpub.com, *W*: www.conceptpub.com

Cataloging in Publication Data--Courtesy: D.K. Agencies (P) Ltd. <docinfo@dkagencies.com>

Rana, Lalita, 1960- **author.**
Paradigmatic shifts in the geographical thought : a methodological examination / Lalita Rana.
pages cm
Includes bibliographical references and index.
ISBN 9789388937269 (HB)
ISBN 9789388937276 (PB)

1. Geography—Philosophy. 2. Geographical perception—History. 3. Geography—Methodology. I. Title.

LCC G70.R36 2019 | DDC 910.01 2323

Dedicated to

The Fraternity of Geography

Preface

The methodology, in recent years, has been increasingly used as a substitute for method in scientific and technical contexts. However, the methodology of a field is not a grab bag of special techniques. The methodology of a discipline, in general, is related to its subject matter, approaches to study and methods of collecting and organizing knowledge. It deals with the position and scope of field within the total system of sciences and with the character and nature of its concepts. Methodology, in this sense, is a body of methods, rules and postulates employed by a discipline. It is the analysis of the procedures of inquiry in a particular field. Methodology thrives on change and evolution. In an active field the concepts, laws, hypotheses or techniques are continuously either refined or discarded. Methodology is the logic of this process. In advancing disciplines, the methodological debate is a sign of health. In Geography, as well, there had been extreme diversity of viewpoints on both philosophy and methodology with constant extension and also a shift in focus. However, Geographical thought, at any point in time, represents the prevailing philosophical viewpoints and major methodological approaches in vogue.

The methodology of geography came under question and heavy debate for the first time during the middle of twentieth century when F. K. Schaefer, an American scholar, published his paper entitled "Exceptionalism in Geography: A Methodological Examination". Publication of this paper brought both the qualitative and quantitative revolution in geographic development. Shortly after the publication of Schaefer's paper was discussed the paradigmatic shifts in a discipline by Thomas S. Kuhn, another American scholar, through his seminal work 'The Structure of Scientific Revolutions' (1962). In this work Kuhn painted a picture of the development of science quite unlike that had gone before. His account of the development of science held that science enjoys periods of stable growth punctuated by revisionary revolutions. Kuhn called the core concepts of an ascendant revolution its 'paradigms' and the term became very popular in the second half of twentieth century. And since 1960s, the concept of a paradigm shift is being used in numerous scientific contexts. Kuhn's approach to the history and philosophy of science focuses on conceptual issues like the practice of Normal Science, influence of historical events, emergence of scientific discoveries, nature of scientific revolutions and progress through scientific revolutions. Kuhn argues that the evolution of scientific discipline does not take place from the straightforward accumulation of facts, but rather from a set of changing intellectual circumstances and possibilities. Chronologically, Kuhn distinguishes between various phases, viz. Pre-paradigm, Normal Science, Crisis, Scientific Revolution and Post-Revolution. A science may go through this cycle repeatedly. According to Kuhn, science is not a well-regulated activity. It is a process of varying tension in which peaceful periods characterized by steady accretion of knowledge are separated by crises which can lead to breaks in continuity within discipline. In order to elucidate this process of development of science, Kuhn prepared a model termed as the 'Paradigm of Science'. In this he considered a paradigm as a theory of scientific tasks and methods which regulates the research in a discipline. The paradigm tells researchers what they should be looking for and which methods are, in a particular case, relevant. Kuhn's paradigm gives a scientific explanation of the growth phases of scientific knowledge. It gives useful guidelines for the understanding of the historical development of a discipline. Distinct parallels may be sought between the historical evolution of scientific disciplines as explained by Thomas Kuhn and the historical evolution of the discipline of geography.

Seeking inspiration from the works of both of the great scholars mentioned above, the present book examines the evolution of geographical thought from the perspective of Thomas Kuhn's classic, 'The Structure of Scientific Revolutions' (1962). Kuhn's work offers profound insights into the evolution of scientific disciplines, and therefore may help us to better understand past and present trends in the development of geography. This historical journey of geography can be understood with the help of the model 'Paradigm of Science' as proposed by Kuhn. The historical facts of geographic development are well known and there is no dearth of literature in this regard. Two publications of mine (2008 & 2014) are on this line only. However, the methodological literature, to understand the philosophical background making possible this disciplinary growth, is too small. According to Kuhn's model, the scientific development of a discipline consists of a series of phases. The purpose here is to explore and screen the growth and development of Geography under sequential methodological changes over time in the light of various paradigmatic shifts with the acknowledgement of crises and revolutionary phases of development in the discipline. Accordingly, the work identifies eight paradigmatic shifts intercepted by eight phases of crises in the discipline. The term 'paradigm' is used here, simply, to mean "an idea, concept, model or technique that is capable of generating scholarly concern for a particular period of time." In other words, an attempt is made to interpret the history of geography in Kuhnian terms. It is argued that Kuhn has been the most influential scientific methodologist as far as geography is concerned. His paradigm concept has been regarded as a useful model for histories of geography. Influenced by Kuhn, several geographers have tried to apply the Kuhnian model to the development of geographical thought (e.g. Harvey & Holly, 1981; Stoddart, 1981; Martin, 1985; Johnston, 1997 etc.). Following these attempts, the present book is written with an intension to go beyond the factual history of the discipline, i.e. to explore the ideas, concepts, philosophies and techniques that have been influencing geographical thinking and research. The work is also distinct in terms of identification of various paradigmatic shifts during the Modern, Pre-Modern and Post-Modern periods. Accordingly, the discussion is organized in five sections, viz. Introductory, Pre-Modern, Modern, Post-Modern and the Conclusion.

The manuscript of the book is self-typed. I am especially indebted to Mr. Bhupinder Bhati, an ex-student of the Department of Geography, Shivaji College for extending the cartographic assistance in preparing maps and diagrams. I record my sincere appreciation to the staff of libraries of RTL and ICSSR for laying at my disposal the material on related topics. This has been substituted by the relevant data and information available at Internet.

Lalita Rana

Contents

IV
POST-MODERN

V
CONCLUSION

List of Tables

List of Figures

I
INTRODUCTION

1
Evolution of Geographical Thought

Geography is the study of location and spatial distribution of various phenomena on the surface of earth. Although it is different from other sciences in its subject matter and methodology, it is closely related to other disciplines at the same time. Geography deals with an enormous range of phenomena, ranging from physical, biotic to human. The four spheres, namely, lithosphere, atmosphere, hydrosphere and biosphere constitute the geographic environment. This makes geography highly interdisciplinary, drawing upon and contributing to numerous other fields of knowledge. The science of geography is a field science and the empirical observations of landscape, physical as well as cultural, make it truly spatial in nature. Therefore, exploration or fieldwork is an important research technique in geography. Fundamentally speaking, geographers are concerned with the territorial space. Geographers vary considerably in their interests; some are concerned with the natural phenomena and others with the human artifacts. But implicit in all these interests is a concern with location and spatial patterns, and a curiosity about terrestrial space. This space is the binding factor. There is no more essentially geographical activity than explaining a spatial distribution. A geographer's task calls for gathering, representation and analysis of the information related to location and distribution of various phenomena. Therefore, the questions pertaining to location, distance, direction, spread and spatial succession constitute the basic queries of a geographer; as also the problems of accessibility, innovation, diffusion, density and other derivations of relative location. Contemporary geography is a spatial science, increasingly focusing on the study of spatial aspects of social and economic phenomena, and thus being pre-eminently anthropocentric. The space of geography is that of the earth's surface, the home of man. The nature of geography is, thus, integrative, interdisciplinary, empirical and anthropocentric.

Various disciplines geography relies on include the earth sciences, pure sciences, life sciences, social sciences, behavioural sciences, and the sciences dealing with data collection, representation, computation and analysis. The subject matter of geography is a clear manifestation of its interdisciplinary nature and reflects the physical, human and biotic aspects of study. Besides, the representation of these elements with the help of maps and diagrams also falls within the ambit of geography. Accordingly, the study of physical, human, biotic and practical aspects constitute the subject matter of geography. The subject matter of geography, as outlined above, is enormous, wide-ranging and is under constant change.

Geography has grown just like a tree, its roots lying in different schools of thought. The discipline as we know today is the contribution of disparate Schools of Thought. The knowledge of geographical nature has infiltrated from the past through present in the form of various fields of learning, philosophies, terminologies, concepts and techniques, as bequeathed to Geography from its various Schools of Thoughtnamely, European (Greek & Roman), Indian, Chinese and Arab in Pre-Modern period; and German, French, American, British and Soviet in Modern period.[1] They refined its methodology from time to time. However, the history of development of geographical thought has not been an affair of continuous and smooth progress.

Scholarly observations of the Eastern and Western civilizations had begun in Pre-historic times, approximately about 800 B.C. and earlier, and the field of geography has highly gained by these in the later stage of its development. Some of the first truly geographical studies occurred more than four thousand years ago in the regions often referred to as "the cradle of civilizations", viz. Mesopotamian, Chinese, Indus and Nile Valleys. These civilizations were first to explore the places and spaces within and outside their homelands.[2] During ancient period (800 B.C.–500 A.D.), there was greater and greater concentration of scholarly research in Europe and Asia. This was particularly under the influence of the Graeco-Roman, Indian and Chinese civilizations. Where the ancient Greek geography was largely an academic field, the Romans and the immediate successors, made far more extensive practical use of the discipline. In the Orient, the geographical studies in India began with the dawn of Indian civilization and these works cover the study of Universe, the Earth and the Indian Subcontinent. The ancient Chinese literature of geographical nature, on the other hand, deals with the field like human, regional, physical and mathematical geography, and the accounts of travels and explorations.

The Medieval Age is a very long period of development in the history of geographical thought, commencing from the fall of Roman Empire (Fifth century A.D.) to 1750s, the beginning of Industrial Revolution. The middle age began with the onset of Dark Age in Europe (500-1100 A.D.), followed by the rise of Arab School of Thought (800-1400 A.D.) and Age of Discovery (1400-1650 A.D.). As far as the West is concerned, little academic progress occurred in geography after the Roman period and initially the middle age was a time of intellectual stagnation. Therefore, this Early Medieval phase is termed as the Dark Age in Europe. This proved to be a long period of decline during which the geographic horizons closed in. The wide-ranging Greek geographical concepts were largely forgotten except among Arab scholars. When the medieval Europe had plunged into darkness, the Arab school rose to mark the end of this recession and return to more scholarly research in geography. The voluminous work of Islamic writers was largely responsible for the foundations of knowledge present in later Western geography. The Age of Exploration began in Europe in the late Medieval Phase and the geographic horizons were again pushed back. Enormous information of geographical nature was gathered. The Age of Discovery was also a renaissance phase in geography. During this period numerous journeys of geographical explorations were commissioned by a number of European nations. These voyages opened up new horizons and stimulated geographic writings. The reintroduction of Ptolemy's[3] writings, which had been lost in Europe but preserved in the Islamic world, provided a systematic framework to tie together and portray geographical information. The great voyages of exploration in sixteenth and seventeenth centuries revived a desire for both accurate geographic detail, and more solid theoretical foundations. The 'Geographia Generalis' by Bernhard Varenius and Gerardus Mercator's world map are prime examples of the new breed of scientific geography.[4]

By eighteenth century, geography came to be recognized more and more as a discrete discipline and became part of university curriculum in Europe. With the rise of the German

School, geography entered the Modern period of its development (1800-1950s). A number of other Modern Schools of Thought emerged thereafter, namely French, American, British and Soviet. Before the beginning of the Modern Period, geography had already bifurcated into two basic branches, viz. physical and human. After that it went on splitting, as continued research became an essential feature of the discipline. From the early nineteenth century to mid twentieth century, geography was developing in several directions under the influence of distinct schools of thought. During the entire nineteenth century, the emphasis of geographical studies kept on shifting between physical and human, with none of the field being prominent. However, towards the close of nineteenth century, the systematic studies on physical geography superseded everything else. The followers of Humboldt and Darwin carried out morphological research and produced studies based on landforms and climate. Long phase of dualistic tendencies also characterized geography, since the beginning of Modern period up to Second World War. During this period, i.e. from late nineteenth century until Second World War, geographers allround the world were seeking to establish the status of geography as an independent discipline, distinct in concepts and procedures, from other disciplines. However, most of the existing controversies in geography got resolved by mid-twentieth century, particularly after the publication of the Paper "Exceptionalism in Geography: A Methodological Examination" by F.K. Schaefer. With this geography is said to have entered its Contemporary (Post-Modern) Phase of development. The important developments of this phase include: the Quantitative Revolution; the use of Behavioural, Humanistic and Marxist Approaches; and the application of Models, Theories and Systems Analysis in geographical studies. From mid twentieth century onwards, the geographic research experienced a new shift in methodology. Three important trends have been observed in the field of geography especially after Second World War: (i) Integration; (ii) Interdependence; and (iii) Utilitarian approach (making geography more and more useful for mankind). The process of integration (unification) has replaced the process of separation (bifurcation) that was evident in the beginning of Modern period. The trend is now for all fields of study to come together around specific human problems. And geography has also turned more and more anthropocentric with time.

In fact, two major revolutions have changed the course of geographic development, viz. the Industrial Revolution in 1750s and the Methodological Revolution in 1950s. Accordingly, the phases of geographic development, now, may be identified as Modern, pre-Modern and post-Modern. Hence, Pre-modernism, Modernism, and Post-modernism may be considered as distinct philosophical systems, looking at the world in three very different ways. Each of these is talked about as periods being defined by the dominant philosophical and methodological system of the time in geographical development. In other words, from the beginning of history up to 1750s, the dominant way of viewing the world was largely consistent with the pre-modern philosophical system. Around the 1650s, pre-modernism started losing its influence as the dominant system and was being replaced by the modernist mind-set. For about 300-years, this was the dominant philosophical system, especially in Western culture. The 1950s also considered the time when the transition from modernism to post-modernism occurred. The differences among these three periods are represented best in their epistmologies[5]. Each of these philosophical approaches has different ways of looking at and understanding the discipline. However, the present geographical thought is an amalgamation of the Western and Eastern philosophies since beginning till present.

Pre-modernism (up to 1750)

The developments during the Ancient and Medieval phase characterize the Pre-modern period. The Ancient schools of thought represented the Greek, Roman, Indian and Chinese traditions, whereas the Arabian civilization generated the medieval tradition.

Ancient philosophy of the Graeco-Roman (Western) world is usually divided into four periods, viz. that of Naturalism, Metaphysical, Ethical and Religious. The main subjects of ancient Western philosophy were related to understanding the fundamental causes and principles of the universe with a purpose of obtaining fixed and certain knowledge about it; questions about things that cannot be perceived by the senses, such as numbers, elements, universals, and gods; and the analysis of patterns of reasoning and argument.

The ancient Eastern philosophy, on the other hand, represents the broad traditions that were popular in India and China. The Indian philosophy (Sanskrit: *Darshanas*) refers to any of several traditions of philosophical thought that originated in the Indian subcontinent, including the Hindu, Buddhist and Jain philosophies. All of these philosophies have a common underlying theme of *Dharma* (religion). They have been formalized and promulgated around 1000 BC. with residual commentaries and reformations continuing up to as late as the twentieth century. Hindu philosophy constitutes an integral part of the culture of South Asia, and is the first of the religious philosophies that were influential throughout the Far East. The great diversity in thought and practice of Hinduism is nurtured by its liberal universalism[6]. Similar to Western philosophy, the Chinese philosophy also covers a broad and complex range of thought, possessing a multitude of schools that address every branch and subject area of philosophy. The great philosophical schools were formulated during the Spring and Autumn Period and Warring States Period (771-221 B.C.); and they came to be known as the Hundred Schools of Thought. However, the most influential of these were Confucianism, Taoism and Buddhism.

In general, the primary epistemology of the pre-modern period was based upon revealed knowledge from authoritative sources. It was believed that the way to this knowledge is through direct revelation. This direct revelation was generally assumed to come from God. The Church, for instance, being the holders and interpreters of revealed knowledge, was the primary authority source in pre-modern times in the European world. Similar lines of thought were observed in the Indian and Chinese Realms during ancient times and in Arab World during medieval period.

Throughout Pre-modern (Ancient & Medieval) Phase of its development, geography was considered as a descriptive field of study. This followed the traditional definition of Geography, as given by Eratosthenes, a Greek scholar living in third century B.C. Eratosthenes coined the term 'geography' (from '*geo*' meaning 'the earth' and '*graphe*' meaning 'the description'). The traditional definition of our discipline, i.e. 'Geography is the description of Earth' has been derived from the literal meaning of this term. Resultantly, the enquiries of geographical nature had begun with the basic question, i.e. "what is where". The emphasis of study of the Pre-modern schools of geographical thought was on this very question.

Modernism (1750-1950s)

The shift in power moved away from the religion during Modern period, and the politics (governments, kings etc.) and universities (scholars, professors) took over as the primary sources of authority. Two new approaches to knowing became dominant in this period. The first was 'empiricism', i.e. knowing through the senses, which gradually evolved into scientific empiricism or modern science later. The second approach was 'reason or logic'. Along with these, the nature of geographical query also shifted to "why what is where" after the Age of Discovery. However, a further refinement was seen in Modern period, when geographers began to seek answer to the questions as "how and why what is where". In an attempt to answer each set of these questions, geography gained a new and more meaningful definition every time.

With the development of the natural sciences in the eighteenth and nineteenth centuries it became apparent that mere description would not do. It required to be followed by explanation. To explain phenomena were described means to recognize them as instances of laws. Another way of saying the same thing is to insist that science is not so much interested in individual facts as in the patterns they exhibit. In geography, the major pattern producing variables are, of course, spatial. Hence, geography came to be conceived as the science concerned with the formulation by the laws governing the spatial distribution of features on the surface of earth since Modern period onwards. This implies that geographers must describe and explain the manner in which things combine 'to fill an area'. These combinations change, of course, from area to area. Different areas contain different features or the same features in different combinations. These differences either in the combinations of features or in their arrangement from place to place underlie the commonsense notion that areas differ. Following the Greek geographers, this viewpoint is called 'chorological'.

This chorological viewpoint presented geography with a problem of great methodological controversy. The geographer's investigations, be a physical, economic, political or any other, are of two types: either 'regional' or 'systematic'. A region contains a special, unique, yet in some ways uniform combination of kinds or categories of phenomena. The detail with which the regional geographer describes, lists, or catalogues these features at the outset of his investigation depends on the size of the region considered. The systematic geography, on the other hand, searches for laws. Spatial relations among two or more selected classes of phenomena must be studied all over the earth's surface in order to obtain a generalization or law. Systematic geography attempts to formulate the rules and laws which are applied in regional geography. However, Systematic geography always obtains data from regional geography, just as the theoretical physicist has to rely on the laboratory for his. Regional geography is like the laboratory in which the theoretical physicists' generalizations must stand the test of use and truth. It seems fair to say, then, in conclusion, that regional and systematic geography are inseparable and equally indispensable aspects of the field.

The difference of opinion about the relative role and importance of regional and systematic geography may be traced to the preference given either one or the other at various periods in the history of the field. Physical geography, e.g. had a phase in the late nineteenth century when it concentrated on systematic work at the expense of regional studies. Therefore, at the beginning of the century, largely as a reaction to this exclusive concentration on systematic studies on the part of the physical geographers, the interest began to shift towards social or human geography. This was the phase when regional descriptive geography appeared to geographers to be the only honourable occupation. Hettner[7], e.g. thought that the core of geography was regional. Hartshorne[8] believed that systematic geography was indispensable to regional work; but the heart of geography is nevertheless regional.

The anti-scientific sprit of historicism was one of the major intellectual forces of the nineteenth century. Through Kant[9] and Hettner it has penetrated geographical thought and powerfully affected its course. Schaefer[10] called it 'exceptionalism' and traced its roots to Kant, to whom history and geography find themselves in an exceptionalist position different from that of the so-called systematic sciences. This grouping of geography with history has tempted many subsequent writers to elaborate the alleged similarity in order to obtain some insight into the nature of geography. This was undoubtedly, one of the historicist variant of the claim to uniqueness with which the geographers have been dealing throughout the modern period. Spurious similarities between history and geography were constructed. If history, according to the historicist, deals with unique events and if geography is like history, then geography, too, deals with unique and must try to 'understand' rather than search for laws. According to Hettner, on the other hand, both history and geography are essentially chorological. History arranges phenomena in time and geography in space. Both, in contrast to the other disciplines, integrate phenomena heterogeneous among themselves. Also these phenomena are unique. No historical event and historical period is like any other. In geography no two phenomena and no two regions are alike. Thus, both fields face the task of explaining the unique. Such explanation is, therefore, unlike all scientific explanations which 'explain' by subsumption under laws.[11] The strength of historicism in geographical thought, ever since Kant and up to mid twentieth century is a matter of record. However, the chorological view in geography was challenged by F.K. Schaefer in mid twentieth century through his powerful paper. This brought another turning point in the history of geographical development.

Post-modernism (1950 onwards)

Post-modernism brought with it a questioning of the previous approaches to knowing. Instead of relying on one approach to knowing, it advocated for an epistemological pluralism which utilizes multiple ways of knowing. Along with the pre-modern ways (revelation) and modern ways (science and reason), Post-modernism also includes many other ways of knowing such as intuition, relational and spiritual. Post-modern approaches seek to deconstruct previous authority sources and power. Contemporary (Post-modern) geography is an all-encompassing discipline that seeks to understand the world and all of its human and natural complexities; not merely where objects are, but how they have changed and come to be. The perspective towards Earth studies also changed from terrestrial to digital during post-modern period. This has been accompanied by a number of methodological changes in the discipline.

Thus, geography took about 2000 years to evolve as a scientific discipline and obtain a more or less clear-cut definition, content, purpose and methodology. Since the basic nature of geography has been shaped by different schools of thought emerging one after the other from ancient to modern times, providing the discipline with its standard literature, the period up to Second World War is identified as the Classical Phase of Development, and the one thereafter as Contemporary Phase. The general nature of geographic developments has been more 'philosophical' during its Classical (Pre-modern and Modern) Phase and more 'explanatory' in the Contemporary (Post-modern) Phase. The Post-modern period reflects more methodological concerns than conceptual.

NOTES

1. Rana, L. (2014).
2. Rana, L. (2008), pp. 47-49.
3. Claudius Ptolemy (A.D. 100–170) was a mathematician, astronomer and geographer. Interestingly, he was Roman by birth, Egyptian by place of residence and Greek as per his literary contribution.
4. Rana, L. (2008), pp. 116-129.
5. Epistemology is the study or a theory of the nature and grounds of knowledge especially with reference to its limits and validity.
6. Universalism is a religious, theological, and philosophical concept with universal applicability. It claims that religion is a universal human quality. Religion in this context is defined as "a set of beliefs concerning the cause, nature, and purpose of the universe, especially when considered as the creation of a superhuman agency or agencies, usually involving devotional and ritual observances, and often containing a moral code governing the conduct of human affairs."
7. Alfred Hettner (1859–1941) was a German geographer, known for his concept of chorology, the study of places and regions. His book *Europe* was published in 1907. According to him, geography is a chorological science or it is a study of regions. Hettner rejected the view that geography could be either general or regional. He proposed it the study of regions—especially in the form of his *Länderkunde* approach—suggested as the main field of geography.
8. Richard Hartshorne (1899–1992) was a prominent American geographer who specialized in economic and political geography and the philosophy of geography. He is particularly known for his 1939 methodological work *The Nature of Geography*.
9. Immanuel Kant (1724-1804) is one of the most influential philosophers in the history of Western philosophy. His contributions to metaphysics, epistemology, ethics, and aesthetics have had a profound impact on almost every philosophical movement that followed him.
10. Fred K. Schaefer (1904-1953), an American geographer, is considered as one of the pioneers of quantitative revolution. He is well known for his article in American periodical, Annals of Association of American Geographers (AAAG) called *Exceptionalism in Geography: A Methodological Examination*. It was both a repudiation of Richard Hartshorne's position in United States, and a call for a scientific approach to geography based upon the search for geographical laws (the ultimate form of a scientific generalization). Schaefer died before his article even appeared in print, and so he was never able to elaborate his argument, nor defend himself from Hartshorne's subsequent critique. But the article became a rallying point for the younger generation of geographers who were intent on reinventing the discipline as a science, or spatial science as it was later dubbed.
11. Schaefer, Fred K. (1953), pp. 226-249.

REFERENCES

Kuhn, T.S. (1962). *The Structure of Scientific Revolutions*, Chicago: University of Chicago Press.

Rana, L. (2008). *Geographical Thought: A Systematic Record of Evolution*, Concept Publishing Company, New Delhi.

Rana, L. (2014). *Ancient Tradition in Geography: The Western & Oriental Perspective*, Classical Publishing Company, New Delhi.

Schaefer, Fred K. (1953). "Exceptionalism in Geography: A Methodological Examination", *Annals of the Association of American Geographers*, Vol. 43 (3): 226-249.

2
Paradigmatic Shifts

A discipline always proceeds within a framework of some conceptual matrices, called paradigms that focus the philosophical perspectives and methodologies in vogue. As stated earlier, the term 'paradigm' was coined by Thomas S. Kuhn[1], one of the most influential philosophers and historians of science of twentieth century, in his classic book, *"The Structure of Scientific Revolutions"*, published with several editions after 1962 (1970, 1996, 2012). Through his publication, Kuhn challenged the then prevailing view of progress in 'normal science'. Normal scientific progress was viewed as 'development-by-accumulation' of accepted facts and theories. In order to elucidate this process of development, Kuhn prepared a model that he termed as the 'Paradigm of Science' (Fig. 2.1).

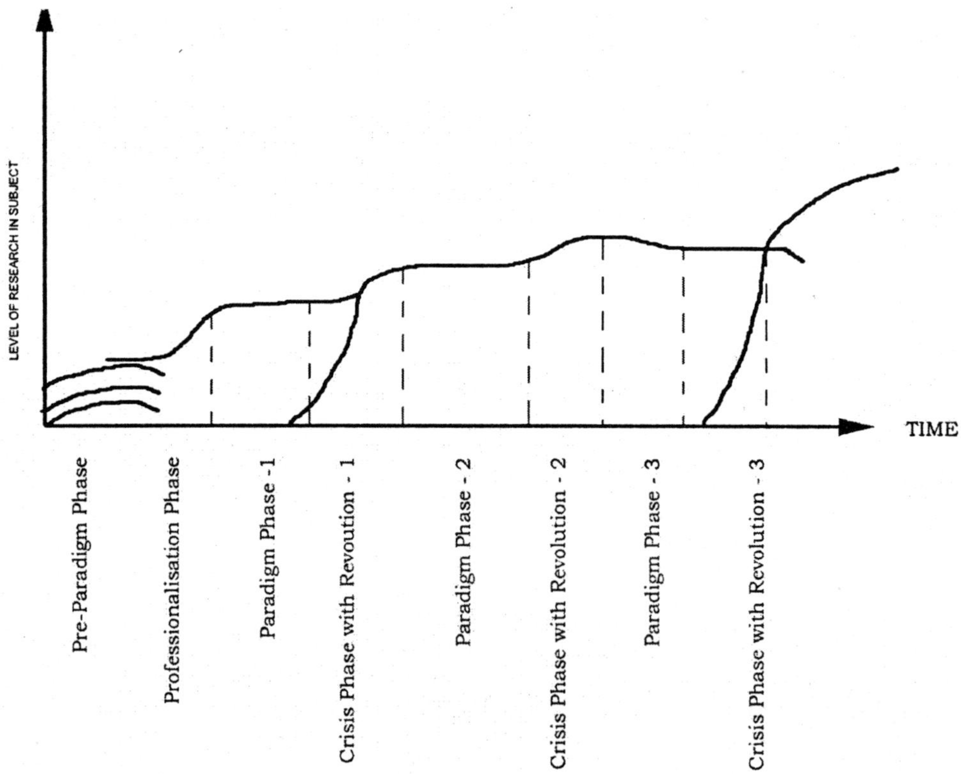

Fig. 2.1: Paradigm of Science

This Model is a simple cycle of progress challenging the world's prevalent conception of science, which he stated was a steady progression of the accumulation of new ideas. Accordingly, the science advances by occasional revolutionary explosions of new knowledge. Each revolution is triggered by introduction of new ways of thought large enough to be called new paradigms. From Kuhn's work came the popular use of terms like "Paradigm," "Paradigm Shift," etc.[2] According to Kuhn, the scientific knowledge progresses and develops like a plateau. There are sudden upheavals and abrupt rise, followed with smooth progress. It is argued that scientific advancement is not evolutionary, but rather is a 'series of peaceful interludes punctuated by intellectually violent revolutions, and in those revolutions one conceptual worldview is replaced by another'. Paradigm Shift is a change from one way of thinking to another. It is a revolution, a transformation, a sort of metamorphosis. It is a change in the basic assumption within the ruling theory of science.[3] In his model, he argued that scientific research and thought are defined by 'paradigms' that consist of formal theories, classic experiments and trusted methods. Paradigm's role has been identified as a vehicle to scientific theory.[4] Paradigms act as exemplars, i.e. concrete examples of research to be followed. Through the theories they embody, paradigms prove to be constitutive of the research activity. A paradigm represents the universally recognized scientific achievements that for a time provide model problems and solutions to a community of practitioners. Kuhn used the term 'paradigm' in two different meanings. In the first one, 'paradigm' designates what the members of a certain scientific community have in common, that is to say, the whole of techniques, patents and values shared by the members of the community. In the second sense, the paradigm is a single element of a whole, acting as a common model or an example; i.e. paradigm means simply 'example' that stands for the explicit rules and thus defines a coherent tradition of investigation. The paradigm is in this sense just an example, a single phenomenon, a singularity, which can be repeated and thus acquires the capability of tacitly modeling the behaviour and the practice of scientists.[5] The functions of a paradigm are to supply puzzles for scientists to solve and to provide the tools for their solution. Kuhn defined 'paradigm' as 'the entire constellation of beliefs, values, techniques, and so on shared by the members of a given community'. According to him, the science is not a well-regulated activity where each generation automatically builds upon the results achieved by earlier workers. No scientific discipline progresses in a normal steady manner. This growth is a process of varying tension in which peaceful periods of steady growth are separated by periods of crisis. In other words, a 'paradigm' is defined as 'universally recognized scientific achievements that for a time provide model problems and solutions to a community of practitioners'[6]. In its established usage, a paradigm is an accepted model or pattern. Three main paradigm types have been identified[7]: (1) the metaphysical paradigms or metaparadigms; (2) the sociological paradigms; and (3) the artifact or construct paradigms. The meta-paradigm represents a total global view of science. The sociological paradigm is based on universally recognized scientific achievements. And the artifact, which is central to Kuhn's formulation, is textbook, an instrument, or classic work. A paradigm, according to Kuhn, sets the framework within which a science proceeds; it describes what is to be observed and scrutinized. In short, a paradigm is a comprehensive model of understanding that provides a field's members with viewpoints and rules on how to look at the field's problems and how to solve them. The paradigm tells researchers what they should be looking for and which methods they should use in this connection. A viewpoint becomes a paradigm if it has a large following and if it has a clearly defined theoretical and methodological base. Eventually, however, there may generate insoluble theoretical problems or experimental anomalies that expose a paradigm's inadequacies or contradict it altogether. This accumulation of difficulties triggers a crisis that can only be resolved by an intellectual revolution that replaces an old paradigm with a new one. A crisis in science arises when confidence is lost in the ability of the paradigm to solve 'anomalies'. A scientific revolution occurs, according to Kuhn, when scientists encounter anomalies that cannot be explained by universally accepted paradigm within which scientific progress has thereto been made. The paradigm helps the scientist to acquire theory, methods and standards together, usually in an inextricable mixture. Therefore, when paradigms change, there are usually significant shifts in the criteria determining the legitimacy both of the problems and of the proposed solutions. Since no paradigm ever solves all the problems it defines and since no two paradigms leave all the same two problems unsolved, paradigm debate is obvious. The paradigm debates are revolutionary. Since new paradigms are born from old ones, they ordinarily incorporate much of the vocabulary and apparatus, both conceptual and manipulative, that the traditional paradigm had previously employed[8]. But, they seldom employ these borrowed elements in quite a traditional way. Within the new paradigm, old terms, concepts, and experiments fall into new relationships with the other.

According to Kuhn, the evolution of a scientific discipline, as practiced by members of a scientific community, may be characterized by the following stages: (i) Pre-paradigm stage, during which a body of phenomena is examined by scientists supporting competing schools of thought, with no common body of belief; (ii) Development of paradigm consensus, or a common body of belief among practicing scientists within the field; (iii) Normal science, in which the paradigm is further articulated to better explain the subject body of phenomena;(iv) Crisis associated with anomalies that are unexplainable within the existing paradigm; (v) The appearance of a new paradigm incommensurable with the old, followed by debates between advocates of the rival paradigms; (vi) Revolution, in which the consensus associated with the old paradigm is replaced by consensus on the new paradigm; (vii) Resumption of normal science based upon the new paradigm; and (viii) Recycling through stages 4 to 7. Thus, according to Kuhnian Model, the scientific development consists of a series of phases[9]. More precisely, they consist of: 'Pre-Paradigm Phase', 'Professionalization Phase', 'Paradigm Phase-1', 'Crisis Phase with Revolution', 'Paradigm Phase-2', 'Crisis Phase with Revolution-2', 'Paradigm Phase-3', and so on (Fig. 2.1). In other words, the Crisis Phase with Revolution is the starting point for the Paradigm Phase-2, which in turn is followed with 'crisis phase' and to again 'paradigm phase'. This sequence of crisis, revolution and paradigm continues throughout the history of science and helps in the advancement and decline of disciplines. Even the period of 'normal science' does not last indefinitely, and sooner or later, it is replaced by a 'crisis phase'. This occurs because more and more problems are accumulated which cannot be solved within the framework of the ruling paradigm.

The first phase of scientific development in a discipline is the Pre-Paradigm period. All new fields emerge at this Pre-Science Stage, where they have begun to focus on a problem area but are not yet capable of solving it or making major advances.[10] Kuhn described the pre-paradigm stage as a state of science at its infancy.[11] This is the phase of 'immature science', lacking consensus. Competing schools of thought possess differing procedures, theories, and even metaphysical presuppositions. This period is characterized by a rather indiscriminate collection of data over a very wide field. In a branch of science that becomes the subject of systematic study, the pre-paradigm period is marked by conflicts between several distinct schools which grow around individual scientists. This phase could be labelled a multi-paradigmatic period, as each school of thought develops its own model solutions.[13] This period is full of communication between various schools of thought and with other scientists and laymen. But none of these schools considers itself to be more 'scientific' than the other. There exists a low level of specialization in the discipline. The pre-paradigm phase exists only once without consensus on any particular theory. This phase is characterized by several incompatible and incomplete theories. Consequently, there is no common body of facts that may be taken for granted. The period, in particular, is regularly marked by frequent debates over legitimate methods, problems and standards of solutions, though these serve rather than to define schools than to produce agreement.[13] Throughout the pre-paradigm period when there is a multiplicity of competing schools, evidence of progress, except within schools, is very hard to find. However, progress is not impossible, and one school may make a breakthrough whereby the shared problems of the competing schools are solved in a particularly impressive manner. This success draws away adherents from the other schools, and a widespread consensus is formed among the practitioners.

From the Pre-Paradigm phase, the scientific development marches into the Professionalization Phase. Professionalization takes place when one of the conflicting schools of thought begins to dominate the others and, thus, a clear answer to the questions raised is given. A particular school of thought may become dominant because it develops new methods or puts questions which come to be regarded as more significant. New researches are thereby undertaken and research makes progress. The phase represents the subject's development from the Pre-Paradigm period to the stage of scientific maturing. One of the conflicting schools of thought often dominates the others in that it seems the best suited to win the discipline's academic esteem.[14] In the development of any science, the first received paradigm is usually felt to account quite successfully for most of the observation and experiments easily accessible to that science's practitioners.[15] Further development, therefore, ordinarily calls for the construction of elaborate equipment, a refinement of concepts, and the development of vocabulary and skills designed for or understood by a specific group of people.

The third phase is the Paradigm Phase-1. It is characterized by a dominating school of thought, which has, often in quite a short space of time, displaced others. A paradigm is established which leads to concentrated research within a clearly distinguishable problem area—an activity described as 'Normal Science'.[16] A mature science, according to Kuhn, experiences alternative phases of Normal Science and Revolutions (Crises). The period of conceptual (paradigmatic) continuity in a discipline is the phase of Normal Science. Normal Science, begins, in which puzzles are solved within the context of the dominant paradigm. As long as there is consensus within the discipline, normal science continues. Over time, progress in normal science may reveal anomalies, facts that are difficult to explain within the context of the existing paradigm. While usually these anomalies are resolved, in some cases they may accumulate to the point where Normal Science becomes difficult and where weaknesses in the old paradigm are revealed.[17] Kuhn claims that Normal Science can succeed in making progress only if there is a strong commitment by the relevant scientific community to their shared theoretical beliefs, values, instruments and techniques, and even metaphysics. This constellation of shared commitments Kuhn calls a 'disciplinary matix'.[18] In Normal Science the key theories, instruments, values and metaphysical assumptions that comprise the disciplinary matrix are kept fixed, permitting a cumulative generation of puzzle-solutions, whereas in a scientific revolution the disciplinary matrix undergoes revision, in order to permit the solution of the more serious anomalous puzzles that disturbed the preceding period of Normal Science.[19] Because commitment to the disciplinary matrix is a prerequisite for the successful Normal Science. Normal Science means research firmly based upon one or more scientific achievements; achievements that some particular scientific community acknowledges for a time as supplying the foundation for its further practice. This phase covers the achievements that are sufficiently unprecedented to attract an enduring group of adherents away from competing modes of scientific activity; and simultaneously, they are sufficiently open-ended to leave all sorts of problems for the group of practitioners to resolve. This suggests some accepted examples of actual science practice—examples which include law, theory, application, and instrumentation together—provide models from which spring particular coherent tradition of scientific research. These are traditions which historian describes under such titles as 'Ptolemaic Astronomy'[20] or 'Aristotelian Dynamics'[21] and so on. Normal Science, the puzzle-solving activity, is a highly cumulative enterprise, eminently successful in its aim, the steady extension of the scope and precision of scientific knowledge.[22] However, the period of Normal Science is sooner or later replaced by a crisis phase. This occurs because more and more problems get accumulated which cannot be solved within the framework of the ruling paradigm. Either more observations shake the underlying theory or a new theory is developed which does not accord with the stipulation of the ruling paradigm.

When enough significant anomalies have accrued against a current paradigm, the scientific discipline is thrown into a state of Crisis. During this crisis, new ideas, or the ones previously discarded, are tried. Eventually, a new paradigm is formed, which gains its own new followers, and an intellectual 'battle' takes place between the followers of the new paradigm and the hold-outs of the old paradigm. As more of these anomalies appear in the discipline it grows weaker, representing the phase of Crisis. It's a crisis because decisions can no longer be made rationally. There occurs stagnation in research, leading to chaos and turmoil. However, the crisis phase has been termed as the temporary dark phase in the development of scientific knowledge. So long as the tools a paradigm supplies continue to prove capable of solving the problems it defines, science moves fastest and penetrates most deeply through confident employment of these tools. The reason is clear. As in the manufacture so in the Science—retooling is an extravagance to be rescued for the occasion that demands it. The significance of crises is the indication they provide that an occasion for retooling has arrived. The crisis phase is characterized by a reassessment of former observational data, new theoretical thinking and free speculation. This involves basic philosophical debates and a thoroughgoing discussion of methodological questions. The crisis phase ends when a new paradigm attracts a growing number of researchers. In case the crisis phase terminates owing to the acceptance of a new paradigm, it becomes the inaugural point of the revolutionary phase. The research during crises very much resembles research during the pre-paradigm period, except that in the former the locus of difference is both smaller and more clearly defined. And all the crises close with the emergence of a new candidate for paradigm and with the subsequent battle over its acceptance. The transition from a paradigm in crisis to a new one from which a new tradition of normal science can emerge is far from a cumulative process, one achieved by an articulation or extension of the old paradigm. Rather it is a reconstruction of the field from new fundamentals, a reconstruction that changes some of the field's most elementary theoretical generalizations as well as many of its paradigm methods and applications. The crisis phase may end when it seems that no significantly better theory can be developed to solve the problems and thus, consequently, research must continue using old paradigm. Alternatively, the crisis phase ends when a new paradigm attracts a growing number of researchers away from the old paradigm.[23] When the transition is complete, the profession will have changed its view of the field, its methods and goals.[24] The process has been described as "picking up the other end of the stick", or the process that involves "handling the same bulk of data as before, but placing them in a new system of relations with one another giving them a different framework." Just because the emergence of a new theory breaks with one tradition of scientific practice and introduces a new one conducted under different rules and within a different universe of discourse, it is likely to occur only when the first tradition is felt to have gone badly astray. Often a new paradigm emerges, at least in embryo, before a crisis has developed for or been explicitly recognized. The crisis involves a complex break in the continuity of research, with a comprehensive reconstruction of the theoretical structure of a research field rather than a steady development and accumulation of knowledge. The crisis phase is characterized by a reassessment of former observational data, new theoretical thinking and free speculation. This involves basic philosophical debates and a thorough-going discussion of methodological questions. All the new discoveries are causes of or contributors to paradigm change.[25]

Crisis is followed by a Scientific Revolution if the existing paradigm is superseded by a rival. The termination point of the crisis phase, owing to the acceptance of a new paradigm, becomes the inaugural point of the next revolutionary phase. This involves a break in the continuity of research, with a thorough going reconstruction of the theoretical structure of research field rather than a steady development and accumulation of knowledge. The understanding of the truth itself and the scientist's perception of the world can take on a new diversion. This simple evolutionary framework is, however, disrupted by the possibility that certain disciplines, especially in the social sciences (e.g. geography), to be multiple-paradigm sciences. In a multiple-paradigm science, many paradigms are competing for hegemony in the field. Each such Revolution produces a consequent shift in the problem available for scientific scrutiny. The most obvious examples of scientific revolutions are those famous episodes in scientific development that have been the major turning points in the scientific development, as those associated with the name of Copernicus, Newton, Lavoisier and Einstein. Each of them necessitated the community's rejection of one-time-honoured scientific theory in favour of another incompatible with it. The resulting transition to a new paradigm is Scientific Revolution.[26] The proliferation of competing articulation, the willingness to try anything, the expression of explicit discontent, and the recourse to philosophy and to debate over fundamentals, all these are symptoms of a transition from normal to extraordinary research. It is upon their existence more than upon the revolution that the notion of Normal Science depends. The Normal-Science tradition that emerges from a scientific revolution is not only incompatible but often actually in commensurable with that which has gone before.[27] The revolution closes with a total victory for one or two opposing camps. After a given discipline has changed from one paradigm to another, this is called, in Kuhn's terminology, a Scientific Revolution or a Paradigm Shift. Overthrow of Ptolemaic cosmology by Copernican heliocentrism; the displacement of Newtonian mechanics by quantum physics and general relativity; and the acceptance of Plate Tectonics as the explanation for large scale geologic changes (1965), for instance, are all examples of major paradigm shifts. Thus, a paradigm shift is a fundamental change in the basic concepts and experimental practices of a scientific discipline. Kuhn contrasted these shifts, which characterize a scientific revolution, to the activity of normal science, which he described as scientific work done within a prevailing framework (paradigm). Thus, Scientific Revolution is the phase in which the underlying assumptions of the field are re-examined and a new paradigm is established. Eventually the old paradigm is sufficiently replaced and becomes the field's new Normal Science. The cycle then begins all over again, because our knowledge about the world is never complete. And in the Post-Revolution period, the new paradigm's dominance is established and so scientists return to Normal Science, solving puzzles within the new paradigm. The exchange of one paradigm for another is not a wholly rational transaction. The new paradigm will generally provide solutions to the problems which the old one found difficult to resolve.

Kuhn's explanation of scientific development in terms of paradigms was not only novel but radical too, insofar as it gives a naturalistic explanation of belief-change. Kuhn's contribution to the philosophy of science marked not only a break with several key positivist doctrines, but also inaugurated a new style of philosophy of science that brought it closer to the history of science. Kuhn's paradigm gives a very scientific explanation of the growth phases of scientific knowledge. Kuhn's model looking into historical development of sciences has its own merits and demerits. But, despite the criticisms, Kuhn's work has been hugely powerful both with in philosophy and outside it. Unquestionable he was on the most influential philosophers and historians of science of the Twentieth century. Whatever the merits or demerits of model, Kuhnian efforts have developed a long awaited new paradigm for the philosophy of science. This model gives useful guidelines for the understanding of the historical development of the subject. The history of geography in the light of Kuhn's paradigm can be easily understood by pursuing the following description.

Paradigm of Geography

The history of development of geographical thought is not a story of continuous progress. It confronted many evolutionary and methodological problems. The periods of intellectual progress were interrupted by the periods of stability or regression. It is being argued that geography is paradigmatic science and has undergone the metamorphosis, the revolution that signals the demise of one and creation of other.[28] Any idea, concept, method or technique that is capable of generating scholarly concern in a discipline, propelling it to 'normal science' stage, may be called a 'paradigm'. As far as geography is concerned, this cause of scholarly concern, for instance, has been the philosophy related largely to cosmology and theology during Pre-modern period; environmentalism, chorology, logic and reasoning during Modern period; and anthropocentricism in Post-Modern period. These philosophies have led to several paradigmatic shifts in the discipline in the past. In the light of Kuhn's Model one can discuss the evolution of paradigms in geography through various stages in succession. Accordingly, eight Paradigmatic Phases interrupted by eight Crisis Phases may be identified during the whole course of development in geography till contemporary times (Fig. 2.2).

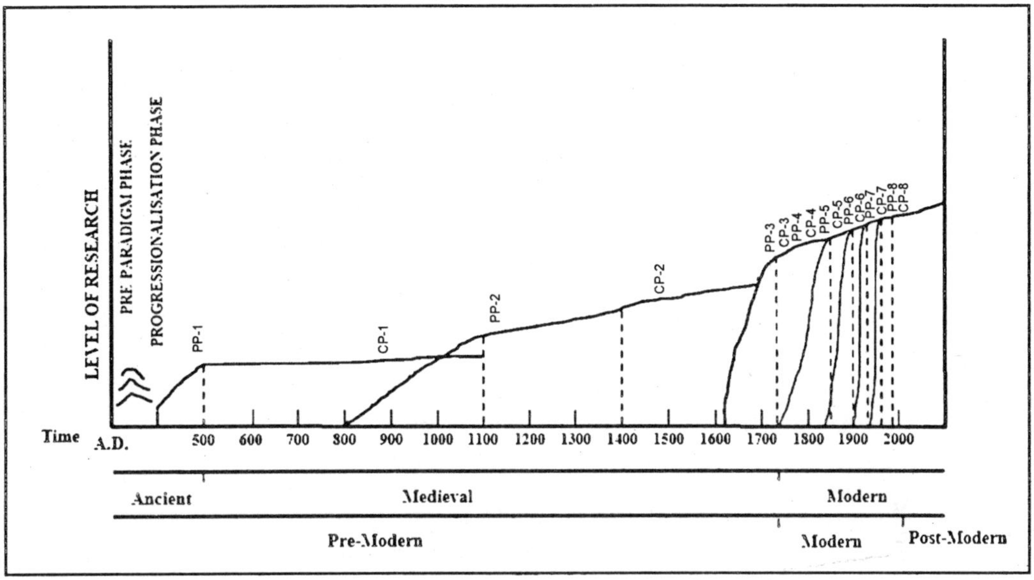

PP—Paradigm Phase
CP—Crisis Phase

Fig. 2.2: The Paradigmatic Shifts in Geographic Development

As evident from the diagram above, Geography, during its Pre-modern period of development, had already passed through two paradigmatic and two crisis phases. It experienced other three paradigmatic shifts intervened by two crisis phases during its Modern period of development. The paradigmatic shifts in the discipline have been relatively faster in the Post-modern period than earlier. More recently, the discipline has encountered four critical phases intervened by three paradigmatic shifts, and it is still undergoing a methodological revolution. Geography after mid of twentieth century to early 1990s has undergone change from the 'mono-discipline' approach, involving 'quantitative or qualitative' methodology to the 'multi-disciplinary' approach.[29] Geography, like other subjects, has passed from descriptive to the model building stage. This process started in the nineteenth century and got accelerated thereafter. Till present, Geography has passed from the descriptive and teleological phase to quantification, radical and materialism stage.[30] Table 2.1 attempts to systematize the development of Geography since Pre-Modern to Post-Modern periods.

Pre-Modern period witnessed the Pre-Paradigm phase of Geography followed by Professionalization and two initial Paradigmatic shifts. These paradigmatic shifts in the discipline are linked to the emergence of Graeco-Roman School in ancient times and Arab School in medieval times. The Dark Age and Age of Discovery brought crises in the discipline during ancient and medieval periods. Emergence of Chorological perspective, Videlian Tradition and Man-Environment based studies were the major developments during Modern period. However, the major methodological concerns of modern period have been the issues related to systematic and regional approach to geography. The German, French, American and British Schools have largely been responsible for these developments. The prevalence of numerous dualistic tendencies has been crucial in

Table 2.1: Methodological Development in Geography

Pre-Modern

- Scholarly observations of the Western and Eastern Civilizations (Pre-Paradigm Phase)
- Ancient Schools of Thought (Professionalization)
- Graeco-Roman School (Paradigm Phase I)
- Dark Age in Europe (Crisis Phase I)
- Medieval Thought and Rise of Arab School (Paradigm Phase II)
- Renaissance and Age of Discovery (Crisis Phase II)

Modern

- Dominance of German School (Paradigm Phase III)
- Darwinism and Dualistic Tendencies (Crisis Phase III)
- Rise of Determinism (Paradigm Phase IV)
- Paradigmatic Shift to Human Studies (Crisis Phase IV)
- Videlian Tradition and Chorological View (Paradigm Phase V)

Post-Modern

- Schaefer-Hartshorne Debate (Crisis Phase V)
- The Nomothetic Turn (Paradigm Phase VI)
- Quantitative Revolution (Crisis Phase VI)
- Emergence of Spatial Science School (Paradigm Phase VII)
- Critical Revolution (Crisis Phase VII)
- Anthropocentric Twist (Paradigm Phase VIII)
- Digital Revolution (Crisis Phase VIII)

bringing about crises in the discipline during this period. The Radicalism, Humanistic trends and Social Relevance of Geography have been instrumental in bringing about the paradigmatic shifts in Geography in its Post-Modern period. The methodology of geography has been under discussion more elaborately now than earlier. Therefore, since 1950s, the discipline has undergone four crises phases.

NOTES

1. Thomas Samuel Kuhn (1922-1996) was the American historian of science noted for his influential works of history and philosophy written in the twentieth century. *The Structure of Scientific Revolutions* was his most important work.
2. "The Kuhn Cycle", Wikipedia, the Free Encyclopedia.
3. Kuhn, Thomas. S. (1970), p. 10.
4. Kuhn, Thomas, S. (1962), p. 108.
5. Agamben, Giorgio (2002).
6. Kuhn, Thomas. S. (1962), p. x.
7. Masterman, M. (1970).

8. Kuhn, Thomas. S. (1962), p. 148.
9. Holt-Jenson, A. (2009), p. 102.
10. The Kuhn Cycle, *op. cit.*
11. Thomas Kuhn, (2004, 2011). *op. cit.*
12. Holt-Jenson, A. (2009), p. 103.
13. Kuhn, Thomas S. (1962), p. 48.
14. Holt-Jenson, A. (2009), p. 103.
15. Kuhn, Thomas. S. (1962). P. 64.
16. Holt-Jenson, A. (2009), p. 103.
17. 'Thomas Kuhn', (2004, 2011). *op. cit.*
18. *Ibid.*
19. Kuhn, Thomas S. (1962), p. 65.
20. In astronomy, the geocentric model (also known as the Ptolemaic system) is a superseded description of the universe with Earth at the centre. Under the geocentric model, the Sun, Moon, stars, and planets all circled Earth. The geocentric model served as the predominant description of the cosmos in many ancient civilizations. Two observations supported the idea that Earth was the centre of the Universe. First, the Sun appears to revolve around Earth once per day. While the Moon and the planets have their own motions, they also appear to revolve around Earth about once per day. The stars appeared to be on a celestial sphere, rotating once each day along an axis through the north and south geographic poles of Earth. Second, Earth does not seem to move from the perspective of an Earth-bound observer; it appears to be solid, stable, and unmoving. The ancient Greeks believed that the motions of the planets were circular and not elliptical, a view that was not challenged in Western culture until the seventeenth century through the synthesis of theories by Copernicus and Kepler. The astronomical predictions of Ptolemy's geocentric model were used to prepare astrological and astronomical charts for over 1500 years. The geocentric model held sway into the early modern age, but from the late sixteenth century onward, it was gradually superseded by the Heliocentric model of Copernicus, Galileo and Kepler.
21. Aristotle taught that the substances making up the Earth were different from the substance making up the heavens. He also taught that dynamics, the branch of physics that deals with motion, was determined primarily by the nature of the substance that was moving. For example, stripped to its essentials, Aristotle believed that a stone fell to the ground because the stone and the ground were similar in substance (in terms of Aristotle's four basic elements, they were mostly "earth"). Likewise, smoke rose away from the Earth because in terms of the four basic elements it was primarily air (and some fire), and therefore the smoke wished to be closer to air and further away from earth and water. By the same token, Aristotle held that the more perfect substance (the "quintessence") that made up the heavens had as its nature to execute perfect (that is, uniform circular) motion. He also believed that objects moved only as long as they were pushed. Therefore, objects on the Earth stopped moving once applied forces were removed, and the heavenly spheres only moved because of the action of the Prime Mover, who continually applied the force to the outer spheres that turned the entire heavens.
22. Kuhn, Thomas S. (1962), p. 52.
23. Holt-Jenson, A. (2009), p. 103.
24. Kuhn, Thomas S. (1962), p. 85.
25. *Ibid.*, p. 66.
26. *Ibid.*, p. 61.
27. *Ibid.*, pp. 90-91.
28. Hussain, M. (1995), p. 306.
29. Kuhn, Thomas S. (1962), p. 18.
30. Malik, Anand (2014).

REFERENCES

Agamben, Giorgio (2002). "What is a Paradigm", http://www.youtube.com, August 2002.

Barnes, B. (1982). *T.S. Kuhn and Social Science*, London: Macmillan.

Braithwaite Richard Bevan, (1953). *Scientific Explanation: A Study of the Function of Theory, Probability and Law in Science*, The Cambridge University Press, London and New York.

Hagget, P. (1979). *Geography: A Modern Synthesis*, New York.

Harvey, D. (1969). *Explanation in Geography*, London.

Holt-Jenson, A. (2009). "Paradigms and Revolutions", *Geography: History and Concepts: A Systematic Guide*, Fourth Edition, Sage, London, pp. 100-124.

Hussain, M. (1995). "Paradigm in Geography", *Evolution of Geographic Thought*, Rawat Publications, New Delhi, pp. 300-340.

Kuhn, Thomas. S. (1962). *The Structure of Scientific Revolutions*, Vol. II, No. 2, Foundations of the Unity of Science, International Encyclopedia of Unified Science.

Kuhn, Thomas S. (1970). *The Structure of Scientific Revolutions*, Second Edition, The University of Chicago Press.

Malik, Anand (2014). "Thomas Kuhn and Changing Paradigm in Geography: Critical Review", *Asian Journal of Multidisciplinary Studies*, Volume 2, Issue 4, April 2014.

Masterman, M. (1970). "*The nature of a paradigm*", in *Proceedings of the 1965 International Colloquium in the Philosophy of Science 4* (3 ed.), Cambridge: Cambridge University Press.

Minshull, R. (1970). *The Changing pattern of Geography*, London.

Schaefer, Fred K. (1953). "Exceptionalism in Geography: A Methodological Examination", *Annals of the Association of American Geographers*, Vol. 43, No. 3.

Shapere, D., (1964). "The Structure of Scientific Revolutions", *Philosophical Review*, 73: 383–94.

Taylor, P.J. (1976). "An Interpretation of The Quantification: debate in British Geography", *Transactions Institute of British Geographers*, New Sr. I, pp. 129-142.

"*Paradigm Shift*", Wikipedia, the Free Encyclopedia.

"*The Structure of Scientific Revolutions*", (Thomas S. Kuhn), Wikipedia, the Free Encyclopedia.

"*Useful notes on The Kuhn's Paradigm*", Wikipedia, the Free Encyclopedia.

II
PRE-MODERN

3
The Earliest Scholarly Concern (Pre-Paradigm Phase)

The scholarly observations of the Western and Eastern civilizations during ancient period mark the Pre-Paradigm phase of geographic development. Since fifth century B.C. the scholars in the West and Orient had begun making meaningful observations of the world around. Although, beginnings of the first paradigmatic phase in geography may be linked to Greeks in the West, the literary and other evidences suggest that that there were schools even before and also parallel to Greeks that have been instrumental in the process of evolution of Geography as a discipline of scholarship during ancient period. The study of written past, begins from the recorded human history in the Old World until the Early Middle Ages in Europe and the Qin Dynasty in China[1]. The period following these events also includes the Imperial era in China and the period of the Middle Kingdoms in India.[2] The span of recorded history altogether is roughly 5,000 years, with Sumerian cuneiform being the oldest form of writing discovered so far[3]. The term 'classical antiquity' is often used for this long period of cultural history.

Western

Towards West the earliest developments took place in western Asia, Egypt and Europe. The geographers credit the beginning of geographical writings to Homer, a Greek scholar, but the earliest known map was made by the Pre-Greek civilizations (particularly Sumerians) about 2700 B.C. The scholars of Babylon had already collected a large number of data on the motions of stars and planets, and they had developed the concept that the position of the celestial bodies had a fundamental effect on human activities. The Graeco-Roman people in Europe had a deep sharing of culture, knowledge, science and technology with many of the preceding civilizations that emerged in Mesopotamian region.

Mesopotamian

The Mesopotamian civilization emerged on the fertile plain of the Tigris and Euphrates rivers, commonly known as the "cradle of civilization"[4]. The region was famous as one of the riverine civilizations as that of Indus Valley in Indian subcontinent and the Yellow River valley in China. Various cities here served as capitals of the succeeding Empires of, for instance, the Sumerians, Babylonians, Assyrians, Egyptians and Phoenicians. Mesopotamian history extends from the emergence of urban societies in Southern Iraq in the 5th millennium BC to the arrival of Alexander the Great in the fourth century BC which is seen as the hallmark of the Hellenization[5] of the Near East.

The primitive inhabitants of this Fertile Crescent were Sumerians, the inhabitants of southern Mesopotamia. Invention of the first language script was their major contribution. Succeeding Sumerians, the Babylonian civilization endured from eighteenth to sixth century BC. It was like Sumerian, i.e. urban in character, although based on agriculture rather than industry. The Babylonians were an inventive and intelligent people who made large advances in the techniques of agriculture and science. However, a considerable amount of Babylonian literature was translated from Sumerian originals. Babylonians also adopted the Sumerian style of writing. The city of Assur, a colony of Babylonia, grew and soon turned out to be a great empire, Assyria, in the northern reaches of Mesopotamia, the southern half being Babylonia. Spanning four countries, namely Syria, Turkey, Iran and Iraq, the Assyrian people also carried on the cultural heritage from Old Babylonian period. Occupying the Mediterranean coast north of Palestine, the region was inhabited by the Phoenician civilization during first millennium B.C. Phoenicians were an enterprising maritime trading culture. Forming a string of cities along the Mediterranean coast, they were amongst the greatest traders of their time and owed a great deal of their prosperity to trade. Fernand Braudel[6], a French historian, remarked that Phoenicia was an early example of a 'world-economy' surrounded by empires. The most important of their early cities were Tyre, Sidon, Berytus (modern Beirut), and Byblos. These coastal cities were hemmed in on the land side by the Lebanon Mountains. The only obvious opportunity for expansion and economic gain was by sea. As traders these great mariners are believed to have been the first civilized culture to reach Britain and the Azores.

The origins of philosophy can be traced back to the early Mesopotamian's wisdom which embodied philosophies of life, particularly in the form of dialectic, dialogues, epic poetry, folklore, hymns, lyrics and prose. Babylonian reasoning and rationality developed beyond empirical observation. Developing the earliest form of logic, the Babylonian thought had a considerable influence on early Greek philosophy and Hellenistic philosophy. By 3200 B.C., the Sumerians had invented cuneiform[7], the earliest known form of writing used in the Middle East for thousands of years. With the invention of writing, the Sumerians were able to record their knowledge and pass it on to others. This gave rise to schools and education, religion, bureaucracy, and division of labour and social class system. An extremely large body of texts in Mesopotamia has survived on cuneiform tablets. Besides, the Phoenicians were also the first society to make extensive use of the alphabet. They are credited with spreading the Phoenician alphabet throughout the Mediterranean world. This proto-alphabet is one of their lasting contributions to civilization. The Phoenician alphabet arose from a need to communicate with the diverse languages of their trading partners that encircled the Mediterranean Sea. The Phoenician alphabet served as the origin of the Hebrew, Aramaic, Greek and Arabic alphabets. The Phoenician alphabet is generally believed to be the ancestor of all modern alphabets. When further refined and spread by the Greeks and Romans, it became the alphabet used today by most western cultures. Through their maritime trade, the Phoenicians spreaded the use of the alphabet to North Africa and Europe where it was adopted by Greeks, who later passed it on to the Romans.

Among the sciences mathematics, astronomy and astrology occupied a conspicuous place in Mesopotamian society. Their system of mathematics was sexagonal (a base 60 numeral system). From the sexagesimal system of mathematics has been derived the modern day usage of the clock with its 60 seconds, 60 minutes and 12 hours, 360 (60 × 6) degrees in a circle and the 12-month calendar which are still in use today. Since the Babylonian civilization replaced the Sumerian, it inherited their number system. But, the Babylonian number system was a modified version; it was a 'positional system'[8] and this advance by the Babylonians was undoubtedly their biggest achievement in mathematics. This mathematical knowledge was helpful in map-making. Among the other mathematical accomplishments, particularly those of Babylonians, was the use of square root, knowledge of the Pythagorean Theorem (well before Pythagoras), and familiarity with the general rules for measuring the areas and distances. This measurement for distances eventually was converted to a time-mile used for measuring the travel of the Sun, therefore, representing time.[9] The Assyrian culture also showed a dramatic growth in science and mathematics. Among their great mathematical inventions was the division of circle into 360 degrees. Assyrians were the first to invent longitude and latitude in geographical navigation. Like their contemporaries, the ancient Egyptians are credited with devising the decimal system and complex mathematical formularizations.

Astronomy was of old standing in Babylonia. The zodiac was a Babylonian invention of great antiquity; and eclipses of the sun and moon could be foretold. There are dozens of cuneiform records of original Mesopotamian eclipse observations. Observatories were attached to the temples, and reports were regularly sent by astronomers to the king. The stars had been numbered and named at an early date. Great attention was naturally paid to the calendar. Babylonian astrology was based on the belief that the entire universe was created in relation to the earth. Thus, the ancients saw it as no accident that the stars and planets were set in a certain divine order at the time of creation. The first evidence of recognition that astronomical phenomena are periodic and of the application of mathematics to their prediction is Babylonian. The Babylonian texts contain catalogues of stars and constellations as well as schemes for predicting heliacal risings and the settings of the planets, lengths of daylight measured by a water-clock, gnomon, shadows, and intercalations. The oldest rectangular astrolabe dates back to Babylonia (1100 BC.). During the 8th and 7th centuries BC, Babylonian astronomers developed a new approach to astronomy. They began studying philosophy dealing with the nature of universe and began employing an internal logic within their predictive planetary systems. This was an important contribution to astronomy and the philosophy of science. This new approach to astronomy was adopted and further developed in Greek and Hellenistic astronomy. The Babylonian development of methods for predicting the motions of the planets is considered to be a major episode in the history of astronomy.

Mesopotamian people, particularly the Sumerians, invented many technologies, most notably the wheel, the most important mechanical invention in history. Other Mesopotamian inventions include metalworking, copper-working, glassmaking, lamp making, textile weaving, flood control, water storage, as well as irrigation. They were also one of the first Bronze Age people in the world. Early on they used copper, bronze and gold, and later they used iron. The Mesopotamian knowledge passed on to the Greeks and Romans; and had a powerful impact on the development of the Western world.

European

Among Europeans, it was Greeks and Romans with whom the geographers credit the beginnings of the discipline. The Romans, in general, carried forward the Greek tradition.

After the decline of Mesopotamian civilizations, the Greeks assimilated technology and knowledge from almost all cultures they encountered—picking and choosing the best. The main feature of Greek thought, therefore, represents the mathematical and astronomical knowledge of the Assyrians, Babylonians, Egyptians and others. The scholarly observations in Europe are believed to have begun roughly with the earliest-recorded Greek poetry of Homer (Seventh century B.C.), and continued through the rise of Christianity up to the fall of the Roman Empire (Fifth century A.D.). The ancient Greeks were divided into three cultural groups, viz. Ionians, Dorians and Aeolians, collectively known as 'Hellenes' (Fig 3.1). However, it is the Ionians, located around the shores of the Aegean Sea in Asia Minor that the story of philosophy takes its beginning. These philosophical developments in ancient Graeco-Roman world are represented by four periods, viz. the period of Naturalism, Metaphysical, Ethical and the Religious.[10] Beginning with the Milesian School and culminating in the Neo-Platonic School, there emerged numerous schools of thought in-between.

The period of Naturalism was dominated by the Milesian, Pythagorean, Ephesian, Eleatic, Pluralists' and Atomists' streams of thought. The initial phase of philosophical development that began in the city of Miletus is called the Milesian (or Ionian) School. Usually the Earlier Ionians (Sixth Century B.C.) are distinguished from the *Later Ionians* (Fifth Century B.C.). The association of Thales, Anaximander and Anaximenes constituted the Early Milesian School; whereas that of Heraclitus, Empedocles and Anaxagoras formed the group of Later Ionians. The teachings of Milesian School, in general, focused on cosmological concerns, particularly the underlying substance of the universe. The School believed that the world originated from a primitive (natural) substance. This primitive substance was water for Thales, "the boundless" (apeiron) for Anaximander and atmospheric vapour (aer) for Anaximenes. These early philosophers believed that the first substance of the universe, be it water, matter or air, that formed the world was alive and self-moving. It produced life and was life at the same time. Because it moved itself, they said it was divine. Like the early Ionians, the Later Ionians were also interested in the problem of the origin and nature of the universe. But, unlike their predecessors, they distinguished the primitive world forming force from the primitive matter of which the world was made.

'Pythagoreanism', revolving around the philosophy of Pythagoras, cultivated the mathematical sciences. The study of mathematics led to the observation that everything could be represented through a number. The realization that all things are numerable, and can be related to each other in a numerical proportion led the Pythagoreans to their emphasis on the value of number in explaining the world order. The Pythagorians termed the universe as *kosmos*, which itself meant order or arrangement. The similarity between the whole and its parts they expressed in terms of some proportion supposed to exist between the whole and its part, and the part and its parts. As the musical scale is defined and limited by these numerical ratios, so every whole is made by the action of Limit (order) upon the Unlimited. The correct proportion between the whole and its parts was called harmony, meaning perfect arrangement. Like the musical harmony the, world harmony is also dependent on number. In making number the first substance of the world, the Pythagoreans transferred these mathematical conceptions to material reality. Consequently, they said that points, lines, and surfaces are the real units from which all bodies in the world are made. Representing the philosophical thought of Heraclitus of Ephesus, the 'Ephesian' school developed as a response to the philosophy of the Ionian predecessors. It

Fig. 3.1: Greek Colonization of Asia Minor

(After Crabben, J.V.D. (2012). "Greek and Phoenician Colonization," *Ancient History Encyclopedia*)

believed everything that exists, including man himself, exists because it is in a continuous process of passage from one state to another. Everything is in a state of continual flux. Because of its unstable nature, they found the Fire as most closely corresponding to becoming. Accordingly, "all things flow, everything runs, as the water of a river. The water in river seems to be the same, but in reality it is in a state of continuous flow." 'Becoming' is itself the first principle of reality, the essence of things. Heraclitus explained change by saying that it is the upward and the downward path of fire by virtue of which the universe came into being. The relative stability of the world, he said, is due to different 'measures' of the ever-lasting fire, some being kindled (burning upward) and some burning out (going downward) in more or less equal proportions. The balance between the upward and downward paths of the different 'measures' of fire forms what Heraclitus called the 'hidden attunement' of the Universe. This change, the transformation of material from one state into another, does not happen by accident, but rather "according to law", within certain limits and within certain time. This law is named 'logos' by Heraclitus. Therefore, the term 'Ephesian School' could be applied to the Greek thought which saw the harmony of world as the resolution of many diverse tensions in the unity of the one reality, which is fire. This concept of unity in diversity is Heraclitus' most significant contribution to philosophy. The 'Eleatic' School, on the contrary, was based on the doctrine of Permanence, changelessness, the most noted thinker of the School being Parmenides. The basic tenet of his philosophy reveals the fundamental approach to the problem of being. He thought of Being as a space-filling mass and Non-Being as empty space. Being is unchanging and unmoving. "To change" means that something becomes what it is not. Since "to be" meant "to exist" for Parmenides, he said that it was impossible for something which is, (i.e. exists) to become what it is not (i.e. not to exist). Similarly, Being is unmoving; "to move" means that something moves into a space where something is not (i.e. does not exist). But Being is the full. Therefore, empty space, or a space where nothing is (a space of Non-Being) logically cannot exist. Parmenides held that only that perception is true which shows us an unchanging Being. Since our senses tell us of a world of change and decay, (i.e. of Non-Being in his view), Parmenides said that they are the cause of intellectual error. They only show us what appears to exist. They tell us nothing about the one, indivisible, unchanging Reality. The only way, he said, we can know the nature of reality is through the structure of thought. Parmenides greatly influenced Socrates, Plato, and Aristotle. Putting to themselves the problem of being (Parmenides) and of 'becoming' (Heraclitus), the 'Pluralists' attempted reconciliation between the two factions and tried to overcome the opposition between 'Being' and 'Non-being'. The chief philosophers of this group were Empedocles and Anaxagoras. Parmenides held that Being was an indefinite, space-filling, unchanging and eternal mass. Empedocles tried to reconcile the world of change with unchanging Being by saying that the world as we know it is composed of four basic material particles which are themselves unchanging and indestructible. He called them "the roots of everything"; they later came to be called 'elements'. According to Empedocles, the four elements were earth, air, fire, and water. The objects of the physical world, he said, are made up of a haphazard combination of these four elements. Empedocles' theory dealing with the four elements endured longest, and fell into decline only with the advent of modern chemistry. It was Empedocles who established that these four ultimate elements make all the structures in the world. According to the different proportions in which these four indestructible and unchangeable elements are combined with each other the difference of the structure is produced. It is in the aggregation and segregation of elements thus arising, that Empedocles found the real process which corresponds to what is popularly termed growth, increase or decrease. Nothing new comes or can come into being; the only change that can occur is a change in the juxtaposition of element with element. The objects of the physical world, he said, are made up of a haphazard combination of these four elements. Although the elements never change their nature, the different ways they can be combined result in the coming-into-being of different physical objects. This theory of the four elements became the standard dogma for the next two thousand years. Another philosophy to try to resolve the paradox of unchanging being and a changing world was that of a school which came to be called the 'Atomists', the school most radical of all. The best known philosophers of this school were Leucippus and Democritus. The atomists believed that the Universe is composed of an infinite number of elements which are characterized by quantitative differences. There exist invisible atoms which are impenetrable, invisible spatial entities differing only in form, weight, size. The nature and organization of matter is the homogeneity of all bodies and that indeterminate matter is divided into an infinite number of molecules (atoms) differing in size and form but endowed with perpetual motion which is derived from their essence. Because atoms are endowed with motion, the School admits a second primordial element, the 'void', that is, infinite space which surrounds the atoms and gives them the possibility of movement. The differences in gravity cause the atoms to whirl into motion, thus giving origin to the formation of things. Every union of atoms indicates a birth, just as every separation of atoms indicates a death. Thus from the primitive void have come the stars and the earth and all beings, including man. The system of Democritus presents to us a world regulated by mechanics (motion) and by the natural laws which act in the picture of cosmic necessity. Motion of individual distinct atoms could produce an orderly cosmos in which atoms are not just randomly scattered, but cluster to form masses of distinct types. The theory that the Universe is composed of an infinite number of elements which are characterized by quantitative differences has undergone many and important modifications, but it has maintained its validity even till date.

The Metaphysical period of Greek philosophy, occupied the entire fourth century before Christ. The problem which claimed the interest of thinkers during this period is no longer the cosmological question, but man in his concreteness, namely in his knowledge, morality and rights. The Period includes the philosophies, especially those of Sophists, Socrates, Plato and Aristotle. Sophistic thought can be considered as a transition from the old cosmological concepts to the new ideas about man. The Sophists (like Protagoras and Gorgias) were the philosophers to introduce the study of man and discuss problems of knowledge, ethics and right. They were interested in human nature and man's actions in the world; and claimed that "Man is the measure of all things." Sophists held that knowledge is essentially empirical and relative to man. The reality was reduced to the subjective experience of man. The Sophists were the first to show complete indifference to the problem of the world of matter and to centre their efforts upon man. But man can be an object of study in his sense knowledge as well as in that more profound one of reason. The Sophists stopped at the first, at the immediacy of sense impressions. The analysis of reason was taken up by Socrates and his disciples. The classic Greek philosophy, in fact, revolves round Socrates who made his most important contribution to Western thought through his method of inquiry. Socrates, like the Sophists, also concentrated his attention on man. But the man taken into his consideration is not the individual in particular; rather, he is the universal. The doctrine of Socrates can be summed up in two words: 'concepts' and 'ethics'. For Socrates, the 'concept' is the common view of a thing. In the rational part of every man there exist some notions which are common to all and hence enjoy universality and necessity, and which form the substratum of true understanding or knowledge. True science is universal; that is, it is common to all men and to all times. True science consists in understanding through concepts, which have the same universal characteristics as science itself. To arrive at an understanding of such concepts, Socrates used the inductive method of dialogue (Socratic Method). In ethics, on the other hand, Socrates taught searching for moral concepts. After the death of Socrates, his disciples opened schools with the intention of continuing the teaching of their master. The greatest of the Socratic schools, however, referred to as the Major Socratic School, was the Academy of Plato, which stayed closer to the original intent of the teachings of Socrates. The thought of Plato was revived in the later Academies, and in particular in the last important movement of Greek thought, Neo-Platonism. Plato's metaphysics divides the world into two distinct aspects: the intelligible world of "forms", and the perceptual world we see around us. The perceptual world consists of imperfect copies of the intelligible forms or ideas. These forms are unchangeable and perfect, and are only comprehensible by the use of the intellect or understanding. The most fundamental distinction in Plato's philosophy is between the many observable objects that appear beautiful (good, just, unified, equal, big) and the one object that is what beauty (goodness, justice, unity) really is, from which those many beautiful (good, just, unified, equal, big) things receive their names and their corresponding characteristics. Nearly every major work of Plato is, in some way, devoted to or dependent on this distinction. Plato distinguishes four degrees of knowledge: (a) apprehension of pure sense images, such as dreams and imaginations; (b) perceptive knowledge of sensible objects, the purpose of which is to form a particular judgment; (c) mathematical knowledge—for instance, the apprehension of the particular shape of the perceived object; (d) philosophical knowledge, which consists in the apprehension of the Ideas. Plato argued that 'genuine' knowledge is knowledge based on Reason, not sense-perceptions or opinion. True knowledge comes from contemplation of the truth which, in turn, impels dialectics. The dialectical method consists of comprehension of scattered particulars in one idea, and in the division of the idea into the processes of generalization and classification. The dialectical method produces consistent thinking by generalizing and particularizing. Conceptual knowledge is the only genuine knowledge. Plato's thought is often compared with that of his most famous student, Aristotle. Aristotle, on the contrary, placed much more value on knowledge gained from the senses, and correspondingly earned the label of empiricist. Aristotle's ideas gained fame as the Peripatetic School. Aristotle's views and approaches dominated early Western science for almost 2000 years.

The Ethical Period extended for duration of about three centuries from the death of Aristotle. It represents the emergence of Stoicism, Epicureanism, Skepticism and Eclecticism. During this time metaphysics was displaced by ethics, of pure speculation by practical thought. This displacement was the unavoidable consequence the sadness of the times resulting from political instability in the region. It developed due to the succession of Alexander the Great first, and then the Romans over the Greek land. The endless wars between the successors of Alexander and the Roman Empire affected not only Greece but all the countries of Mediterranean basin. Solutions to the existing problems of life were found in three possible means, viz. to appeal to reason; to appeal to pleasures; and to deceive oneself in the belief that every appeal is useless. These are the characteristics of the three streams of thought which philosophy followed in its ethical development; i.e. Stoicism (the appeal to reason); Epicureanism (the appeal to pleasure); and Skepticism (the renunciation of every solution). These three streams of thought formed the major schools of thought during this period. The philosophy of 'Stoicism'[11] is in many ways similar to the Taoist philosophy of China. Both teach one to attune with his inner nature, which the Romans called 'Reason', the Chinese 'Tao', and the Greeks 'Logos'. Stoics were concerned with the active relationship between cosmic determinism and human freedom, and the belief that it is virtuous to maintain a will that is in accord with nature. It encouraged simple living and contentment with ones present state of being. Because of this, the Stoics presented their philosophy as a way of life, and they thought that the best indication of an individual's philosophy was not what a person said but how he behaved. The stoic philosophy was, in turn, divided into three parts: Logic (a very wide subject including rhetoric, grammar, and the theories of perception and thought), Physics (not just science, but the divine nature of the universe as well), and Ethics, the end goal of which was to achieve happiness through the right way of living according to Nature. The Stoics presented, in popular form, the theories of life that generally opposed materialism. The Stoics held that the soul is a "tabula rasa" (blank sheet) and that only sensations are true. Ideas are reduced to images which are merely the coordinated result of many similar sensations preserved in memory and expressed in speech. In the Epicurean view[12], the highest pleasure (tranquility and freedom from fear) was obtained by knowledge, friendship, and living a virtuous and temperate life. Forms of Christianity and the Indian religions (including *yoga*) promote engagement in a transformational process of refraining from actions of body, speech, and mind. The founders and earliest practitioners of these religions (e.g. Buddhism and Jainism) lived extremely simple lifestyles refraining from sensual pleasures and the accumulation of material wealth. Skepticis[13], on the other hand, held that the wise man can be assured of the greatest degree of quietude and happiness by renouncing all desires. Moreover, they held that it is impossible to reach truth with absolute certainty because sensations vary from individual to individual. Neither the Stoic criterion nor the Epicurean evidence can change this natural behaviour of sensation. Since complete renunciation would be impossible to maintain in practice, the Skeptics had recourse to probabilism. Skepticism is the philosophy of doubt. Since we have no means of knowing the true nature of things, we have a degree of probability, a perceptual clearness. This position led to 'Eclecticism', a form of Skepticism. Eclectic thinking weakened dogmatism and encouraged intellectualism. Thus, Epicureanism viewed the world as a machine; Stoicism conceived the world as an intelligent teleological system and subordinated man to the universal will, and Eclecticism turned its attention to what seemed good in all systems, pieced them together, and presented a world view from the materials at hand.

During the Religious or the last period of development (extending to the fifth century A.D.), the Greek philosophy sought refuge in religion, with an intermingling of Greek, Egyptian, Chaldean, and Jewish religions. This period is represented by three main Schools, viz. the Judaic-Alexandrian, the Neo-Pythagorean the Neo-Platonic. The 'Judaic-Alexandrian School'[14] attempted to coordinate Oriental religion with Greek speculation. The major philosophic content of the School is the Idea of God, conceived as a reality that transcends the world and our limited intelligence. The abyss between God and the world is filled by the logos, the divine reason which comprehends the Ideas, which conceived in a Platonic way, are models of things. All things, then, are reflections of the divine ideas, and man is also a reflection of these same ideas. All these reflections are received in matter, which is the source of evil. The purpose of life is to free oneself from matter and to return to God. Such liberation is accomplished in the intellective way, by raising ourselves above the senses to the contemplation of the Ideas. The highest contemplation is ecstasy, a super rational act which unites us directly with God. The 'Neo-Pythagorean School' attempted to construct a world religion upon Pythagorean doctrines. The teaching of Pythagoras aimed at ethico-religious reform. Beginning with the middle of the first century B.C., a more systematic attempt was made to restore the speculative philosophy of the Pythagoreans and combine it with the practice of astrology and sorcery. The neo-Pythagoreans were methodical eclectics; they admitted into their speculative system not only the traditional teachings of the Pythagorean School but also elements of Platonism, Aristoteleanism, and Stoicism. Besides, they derived from Oriental religions with which they were in contact at Rome as well as at Alexandria, a highly spiritual notion of God. Their cosmology is dominated to a great extent by the belief that the stars are deities and that the powers of air, earth, and sky are demons. Neo-Platonism[15] conceived God as the source and goal of everything. The School tried to resolve this question by overcoming dualism with monism. God, accordingly, is not only the supreme inconceivable reality but also the principle of all realities. The invisible world as well as the visible world, man included, is nothing other than a derivation or emanation from God. God is true happiness for man; but as man cannot reach full knowledge of God by reason, so also he cannot possess happiness of himself; his intellectual knowledge is not sufficient. Man needs a superior help in order to reach God and to possess Him. This superior help comes from God who, beyond all forces of reason, manifests Himself to man, and makes him happy. The reconstruction of Greek philosophy culminated in Neo-Platonism, and Plato's system became the framework for a religious world view.

While, on the one hand, the natural, metaphysical, ethical and religious philosophies inspired the thinking of Western scholars in ancient period, the interest in Natural Sciences also continued to survive. Along with philosophy, the theoretical interest of the time also concentrated on the physical sciences during Greek Age. The knowledge of new lands, new phenomena, and new facts made possible through the expeditions of Alexander also fostered the rise of science. The principal centre of this culture was Alexandria, just as Athens had been the great centre of philosophical speculation. To this centre came scientists from the entire civilized world. Alexandria attained its greatest splendour during the third and second centuries B.C. (through Euclid, Archimedes and Hipparcus) and in the second century A.D. (through Ptolemy). The Mathematics, Physics and Astronomy were the principal sciences developed during this epoch. Besides Mathematics and Physics, the studies on Natural Sciences also covered Zoology and Botony as fostered by Aristotle and Theophratus respectively. Thus, Geography began to be cultivated largely in its astronomo-mathematical aspect during Graeco-Roman period in ancient times.

Indian

India and China were the major centres of learning in ancient times in the East. Occupying a strategic location in Asia, Indian history is at crossroads of cultures from China to Europe. The studies of geographical nature here began with the dawn of Indian civilization in ancient times.[16] The Indian contribution to geography came largely from the astronomers and astrophysicists like Bhaskaracharya, Varahamihira, Aryabhatta etc. Valuable geographical information is contained in Hindu mythology, philosophy, epics, history and sacred laws. Apart from religious records, the travellers' accounts give the description of different regions of the world. The accounts of these travellers reveal that India had close links with the neighbouring lands and Indian scholars were familiar with the geographical conditions of China, Southeast Asia, Central Asia and Mesopotamia. The glorious heritage of India has been shaped and fostered from the Vedic times,[17] and the Vedic age had, in many ways, a decisive influence on Indian culture. According to Winternitz[18], the Vedas belonged to a period stretching from 2500 B.C. to 7000 B.C., proving that the Indian civilization is the oldest. The ancient *Rishis* and sages were our first scholars. All current Scientific, Spiritual techniques and teachings are based on the knowledge passed on from these great scholars. Ancient Indian contribution can be traced back to what is usually referred to the "golden age" of classical Hindu culture, beginning around 500 B.C. with the sixteen monarchies and 'republics' known as the 'Mahajanapadas'. It stretched across the Indo-Gangetic plains from modern-day Afghanistan to Bangladesh. The births of Mahavira and Buddha in the sixth century BC mark the beginning of well-recorded Indian history. The period between 320–550 (Gupta Empire) is known as the Classical Age, when most of North India was reunited under the Gupta Empire. This was a period of relative peace, law and order, and extensive achievements in religion, education, mathematics, arts, grammar, composition, logic, metaphysics, mathematics, medicine, astronomy, Sanskrit literature and drama. Trading with the ancient world, the Indians were in contact with the Phoenicians, Romans, Greeks, Arabs, Syrians, Jews and the Chinese. India is estimated to have had the largest economy of the world between the 1st and 15th centuries A.D., controlling between one-third and one-quarter of the world's wealth up to the time of the Mughals, from whence it rapidly declined during British rule.

The different systems of Indian philosophy (*darshanas*) represent the special perception of Indian thought. The major streams of these thought are classified into two broad categories, viz. orthodox (*astika*) and heterodox (*nastika*), conveying a division of thinkers into 'theist' and 'atheist'. Orthodox systems are those which accept the authority of the Vedas, while the heterodox systems are those which reject it. To the latter group belong the three systems of *Charvaka, Buddhism* and *Jainism*. The other six systems (*Shaddarshanas*) belong to the former group. These are *Samkhya, Nyaya, Vaisesika, Yoga, Purva Mimamsa* and *Uttara Mimamsa* or *Vedanta*. They generally deal with four topics: Existence and nature of *Brahman*, Nature of the individual soul (*jiva*); Creation of the world (*jagat*); and liberation (*moksha*). *Samkhya*, the oldest of all, is a dualistic philosophy that postulates everything in reality stems from *purusha* (self, *atma* or soul) and *prakriti* (matter, creative agency or energy). The philosophical system of *Yoga*, as expounded by Patanjali[19], is closely allied with the Samkhya School. The Nyaya School is based on the *Nyaya Sutras*, written by Aksapada Gautama, in the second century BC. The most important contribution made by this school is its methodology. This methodology is based on a system of logic that has subsequently been adopted by the majority of the Indian schools. This is comparable to the relationship between Western science and philosophy, which was derived largely from Aristotelian logic. Nevertheless, Nyaya was seen by its followers as more than logical in its own right. They believed that obtaining valid knowledge was the only way to gain release from suffering, and they took great pains to identify valid sources of knowledge and distinguish these from mere false opinions. According to Nyaya, there are four sources of knowledge: perception, inference, comparison, and testimony. Knowledge obtained through each of these is either valid or invalid. Nyaya developed several criteria of validity. In this sense, Nyaya is the closest Indian equivalent to analytic philosophy. The Vaisesika School was founded by Kanada[20] and it postulates an atomic pluralism, as discussed by ancient Greeks. All objects in the physical universe are reducible to certain types of atoms, and *Brahman* is regarded as the fundamental force that causes consciousness in these atoms. The main objective of the Purva Mimamsa School was to establish the authority of Vedas. Consequently, the school's most valuable contribution to Hinduism was its formulation of the rules of Vedic interpretation. Its adherents believe that one must have unquestionable faith in the Vedas and perform the *yajñas*, or fire-sacrifices, regularly. They believe in the power of the mantras and *yajñas* to sustain all the activities of the universe. In keeping with this belief, they place great emphasis on *dharma,* which consists of the performance of Vedic rituals. The Mimamsa accepted the logical and philosophical teachings of the other schools, but felt they did not sufficiently emphasize attention to right action. They believed that the other schools of thought that aimed for release (*moksha*) are not allowed for complete freedom from desire and selfishness, because the very striving for liberation stemmed from a simple desire to be free. According to Mimamsa thought, only by acting in accordance with the prescriptions of the Vedas may one attain salvation. The Vedanta, or Later Mimamsa School, concentrates on the philosophical teachings of the Upanishads rather than the ritualistic injunctions of the Brahmanas.

Ancient Indians possessed great knowledge in all areas of human development. Science and Technology, particularly, covered all the major branches of human knowledge and activities, including mathematics, astronomy, physics, chemistry, medical science, cosmology, astrology, astronomy, fine arts, architecture, warfare, sports and games. In fact several inventions and discoveries, that are believed to have originated from the Western world have, actually been carried out by our ancestors. For instance, the observations regarding the sphericity of Earth and the gravitational pull are believed to have been made first by the ancient Indian scholars. From the times of Vedic period, the great culture of India has presented humanity with a gold mine of discoveries and inventions, ranging from aviation technology, to medicine and botanical science, atomic theory, plastic surgery, Yoga etc. The historian Will Durant has said, "Mother India is in many ways, the mother of us all." (The Story of Civilization; Our Oriental Heritage). The knowledge contained in these scriptures covers the views of ancient Indian scholars on the study of the Universe, the Earth and the Indian Subcontinent. The ancient Indian literature deals with many problems pertaining to cosmology, cosmogony and cosmography. In the Rigveda, for instance, there is a description of various heavenly bodies including the Sun, the Moon, different planets and constellations. The principle of gravity is known to have discovered by Sir Isaac Newton, but the ancient Indian scholar, Bhaskaracharya II had talked about gravity much before Newton. The Treatise of the great Indian astronomer Aryabhatta, *Aryabhateeyam,* contains many theorems of motion of bodies. The concept and symbol of 'zero' is also believed to be the Indian contribution to the world. It was introduced in Europe about 1000 A.D. Observations about the earth are related to its origin, sphericity, eclipses, size and dimensions, latitudes-longitudes and local time, cardinal points, earthquakes and volcanoes, atmosphere and seasons, and physical divisions of this planet. The estimates regarding the earth's volume, circumference and mass, as given in 'Suryasidhanta', the treatise of Indian astronomy of ancient period, are nearly correct and reliable. As far as the origin of the earth is concerned, they believed in the solidification of earth from gaseous matter. The earth's crust, according to them, is made of hard rocks (*sila*), clayey material (*bhumih*) and sandy material (*asma*). The Puranas mention the earth to be apparently floating on the water like a sailing boat on the river. They were also aware of the fact that there is more land surface in the Northern Hemisphere. The concept of *Prithvi* (Earth) was the most basic in the study of geography. It has been profusely used in the Vedas and Puranas. The study of the earth or *Prithvi* has been called '*Bhugol*' or Geography. Thus, the use of the term *Bhugol* for the discipline of Geography is the most appropriate; '*bhu*' meaning 'the Earth' and '*gol*' meaning 'round', i.e. 'the study of the round earth'. It clearly suggests that the ancient Indians endorsed the earth being a sphere, and not a flat disc as believed by some of their parallel civilizations. This fact was further affirmed by the periodical occurrence of eclipses and the rise and setting of Sun. The estimates made by Aryabhatta in this regard were very close to those established in the modern period. It was also well known to the ancient Indian scholars that the earth is an oblate spheroid slightly flattened at the poles. Also, its equatorial diameter is greater than its polar diameter. Besides proving Earth's sphericity, the Indian astronomers also formulated ideas about gravity and gravitation. Further, it is widely believed that Christopher Columbus was the first person to discover that the Earth is round, during his voyages around 1500 A.D. However, in 499 A.D., Aryabhatta had already defined the shape of the Earth in his work `*Aryabhateeyam*', as: *Mrujjalashikhivaayumayo Bhoogola: sarvatho vruttha*: [Meaning: Earth which is made of soil, water, fire and air is circular when viewed from all sides. i.e. The Earth is spherical][21]. The ancient Indian scholars were aware of the causes of occurrences of eclipses (*grahanas*). Varahamihira considered the effects of eclipse month-wise and emphasized that eclipse in *Pusa* (December) leads to famine and its occurrence in April and May results in good rainfall, while and eclipse in *Phagun* (March) and *Asadh* (June) is unlucky. It was believed that the looking at eclipses is bad, because seeing the eclipse (particularly Solar) is harmful for the naked eyes. The facts related to the size and dimension of the earth were quite near to accuracy. *Akshansh* and *Deshantar* are the terms used for 'latitudes' and 'longitudes' respectively in the ancient Indian literature. Puranas have a reference of three imaginary lines of latitudes passing through Equatorial belt, North Pole and South Pole. Accordingly, three major regions have been identified in the Literature, viz. Equatorial (*Nirakshadesha*), Northern Polar (*Meru*) and Southern Polar (*Bhadvanala*). The North Pole has been called as *Zenith* and the South Pole as *Nadir*. The South Pole was truly considered as the antipode of the North Pole, i.e. diametrically opposite to it. However, the world was not believed to exist beyond Equator, as the region here was compared to hell of the earth. The Eastern part, on the other hand, was believed to be 'the land of Gods'. This thinking is in consonance with that of the Europeans in the Early Medieval period, when the Dark Ages prevailed and the East in 'T-in-O' Maps was assumed to be the place of *Adam* and *Eve*. The ancient Indian scholars have also drawn Prime Meridian. These imaginary lines, the position of Sun and various stars have helped them to determine local time at various places. In Rigveda, there is formulated the idea of four main directions, viz. *Purva* (East), *Paschim* (West), *Uttar* (North) and *Dakshin* (South). By adding Zenith (*Meru*) and Nadir (*Bhadvanala*) it was raised to six. But, afterwards, ten directions have been frequently mentioned in the Puranic literature. The designation of these directions in the Puranas is significant in the sense that it bears concept of the Gods dominating in each of these directions. These ten directions are *Purva* (East), *Agneyay* (Southeast), *Dakshina* (South), *Nairitya* (Southwest), *Paschim* (West), *Vayavya* (North-west), *Uttar* (North), *Isana* (Northeast), *Urdhava* (Zenith) and *Adhoh* (Nadir). The knowledge regarding the earthquakes, atmosphere, weather, climate and seasons in this period is excellent. For 'earthquakes' the term *bhukamp* has been used in Puranas. It was assumed that the deities of Air, Fire and Water cause the earthquakes. The ancient Indian scholars have identified the vacuum between the earth and the heaven as *Antariksha* and believed in the occurrence of various weather phenomena here, as rain, winds, clouds, lightening, fog and frost etc. Ramayana furnishes a lot of this kind of information. Bhaskaracharya has conceived the thickness of this *Antariksha* around the earth to be 12 *yojans* (or 96 kms). As far as the knowledge about the seasons is concerned, it is based largely on the studies in

India. Rigveda mentions five seasons. In Valmiki Ramayan, however, six seasons have been identified. They are: *Basant* (Spring), *Grishma* (Summer), *Prourit/Varsha* (Rainy), *Sharad* (Autumn), *Hemant* (Winter) and *Shishir* (severe Winter). The known world during the Puranic period was divided into seven Dwipas or 'regions'. These divisions exclude the parts of world discovered later i.e. towards the close of medieval period, as American Continents, Greenland, England and Antarctica. These seven regions were known as Jambu Dwipa, Krauncha Dwipa, Kusha Dwipa, Plaksha Dwipa, Pushkara Dwipa, Shaka Dwipa and Shalmali Dwipa.[22]

Geographical knowledge about the Indian Subcontinent is associated with its physical identification, people and culture, and relief and drainage. In Vedic and Puranic literature, the entire country from Himalayas to Kanyakumari has been referred to as *Bharatvarsha by* Aryans. This name has both geographical and historical significance. The name is derived after the name of *Bharat*, the ruling king. Bharata according to Hindu legend was the first king to conquer all worlds as known to the Vedic Aryans, uniting it into a single entity. Bharata's paramount position as founder of modern India. According to the Mahabharata, Bharata's empire covered all of the Indian subcontinent, Afghanistan and Persia. This Bharatvarsha, in ancient times was divided into *Nav-Khandas* (nine divisions)-eight shown as part of Greater India and the nineth surrounded by the sea (i.e. present day Sri-Lanka). Certain parts of the country are very distinctively mentioned in the ancient Indian literature. They are, e.g. *Sapta-Sindhu* (Punjab Plains), *Aryavarta* (the Aryan domain) and the region of Indus valley or the Upper Gangetic Plains. Culturally, caste had been the strongest element in ancient Hindu way of life. Bharatvarsha was inhabited by four castes, viz. *Brahmins, Kshtriyas, Vaishyas* and *Shudras*. The Vedas, Epics and Puranas make mention of a series of mountains in Bharatvarsha. They are, for instance, Himalayas (*Himavat*), existing like a bow in its northern part and divided into *Antagiri* (Inner Himalayas) and *Bahyagiri* (Outer Himalayas); Kailash Parbat, the abode of *Apsaras* (nymphs) and *Devas* (Deities) and rich in diamonds, minerals and other precious stones; Vindhayans, the extensive mountains with hundreds of peaks, variegated trees and creepers; Mahendra Mali, the Eastern Ghats; Sahyadri, the Western Ghats; Rika, the mountain range from Ken to Ton rivers north of Vindhayans; and Suktiman, the mountains of Khandera, Ajanta and Golkunda. The descriptions are also available for a number of Himalayan and other Inland rivers. Rigveda has mentioned various rivers originating from Himalayas, viz. Ganga, Yamuna, Brahmaputra, Saraswati, Satudri (Sutlej), Asikni (Chenab), Vitasta (Jhelum), Arjikeya (upper part of Indus), Susoma (Savan), Sindhu (Indus), Kubha (Kabul), Gomati (Gomala), Krumu (Kurrum) etc. Among the inland rivers the important ones that get a mention are Narmada, Tapti (Tapi), Godavari, Krishna, Cauvery and Tungbhadra. However, the most elaborate descriptions exist about Ganga and Brahmaputra. The religious flavour is very strong in these descriptions, as the rivers have been considered sacred, to be worshipped as Goddesses in the Hindu mythology. Although, the subsequent invaders ruthlessly destroyed the architectural achievements of this country and imposed their despotic rule, they also transmitted Indian sciences to the outside world. Along with Mahmud of Ghazni, for instance, came scholars like Al Beruni who studied Indian texts and translated them into Arabic.

Chinese

China was another major centre of learning in the ancient world and the record of geographical work here is highly impressive. The ancient Chinese scholars and travellers possessed a rich heritage of geographical knowledge. This has been passed on to the succeeding generations in spite of physical isolation from the West and the communication (language) barrier. Known as the People's Republic of China (PRC), China refers to a long-standing civilization comprising successive cultures dating back more than 5,000 years. In fact, the developments that began here in ancient times continued up to the Age of Discovery and later. With one of the world's longest periods of mostly uninterrupted civilization and the world's longest continuously used written language system, China's history has been largely characterized by repeated divisions and reunifications amid alternating periods of peace and war, and violent imperial dynastic change (Table 3.1). The

Table 3.1: The Imperial Dynastic Change in China

ANCIENT

3 Sovereigns and 5 Emperors

- Xia Dynasty (2100–1600 B.C.)
- Shang Dynasty (1600–1046 B.C.)
- Zhou Dynasty (1045–256 B.C.)
- Spring and Autumn Period (770-470 B.C.)
- Warring States Period (475-221 B.C.)

IMPERIAL

- Qin Dynasty (221 B.C.–206 B.C.)
- Han Dynasty (206 B.C.–220 A.P.)
- Three Kingdoms (220–280)
- Wei, Shu & Wu
- Jin Dynasty (265–420)
- 16 Kingdoms (304–439)
- Southern & Northern Dynasties (420–589)
- Sui Dynasty (581–618)
- Tang Dynasty (618–907)
- Second Zhou (690–705))
- 5 Dynasties & 10 Kingdoms (907–960)
- Liao Dynasty (907–1125)
- Song Dynasty (960–1279)
- Yuan Dynasty (1271–1368)
- Ming Dynasty (1368–1644)
- Qing Dynasty (1644–1911)

MODERN

- Republic of China (1912–1949)
- People's Republic of China (1949–*present*)

country's territorial extent expanded outwards from a core area in the North China Plain, and varied according to its changing fortunes to include multiple regions of East, Northeast, and Central Asia. The term "China Proper" is used to describe this core territory.

For centuries, Imperial China was also East Asia's dominant cultural influence, with an impact lasting to the present day. As in Egypt, Mesopotamia, and along the Indus River, Chinese civilization began within a major river valley, the Yellow River Valley. Around 4000 B.C. the modern China, with a huge geographical expanse, contained an almost infinite number of ethnic groups and languages. Most Chinese dynasties were based in the historical heartland of China, the China Proper. China Proper is generally thought to be bounded by the Great Wall and the edge of the Tibetan Plateau.[23] The conventional view of Chinese history is that of a country alternating between periods of political unity and disunity and occasionally becoming dominated by foreign peoples, most of whom were assimilated into the Han Chinese population.[24] Cultural and political influences from many parts of Asia, carried by successive waves of immigration, expansion, and assimilation, merged to create the Chinese culture.

The earliest comprehensive Chinese history is traced from about 2800 B.C., with an account of three semi-mystical and legendary individuals, viz. Fu Hsi, Shen Nung and the Yellow Emperor or Huang-di. They taught the Chinese the arts of civilization around 2800-2600 B.C.[25]. Emperor Fu-Hsi (2852-2738 B.C.) was the legendary inventor of the Chinese script; Shen Nung taught the art of agriculture and discovered the medicinal properties of plants; and The Yellow Emperor or Huang-di has been credited with the invention of the principles of Traditional Chinese medicine. Huangdi's reign is seen to provide a useful and validated reference point, from whence a consistent historical narrative has been developed. The Chinese believed that the Three Cultural Heroes were followed by the Three Sage Kings, Yao (around 2350 B.C.), Shun (around 2250 B.C.), and Da Yu (around 2205 B.C.). These Sage Kings ruled with perfect wisdom, clarity, and virtue. In the Chinese model of history, human events follow discernible cycles in which times of great virtue and wisdom are followed by times of decadence and decline. Chinese historians believed the Sage Kings' rule to be the most virtuous period in Chinese history. According to the Chinese historians, the last of the sage kings, Yu, founded a dynasty of kings, the Hsia (Xia). Since its references have traditionally been believed to be legendary, the Shang was the first confirmed historical dynasty who settled along the Huang He river from eastern China. In fact, the earliest discovered written record of China's past dates from the Shang Dynasty and takes the form of inscriptions of divination records on the bones or shells of animals—the so-called oracle bones. Thereafter, the information on dynastic change is more or less continuously available. The Shang dynasty was based on agriculture, augmented by hunting and animal husbandry. Two important events of the period were the development of a writing system, as revealed in archaic Chinese inscriptions found on tortoise shells and oracle bones, and the use of bronze metallurgy. Historians often refer to the period from Qin Dynasty to the end of Qing Dynasty as Imperial China. The major contributions of the Qin include the concept of a centralized government, the unification of the legal code, written language, measurement, and currency of China. The Qin Dynasty is also known for beginning the 'Great Wall of China', which was later augmented and enhanced during the Ming Dynasty. The Han Dynasty was the first one to embrace the philosophy of Confucianism, which became the ideological underpinning of all regimes until the end of imperial China. Under the Han Dynasty, China made great advances in many areas of the arts and sciences. The Emperor consolidated and extended the Chinese empire. This enabled the first opening of trading connections between China and the West, the 'Silk Road'[26]. An increasing number of nomadic people in Northern China adopted Confucianism as personal life guidance and state ideology while becoming gradually assimilated into the Han Chinese civilization. During this rivalry between Northern and Southern China, Buddhism propagated throughout China for the first time, despite facing opposition from Taoist followers. The Tang and the Han are often referred to as the most prosperous periods of Chinese history. The Tang, like the Han, kept the trade routes open to the west and south and there was extensive trade with distant foreign countries and many foreign merchants settled in China. By the mid to late thirteenth century the Chinese had adopted the dogma of Neo-Confucian philosophy formulated by Zhu Xi. There were enormous literary works compiled during the Song Dynasty, such as the historical work of the *Zizhi Tongjian*. During the era after the war, later called the *Pax Mongolica*, adventurous Westerners such as Marco Polo travelled all the way to China and brought the first reports of its wonders to Europe. Before the Mongol invasion (in 1279), Chinese dynasties reportedly had approximately 120 million inhabitants; after the conquest was completed in 1279, the 1300 census reported roughly 60 million people. Chinese merchants explored all of the Indian Ocean, reaching East Africa with the voyages of Zheng He. During the Ming dynasty the last construction on the Great Wall was undertaken to protect China from foreign invasions. For a long period of time, especially between the 7th and 14th centuries, China was one of the most advanced civilizations in the world in technology, literature, and art, although change was very gradual and mitigated by Imperial powers. By the nineteenth century, however, China's political, economic, and military influence declined relative to the growing regional power of Japan and the influence of Western powers. The imperial system in China ended with the establishment of the Republic of China (ROC) in 1912. The following three decades were a period of disunion—the Warlord Era, the Sino-Japanese War, and the Chinese Civil War. The latter ended in 1949 with the Communist Party of China (CPC) in control of mainland China. The CPC established a communist state—the People's Republic of China—that laid claim to be the successor state of the Republic of China. After its victory in the Chinese Civil War, the Communist Party of China established the People's Republic of China (PRC) in 1949. The region is known by this name till date.

The Chinese Thought has been mainly represented by Confucianism, Taoism and Buddhism. Developed from the teachings of the early Chinese philosopher Confucius, Confucianism is a moral code or philosophic world view that has had tremendous influence on Chinese history and culture down to the twenty first century. Besides Chinese, the cultures most strongly influenced by Confucianism include Japanese, Korean and Vietnamese. Like Socrates, Confucius taught that vice sprang from ignorance and that knowledge led unfailingly to virtue. Unlike most Western philosophers, Confucius did not rely on deductive reasoning to convince his listeners. Instead, he used figures of rhetoric such as analogy and aphorism to explain his ideas. As a foundation for the life of perfect goodness, Confucius insisted chiefly on the four virtues of sincerity, benevolence, filial piety, and propriety. Confucius was concerned with people's individual development, which he maintained took place within the context of human relationships. Confucius' disciples continued his philosophical school after his death. The works of Confucius were translated into European languages. Confucius' teachings and his conversations and exchanges with his disciples are recorded in the '*Lunyu*' or '*Analects*'. Confucianism, as passed down to the twentieth and twenty first century developed into the school of the Neo-Confucianism, led by Zhu Xi, who gave Confucianism renewed vigour in the Song and later dynasties. Neo-Confucianism combined Taoist and Buddhist ideas to create a more complete metaphysics than had ever existed before.

Laozi is traditionally regarded as the founder of Taoism (or Daoism). His book, Daodejing (*Tao-te ching*), commonly translated as the "Classic of the Way and Virtue" is next to the Bible in importance, and is considered as the most translated work in world literature. Taoism refers to a variety of related philosophical and religious traditions that have influenced East Asia the West for over two centuries. Its influence on Chinese culture is pervasive, and it reaches beyond China. In about the sixth century B.C., Taoism became 'the way'. Like Confucianism, it has influenced every aspect of Chinese culture. Until the arrival of Buddhism in China, Taoism was the chief religious and philosophical system. Despite the spread of Confucianism and Buddhism, Taoism has continued to influence the Chinese thought and culture up to the present day. Taoist thought generally focuses on nature, men-cosmos correspondence, health, longevity, *wu wei* (action through inaction), liberty, and spontaneity. Fundamental to Taoism is the belief that the Tao is the eternal reality and source of all existence. The goal of individual lives is to avoid earthly things and to live in harmony with the Tao and with nature in hopes of achieving immortality. Taoism focuses on what is perceived to be the undeniable fact that human attempts to make the world better, actually make the world worse. Therefore, it is better to strive for harmony. The dominant interpretation in traditional China is that Dao represents the source of the original, undifferentiated, essential *qi*-energy, the "One," which in turn produces the *yin* and *yang* cosmic forces. While the *yang* energy rises to form heaven, *yin* solidifies to become earth. A further blending of the two generates a "harmonious" *qi*-energy that informs human beings. The Laozi calls it Dao, to signal its presence as the source of the created order.

Buddhism is a philosophy based on the teachings of Gautama Buddha, who lived on the Indian subcontinent from the mid-6th to the early fifth century BC. Although Buddhism originated in India, it has had the most lasting impact on China. Buddhism, just like Confucianism and Taoism, guides people to behave, to be honest and responsible. It promotes harmony and peaceful mind, sharing and compassion. Buddhism emphasizes in 'awakening of mind'. Through learning, one is believed to develop intellectual capacity to the fullest so as to understand, to love and be kind to other beings. Buddhism does not believe in God. Its ability to co-exist with any other religions makes it being developed into one of the largest religions in China. The basic Buddhist concepts base on Four Noble Truths: (1) Life is suffering mentally and physically in the forms of sickness, injuries, aging, death, tiredness, anger, loneliness, frustration, fear and anxiety etc.; (2) All these sufferings are caused by craving. A self-centred person with continuous wanting from others will cause mental unhappiness. In turn, it will cause physical exhaust and fatigue in life; (3) All sufferings can be overcome and avoided. When one gives up endless wanting and endures problems that life evolves without fear, hatred and angre, happiness and freedom will then be obtained. By overcoming the mentality of self-centre and selfishness, one can spend time in meeting others needs and feels life more fulfilled; and (4) There is a Noble Eigthfold path leading to overcome the suffering. The Eightfold path includes Perfect Understanding, Perfect Thought, Perfect Speech, Perfect Action, Perfect Livelihood, Perfect Effort, Perfect Mindfulness and Perfect Concentration. This is said to lead one into the ultimate happiness in life and even afterlife. Buddhism also promotes code of conduct in life. Buddhist teaching introduces the way and concepts of meditation, rebirth, intellectual development and comparisons. It leads to self-understanding, instead of blind believing.

A number of Chinese philosophers, travellers and scholars contributed to geographical knowledge since ancient period, beginning from Laozi and Confucius and followed by

Zhang Qian, Sima Qian, Ban Gu, Chang Heng, Pei Xiu, Huan-Tsang, I Ching, Jia Dan, Sima Guang, Su Song, Shen Kuo, Fan Chengda, Chou Ta-Kuan, Zheng He and Xu Xiake in order. Of these, Zhang He is known for his scholarly works; Confucius, Laozi and I Ching for their philosophical contribution, and most of the others for their travels. The contribution of these scholars goes to the fields of Astronomy & Astrology, Human and Regional Geography, Physical Geography, Mathematical & Practical Geography, and Accounts of Travels and Explorations.

The Chinese mathematics had developed the Decimal System, which was introduced to the Arab world in about A.D. 800 from the Indians[27]. The Chinese were expert in map-making on the latitudinal-longitudinal grid, and this art is believed to have developed much earlier than in Greece or Rome. Also commendable are their observations on weather and climate, since they are known to have prepared a number of weather reports dating back several centuries before Christ. The discovery of the rest of the world by Chinese travellers is an aspect of the history of geography that is often overlooked in Western writings.

In China, the earliest known geographical writing dates back to fifth century B.C., during the Warring States period (481 B.C.-221 B.C.). This was the 'Yu Gong' ('Tribute of Yu') chapter of the book *Shu Jing* (*Classic of History*).[28] The book describes the traditional nine provinces, their kinds of soil, their characteristic products and economic goods, their tributary goods, their trades and vocations, their state revenues and agricultural systems, and the various rivers and lakes listed and placed accordingly.[29] Although the Chinese geographical writing in the time of Herodotus and Strabo were of lesser quality and contained less systematic approach, this would change from the third century onwards, as Chinese methods of documenting geography became more complex than found in Europe (until the thirteenth century). The earliest extant maps found in archeological sites of China date back to the fourth century B.C. and were made in the ancient State of Qin.[30] The earliest known reference to the application of a geometric grid and mathematically graduated scale to a map was contained in the writings of the cartographer Pei Xiu (224–271).[31] From the first century A.D. onwards, official Chinese historical texts contained a geographical section, which was often an enormous compilation of changes in place-names and local administrative divisions controlled by the ruling dynasty, descriptions of mountain ranges, river systems, taxable products etc.[32] The ancient Chinese historian Ban Gu (32–92) started the trend of the gazetteer in China, which became prominent in the Southern and Northern Dynasties period and Sui Dynasty. Local gazetteers would feature a wealth of geographic information, although its cartographic aspects were not as highly professional as the maps created by professional cartographers.[33] From the time of the fifth century B.C. onwards, Chinese geographical writing provided more concrete information and less legendary element. In later periods of the Song Dynasty (960-1279 A.D.) and Ming Dynasty (1368-1644 A.D.) there were much more systematic and professional approaches to geographic literature. The Song Dynasty poet, scholar, and government official Fan Chengda (1126-1193) wrote the geographical treatise known as the *Gui Hai Yu Heng Chi*.[34] It focused primarily on the topography of the land, along with the agricultural, economic and commercial products of each region in China's southern provinces. The polymath Shen Kuo (1031-1095) devoted a significant amount of his written work to geomorphology, basing evidence on marine fossils found far inland, along with bamboo fossils found underground in a region far from where bamboo was suitable to grow. The Ming Dynasty geographer Xu Xiake (1587-1641) travelled throughout the provinces of China to write his enormous geographical and topographical treatise, documenting various details of his travels, such as the locations of small gorges, or mineral beds such as mica schists.[35] Xu's work was systematic, providing accurate details of measurement, and his work read more like a twentieth century field surveyor than an early seventeenth century scholar.

The Chinese were also concerned with documenting geographical information of foreign regions far outside of China. Although Chinese had been writing of civilizations of the Middle East, India, and Central Asia since the traveller Zhang Qian (second century B.C.), later Chinese would provide more concrete and valid information on the topography and geographical aspects of foreign regions. Chinese geographers such as Jia Dan (730–805) wrote accurate descriptions of places far abroad. In his work written between 785 and 805 A.D., he described the sea route going into the mouth of the Persian Gulf, and that the medieval Iranians (whom he called the people of the Luo-He-Yi country, i.e. Persia) had erected 'ornamental pillars' in the sea that acted as lighthouse beacons for ships that might go astray. Confirming Jia's reports about lighthouses in the Persian Gulf, Arabic writers a century after Jia wrote of the same structures, writers such as al-Mas'udi and al-Muqaddasi. The geography of medieval Cambodia (the Khmer Empire) was documented in the book *Zhen-La Feng Tu Ji* of 1297 A.D., written by Zhou Daguan. China was one of the earliest centres of human civilization and Chinese civilization was also one of the few to invent writing independently. The Chinese script is the only logographic script still used in the world.

Further research in these schools viz. European, Indian and Chinese, bought high level of professionalization in various fields of learning. All these developments set the ground for geography to emerge as a discipline of scholarship with time.

NOTES

1. Crawford, O.G.S. (1927).
2. Robinson, C.A. (1951).
3. Breasted, J. H. (1916).
4. Rana, L. (2014). *op. cit.*, p. 25.
5. Hellenization is the historical spread of ancient Greek culture, religion and, to a lesser extent, language, over foreign peoples conquered by Greeks or brought into their sphere of influence, particularly during the period following the campaigns of Alexander the Great in the fourth century B.C.
6. Fernand Braudel (1902–1985) was the foremost French historian of the postwar era and a leader of the Annales School. His scholarship focused on three great projects, each representing several decades of intense study: "The Mediterranean" (1923–1949, then 1949–1966), "Civilization and Capitalism" (1955–1979), and the unfinished "Identity of France" (1970–1985). His reputation stems in part from his writings, but even more from his success in making the Annales School the most important engine of historical research in France and much of the world after 1950. As the dominant leader of the Annales School of historiography in the 1950s and 1960s, he exerted enormous influence on historical writing in France and other countries.
7. The cuneiform script is one of the earliest known forms of written expression. Created by the Sumerians in the late 4th millennium B.C., cuneiform writing began as a system of pictographs. Over time, the pictorial representations became simplified and more abstract. Cuneiforms were written on clay tablets, on which symbols were drawn with a blunt reed called a stylus. The impressions left by the stylus were wedge shaped, thusgiving rise to the name cuneiform, wedge writing.

8. A positional (numeral) system is a system for representation of numbers by an ordered set of numerals symbols (called digits) in which the value of a numeral symbol depends on its position. For each position a unique symbol or a limited set of symbols is used. The Babylonian Number System contained 59 symbols built from a 'unit' symbol and a 'ten' symbol. It was a positional base 60 system and it had some vestiges of a base 10 system within it. This is because the 59 numbers, which go into one of the places of the system, were built from a 'unit' symbol and s 'ten' symbol.
9. Eves, 1969, Chapter 2.
10. Durant Will, 1926; Bertrand Russell, 1967; & Nikolaos Bakalis, 2005.
11. Founded by Zeno of Citium.
12. Epicurianism is a system of philosophy based upon the teachings of Epicurus who was an atomic materialist, following the steps of Democritus.
13. Founded by Pyrrho of Elis.
14. Largely represented by Philo of Alexandria.
15. Plotinus (204-270 C.E.) is considered to be the founder of Neo-Platonism. Taking his lead from his reading of Plato, Plotinus developed a complex spiritual cosmology involving three foundational elements: the One, the Intelligence, and the Soul. It is from the productive unity of these three Beings that all existence emanates, according to Plotinus.
16. Crawford, O.G.S., (1927).
17. The Vedic Age (Inernet).
18. Moriz Winteritz (1863-1937) was a German Indologist. He was an outstanding scholar who wrote on religion, folk-lore, the epics and Sanskrit literature. He is best remembered for his "History of Indian Literature," a work in three volumes, which is considered indispensable for teachers and students alike. Winternitz was also one of the co-sponsors of a critical edition of the *Mahabharata,* a project that held his attention throughout his life.
19. Guruprasad, P.R., (2011).
20. As the founder of "Vaisheshik Darshan"—one of six *darshanas* of India—Acharya Kanad was a genius in philosophy. He is believed to have been born in Prabhas Kshetra near Dwarika in Gujarat. He was the pioneer expounder of realism, law of causation and the atomic theory. He has classified all the objects of creation into nine elements, namely: earth, water, light, wind, ether, time, space, mind and soul. He believed that "Every object of creation is made of atoms which in turn connect with each other to form molecules." His statement ushered in the Atomic Theory for the first time ever in the world nearly 2500 years before John Dalton. Kanad has also described the dimension and motion of atoms and their chemical reactions with each other. The eminent historian, T.N. Colebrook, has said, "Compared to the scientists of Europe, Kanad and other Indian scientists were the global masters of this field."
21. Bourbaki, Nicolas, (1998), p. 126.
22. Rana, L. (2014), pp. 199-276.
23. The 2150 miles long Great Wall of China is a Chinese fortification built in order to protect the various dynasties from raids by Hunnic, Mongol, Turkic, and other nomadic tribes coming from areas in modern-day Mongolia and Manchuria. The Great Wall of China was originally a project of Qin dynasty designed to keep out the nomadic invaders from the north. Some of the wall was built during the Qin dynasty, but most of it that we see today was constructed during the Ming dynasty.
24. Han Chinese are an ethnic group native to China and, by most modern definitions, the largest single ethnic group in the world. Han Chinese constitute about 92 per cent of the population of the People's Republic of China (mainland China), 98 per cent of the population of the Republic of China (Taiwan), 75 per cent of the population of Singapore, and about 20 per cent of the entire global human population. An alternate name that many Chinese peoples use to refer to themselves is "Descendants of the Dragon". Many Han and other Chinese also call themselves "Descendants of the Yan Di (Yan Emperor) and Huang Di (Yellow Emperor)".
25. Rana, L. (2014), *op. cit.*
26. Silk Route was the most well known trading route of ancient Chinese civilization. It linked

Europe and Asia through trade. Through this the trade, mainly in silk, grew under the Han Dynasty in the first and second centuries A.D. In fact, the Silk Route or Silk Road was an interconnected series of routes through Southern Asia traversed by caravan and ocean vessel. It extended over 7,000 km. (5000 miles).

27. Mohanty, J.N., (2000).
28. *The Classic of History* is a compilation of documentary records related to events in ancient history of China. It is also commonly known as the *Shàngshū*. The book consists of 58 chapters (including eight sub-sections. The first five chapters of the book purport to preserve the sayings and recall the deeds of such illustrious emperors as Yao and Shun, who reigned during legendary age; the next 4 are devoted to the Xia Dynasty, the historicity of which has not been definitively established; the next 17 chapters deal with the Shang Dynasty and its collapse.
29. Kurt A. Raaflaub & Richard J.A. Talbert (2009), p. 147.
30. Needham, (1986), Vol. 3, p. 500.
31. Needham, (1986), Vol. 3, p. 501.
32. Needham, (1986), Vol. 3, p. 512.
33. Hsu, (1993), p. 98.
34. Needham, (1986), Vol. 3, p. 517.
35. *Ibid.*

REFERENCES

Bertrand Russell, (1967). *A History of Western Philosophy*, Simon and Schuster, New York, ISBN 9780671201586.

Bourbaki, Nicolas (1998). *Elements of the History of Mathematics*, Berlin, Heidelberg, & New York: Springer-Verlag, ISBN 3-540-64767-8.

Breasted, J.H. (1916). *Ancient times, a history of the early world: an introduction to the study of ancient history and the career of early man*. Boston: Gin and Company.

Crawford, O.G.S.(1927). *Antiquity*, Glou Cester, Eng.

Durant, Will (1926). *The Story of Philosophy*. Simon & Schuster, New York, ISBN 0-671-69500-2.

Eves, Howard (1969). *An Introduction to the History of Mathematics*, Olt, Rinehart and Winston.

Guruprasad, P.R. (2011). "Scientific Reasoning in Ancient India" (*Internet*).

Hsu, Mei-ling (1993). "The Qin Maps: A Clue to Later Chinese Cartographic Development", *Imago Mundi*, Volume 45, pp. 90-100.

Kurt, A. Raaflaub and Richard, J.A. Talbert (2009). *Geography and Ethnography: Perceptions of the World in Pre-Modern Societies*, John Wiley & Sons, ISBN 1405191465.

Mohanty, J.N. (2000). *Classical Indian Philosophy of Induction*, Rowman & Littlefield Publishers.

Needham, J. (1986). *Science and Civilization in China*, Vol. 3. Mathematics and the Sciences of the Heavens and the Earth, Taipei: Caves Books Ltd.

Nokolaos Bakalis (2005). *Handbook of Greek Philosophy: From Thales to the Stoics Analysis and Fragments*, Tafford Publishing, ISBN 1-4120-4823-5.

Rana, L. (2014). *Ancient Tradition in Geography: The Western & Oriental Perspective*, Classical Publishing Company, Karampura, New Delhi.

Robinson, C.A. (1951). *Ancient History from Prehistoric Times to the Death of Justinian*, New York: Macmillan.

4
Ancient Schools of Thought (Professionalization)

The professionalization in the discipline had started becoming evident in the attempts of European, Indian and Chinese scholars during ancient period itself. In spite of physical isolation, fascinating parallels were exhibited in the streams, concepts and methods of study between that of oriental and European civilizations. Numerous philosophers, scholars, mathematicians, administrators and explorers whose attempts constitute the foundation of geography are listed in Table 4.1. The main subject of ancient philosophy was to understand the fundamental causes and principles of the Universe. Therefore, the important development of this period was the accumulation of knowledge, particularly from the fields like cosmology, astronomy, mathematics, meteorology, physics, astrology, geology, chemistry etc.

Table 4.1: The Contributors of Antiquity

- Acharya Kapil (3000 B.C.)
- Acharya Bharadwaj (800 B.C.)
- Baudhāyana, (800 B.C.)
- Homer (Eighth Century B.C.)
- Thales (639-544 B.C.)
- Anaximander (610-547 B.C.)
- Pherecydes (600-550 B.C.)
- Acharya Charak (600 B.C./300 B.C.)
- Acharya Kanad (600 B.C.)
- Acharya Sushrut (600 B.C.)
- Laozi (Sixth Century B.C.)
- Anaximenes of Miletus (585 -525 B.C.)
- Pythagoras (582-507 B.C.)
- Xenophanes (570-480 B.C.)
- Gautama Buddha (563 to 483 B.C.)
- Confucius (551-479 B.C.)
- Hecateus (550-476 B.C.)
- Heraclitus (535-475 B.C.)
- Hanno The Navigator (500 B.C.)
- Parmenides (510 B.C.)
- Leucippus (500-450 B.C.)
- Anaxagoras (500-428 B.C.)
- Protagoras (490-420 B.C.)
- Empedocles (490–430 B.C.)
- Zeno of Elea (490-430 B.C.)
- Gorgias (487-376 B.C.)
- Herodotus (485-425 B.C.)
- Philolaus (480-385 B.C.)
- Socrates (470-399 B.C.)
- Melissus of Samos (470 B.C.)
- Prodicus (465 -415 B.C.)
- Hippias (460-399 B.C.)
- Democritus (460-370 B.C.)
- Xenophon (430-354 B.C.)
- Plato (428-348 B.C.)
- Archytas (428-347 B.C.)
- Eudoxus of Cnidus (408–347 B.C.)
- Pānini (Fourth century B.C.).
- Aristotle (384-322B.C.)
- Pytheas (380-310 B.C)
- Theophratus (372-287 B.C.)
- Eudemus of Rhodes (370-300 B.C.)
- Pyrrho (360-270 B.C.)
- Alexander The Great (356-323 B.C.)
- Epicurus (342-270 B.C.)
- Zeno of Citium (334-262 B.C.)
- Euclid (325-265 B.C.)
- Archimedes of Syracuse (287–212 B.C.)

(Contd.)

Table 4.1 (*Contd.*)

• Chrysippus of Soli (280–207 B.C.)	• Ȃryabhatta I (476–550 A.D.)
• Eratosthenes (276-194 B.C.)	• Varahamihir (499-587 A.D.)
• Apollonius of Perga (262–190 B.C.)	• Brahamgupta (598-668 A.D.)
• Polybius (203-122 B.C.)	• Bhāskara I (600-680 A.D.)
• Crate of Mallus (Second century B.C.)	• Huan-Tsang or Yuan Chang (602–664 A.D.)
• Hipparchus (190-120 B.C.)	• I Ching or Yi Jing (635-713 A.D.)
• Chang-Chien (172-114 B.C.)	• Jia Dan (730–805 A.D.)
• Sima Qian (145 B.C.-90 B.C.)	• Madhav (Eightth century)
• Posidonius (135-51 B.C.)	• Adi Shankara (788 A.D.–820 A.D.)
• Marcus Tullius Cicero (106-43 B.C.)	• Aryabhatta II (about 920 A.D.)
• Strabo (64 B.C.-20 A.D.)	• Sridharacharya (991 A.D.)
• Philo of Alexandria (20 B.C.-A.D. 50)	• Su Song (1020-1101 A.D.)
• Ban Gu (32–92)	• Shen Kuo (1031-1095 A.D.)
• Chang Heng (78–139)	• Brahmadeva (1060-1130 A.D.)
• Nagarjuna (100 A.D.)	• Bhāskara II or Bhaskaracharya II (1114-1183 A.D.)
• Ptolemy (90-168 A.D.)	• Fan Chengda (1126-1193 A.D.)
• Claudius Galen (129-200 A.D.)	• Jayadeva (1200 A.D.)
• Ammonius Saccas (176-242 A.D.)	• Chou Ta-Kuan or Zhou Daguan (Thirteenth century)
• Pei Xiu (224–271)	• Zheng He or Cheng Ho (1371–1433 A.D.)
• Porphyry (234-305 A.D.)	• Xu Xiake (1587-1641 A.D.)
• Proclus (411-485 A.D.)	

Geography, as a discipline of scholarship, has filtered through several fields of learning (Table 4.2). The scholars like Anaxagorus, Strabo, Claudius Galen, Socrates, Shen Kuo, Thales, Anaximander, Democritus, Aryabhatta I & II, Brahamgupta, Bhaskara, Ptolemy, Thales, Anaximander, Euclid, Heraclitus, Chang Heng, Aristotle, Posidonius and Hippias were polymaths whereas the others had shown interest in certain specialized fields of study or travels. These disciplines lie at the roots of geographic development.

Table 4.2: Fields of Study as Contributed by Early Scholars

Algebra:	Bhaskaracharya.
Astrology:	Ptolemy and Varahamihir.
Astronomy:	Thales, Anaximander, Anaxagorus, Democritus, Eudoxus, Appolonius, Hipparchus, Ptolemy, Aryabhatta I & II, Brahamgupta, Bhaskara I, Varahamihir, Brahamdeva, Chang Heng and Shen Kuo.
Botany:	Theophratus.
Cartography:	Chang Heng, Pei Xiu, Jia Dan and Su Song.
Chemistry:	Nagarjun.
Cosmology:	Anaximander, Anaximenes, Heraclitus, Parmenides, Anaxagorus and Acharya Kapil.
Engineering:	Acharchya Bhardwaj (Aviatiation Technology) and Chang Heng.
Exploration:	Hanno the Navigator, Pytheas, Alexander the Great, Chang Chien, Huan-Tsang, I-Ching, Zhou Daguan, Zheng He and Xu Xiake.
Geometry:	Euclid.
Geology:	Chang Heng.
Historiography and History:	Sima Qian, Homer, Pherecydes, Hecateus, Herodotus, Xenophon, Eudemus, Polybius, Strabo, Pliny the Elder Ban Gu.
Literature, and Poetry etc.:	Valmiki, Homer, Hecateus, Fan Chengda, Thales, Pherecydes, Protogoras, Empidocles, Gorgias, Prodicus, Cicero, Zeno of Elea, Zeno of Elea, Socrates, Crate of Mallus and Panini.

(*Contd.*)

Table 4.2 (*Contd.*)

Mathematics:	Thales, Pythagoras, Democritus, Archytas, Eudoxus, Euclid, Archimedes, Eratosthenes, Hipparchus, Ptolemy, Shen Kuo, Baudhayana, Aryabhatta I & II, Brahamgupta, Bhaskara I, Sridharacharya and Brahamdeva.
Medicine:	Claudius Galen, Acharya Charak and Acharya Sushrut.
Meteorology:	Anaxagorus, Chang Heng and Shen Kuo.
Philosophy:	Xenophanes, Heraclitus, Leucippus, Anaxagorus, Philolaus, Mellisus, Plato, Pyrrho, Epicurus, Zeno of Citium, Chrysipus, Strabo, Philo, Claudius Galen, Ammonius Saccas, Porphyry, Proclus, Socrates, Cicero, Gautam Buddha, Adi Shankara, Laozi, Confucius and Shen Kuo.
Physics:	Democritus, Archimedes and Acharya Kanad.
Political Science:	Empidocles.

Among these, the fields of philosophy, astronomy, mathematics and science have been commonly contributed by Graeco-Romans, Indian and Chinese; whereas history and cartography have especially interested the Greeks and Chinese, as also the travels and exploration.

Philosophy

The term philosophy reflects the wisdom of a discipline. Precisely, it relates itself with the Metaphysics, Logic, Epistemology and Ethics.[1] Metaphysics is concerned with explaining the fundamental nature of being and the world. Among ancient Greeks, the earliest speculators (Thales, Anaximenes and Anaximander) were philosophers of nature. The Pythagoreans succeeded these with mathematical abstractions. The level of pure thought was reached partly in the Eleatic philosophers (such as Parmenides and Anaxagoras), but more completely in the work of Socrates. In its universal character, metaphysics resembles dialectics and sophistry. For Greeks, the subject of metaphysics dealt with the first principles of scientific knowledge and the ultimate conditions of all existence. Most of the schools of Indian philosophy are metaphysical. Samkhya, for instance, is a dualistic philosophy that postulates that everything in reality stems from '*Purusha*' (self, *atma* or soul) and '*Prakriti*' (matter, creative agency or energy). *Prakriti* consists of three dispositions known as qualities (*gunas*): activity (*rajas*), inactivity (*tamas*) and steadiness (*sattva*) which arises when the two other *gunas* are held in equilibrium. Because of the intertwined relationship between the soul and these dispositions, an imbalance in disposition causes the world to evolve. Closely allied with the Samkhya, the Yoga accepts the Samkhya psychology and metaphysics, but is more theistic than Samkhya. As explained by Heinrich Zimmer[2], Samkhya provides a basic theoretical exposition of human nature, enumerating and defining its elements, analyzing their manner of co-operation in a state of bondage (*bandha*), and describing their state of disentanglement or separation in release (*moksh*), while Yoga treats specifically of the dynamics of the process for the disentanglement, and outlines practical techniques for the gaining of release, or 'isolation-integration' (*kaivalya*)." The Vaisheshika school postulates an atomic pluralism. All objects in the physical universe are reducible to certain types of atoms, and Brahman is regarded as the fundamental force that causes consciousness in these atoms. Establishing the authority of Vedas, the Purva Mimamsa School made most valuable contribution to Hinduism by formulating of the rules of Vedic interpretation. Its adherents believe in the power of the *mantras* and *yajñas* to sustain all the activity of the universe. According to Mimamsa thought, only by acting in accordance with the prescriptions of the Vedas may one attain salvation. Lastly, the Vedanta or Later Mimamsa School, concentrates on the philosophical teachings of the Upanishads rather than the ritualistic injunctions of the Brahmanas. Logic is the art and science of reasoning which seeks to identify and understand the principles of demonstration and inference. Dating back to Plato and Aristotle, the logic remains integral to fields such as mathematics, computer science and linguistics. The ancient Greeks regarded philosophy as related with the universal. Aristotle, however, found the universal in particular things, and called it the essence of things; while Plato found that the universal exists apart from particular things. For Aristotle, therefore, philosophic method implies the ascent from the study of particular phenomena to the knowledge of essence, while for Plato philosophic method means the descent from knowledge of universal ideas to a contemplation of particular imitations of those ideas. In this sense, Aristotle's method is largely inductive, while Plato is essentially deductive. Among ancient Indian schools Nyaya is the closest Indian equivalent to analytic philosophy. Its methodology, comparable to the Western science, was based on a system of logic as that of Aristotalian logic. Its followers believed that obtaining valid knowledge was the only way to gain release from suffering, and they took great pains to identify valid sources of knowledge. According to Nyaya, there are exactly four sources of knowledge: perception, inference, comparison and testimony. Knowledge obtained through each of these is either valid or invalid. Nyaya developed several criteria of validity. Epistemology is the branch of philosophy concerned with the nature and scope (limitations) of knowledge. Much of the debate in this field has focused on analyzing the nature of knowledge. It also deals with the means of production of knowledge, as well as skepticism about different knowledge claims.[3] There are two kinds of knowing, the one is called 'legitimate' (genuine) and the other 'bastard' (obscure). The 'bastard' knowledge is concerned with the perception through the senses; therefore, it is insufficient and subjective, whereas the 'legitimate' knowledge can be achieved through the intellect. Lastly, Ethics is a branch of philosophy which seeks to address questions about morality. It is conceived as an art of caring for the soul analogous to medicine's care for the body.[4] The ancient Chinese philosophy had been more ethical than others in nature. Questions about the nature and existence of God which have profoundly influenced Western and Indian philosophies, have not been important in Chinese philosophies. In China, Taoism explained the path to contentment; Confucius taught people about the importance of being loyal, unselfish and courteous; and Buddhism focused on achieving absolute peace through an Eightfold Path including rightness of views, intention, speech, action, livelihood, effort, mindfulness and concentration. Among the commonalties of Chinese philosophies are: Epistemological optimism, i.e. human possibility of knowledge; the tendency not to view man as part of nature; the tendency not to invoke a unified and personified supernatural power; the belief that the purpose of philosophy is primarily to serve as an ethical and practical guide; and the political focus.

Cosmology and Cosmogony

The Universe and its origin remained a common point of speculation among all the ancient civilizations. The traditional theories generally depict an original unity which begins to become separated off into a series of opposites. The early philosophers sought to explain the cosmos on the basis of their observation of the regular behaviour of materials that make up the cosmos and all its phenomena, without reference to extra-natural causes or the personified gods as by traditional religions. Anaxagoras, for instance, proposed a rationalistic governing principle that remained free from the mythical or theological characteristics of prior cosmogonies. The Cosmological theory of Anaxagoras suggests that all things have existed from the beginning. But, originally they existed in infinitesimally small fragments of themselves, endless in number and inextricably combined. At some point, the unity is spurred into a vortex motion at a force and a speed. The origin of the world is depicted through this process of motion and separation from the unified mixture. The rotary motion caused the ingredients in the mass to shift. This shifting produces what Anaxagoras calls 'separating off'. The continued spreading rotation produced more and more separation. Anaxagoras proceeded to give an account of the stages in the process from original chaos to present arrangements. The division into cold mist and warm ether first broke the spell of confusion. With increasing cold, the former gave rise to water, earth and stones. The seeds of life which continued floating in the air were carried down with the rains and produced vegetation. Animals, including man, sprang from the warm and moist clay. In this way, Anaxagoras gave a complete account of the universe: of the heavens, the earth, and geological and meteorological phenomena.[5] Anaximenes, like his predecessors, also gave an account of how our world came to be out of previously existing matter. According to Anaximenes earth was formed from air by a felting process. It was a flat disk and there was only one substance, in this case air, from which the whole world and everything in it are composed. The several stuffs: wind, cloud, water etc., are only modifications of the real substance that is always and everywhere present. Anaximenes' notion of successive change of matter by rarefaction and condensation was influential in later theories. Philolaus' ideas about the cosmology of the universe, on the other hand, were drastically different. His thinking revolved around a hypothetical astronomical object he called the Central Fire. He supposed that a sphere of the fixed stars, the five planets, the Sun, Moon and Earth, all moved round his Central Fire. All these made up only nine revolving bodies and he conceived, in accordance with his number theory, a tenth one which he called Counter-Earth. He supposed the Sun to be a disk of glass which reflects the light of the universe. He made the lunar month consist of 29½ days, the lunar year of 354, and the solar year of 365½ days. One of the most widely accepted interpretation of the cosmos is that of Empedocles who posits two cosmic forces which work upon the elements in both creative and destructive ways (Fig. 4.1). These he personifies as Love (*Philia*)—a force of attraction and combination—and Strife (*Neikos*)—a force of repulsion and separation. These two forces are engaged in an eternal battle for domination of the cosmos and that they each prevail in turn in an endless cosmic cycle. Beginning clockwise from top (Fig. 4.1), one finds Love to be completely dominant and drawings all the elements fully together into a Sphere in which, although the elements are not fused together into a single mass, each is indistinguishable from the others. The Sphere then, is a-cosmic state during which no matter can exist, and no life is possible. Then as Love's power gradually weakens and Strife begins to grow in power, he gradually separates out the elements from the Sphere until there is enough separation for matter to come into existence, for the world to be created and for all life to be born. When Strife has achieved total domination we again get, an a-cosmic state in which the elements are separated completely and the world and all life is destroyed in a Whirl. Then Love begins to increase in power and to draw the elements together again, and as she does so the world is again created and life is again born. When Love has achieved full dominance we return once more to the sphere. As

Empedocles puts it: The cosmos exists in a state of constant flux then, beneath which there is a certain sort of stability in the eternity of the elements. The world is in a constant state of organic evolution. Following Empedocles' cosmology, since the time of the sphere, Strife has gained more sway; and the actual world is full of contrasts and oppositions, due to the combined action of both principles. Empedocles attempted to explain the separation of elements, the formation of earth and sea, of sun and moon, of atmosphere.

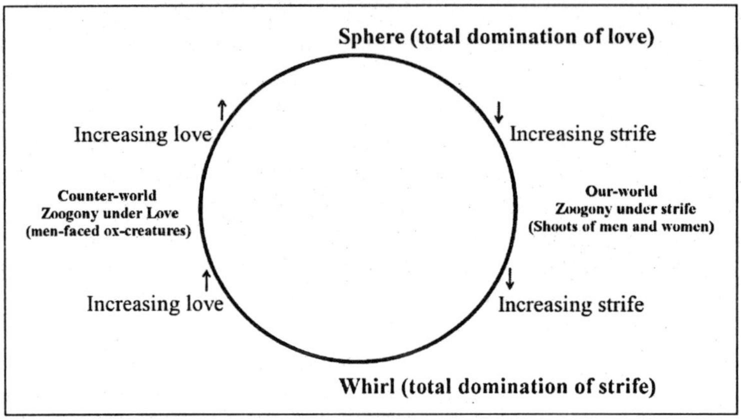

Fig. 4.1: Empedocles Cosmic Cycle

Besides, Heraclitus also offers his most fundamental statement on cosmology.[6] Heraclitus envisages a law like transformation of stuff from fire to water to earth; the transformation is reversible, and in it the same relative quantities of stuff are preserved. There is, then, a kind of conservation of matter, or at least overall quantity of matter. What would make the world to be continuous would be the fact that when one portion of fire turns into water, an equivalent portion of water turns into fire. The overall equilibrium is preserved, even if the water that is now in the sea is not the same water as was in it before. This picture bears a similarity to the image of the river, which remains the same despite its changing material contents. In this view of the world, the mutual transformations of matter are not an accidental feature, but the very essence of nature. Without change, there would be no world. Likewise, Acharya Kapil, recognized as the founder of Samkhya philosophy and the Father of Cosmology in India, asserted that *Prakriti,* with the inspiration of *Purusha,* is the mother of cosmic creation and all energies. The ancient Indian scholars dealt with many problems pertaining to Cosmology, Cosmogony and Cosmography, respectively the study of the science, origin and description of the Universe (*Brahamand*). Four types of theories have been advanced about the origin of the Universe.[7] These are the theories of: Artistic Origin; Mechanical Origin; Instrumental Origin; and Philosophical Origin. According to the Theory of Artistic Origin, the universe has been compared to a house and Rigveda refers various stages in the construction of this universal house. It mentions a number of Gods who who contributed their skills to the construction and completion of the Universe. They wove various materials into a pattern, and shaped the universe by blasting and smelting. The Theory of Mechanical Origin suggests the sacrifice (or disintegration) of a primeval body known as *Adi-Purush.* He was conceived as the soul and nucleus of the

Universe and an embodiment of the supreme spirit. The 'sky', the 'wind', the 'moon' and the 'sun'—all these elements were believed to be the results of dismemberment of *Adi-Purushas* a part of sacrifice ceremony. The scholars having faith in the Theory of Philosophical Origin believed that there was nothing earlier, i.e. no atmosphere, no sky, no days and nights etc. The space was empty (a vacuum). Within this a unit was born on its own, due to inherent heat and it hardened with time. This led to the formation of the Universe later. The Theory of Instrumental Origin, on the other hand, talks about the occurrence of two parent bodies, from which the Universe was created. Agni (God of Fire), Indra (God of Rain), Rudra (God of Storm) and other gods are mentioned as having being instrumental in the creation of the *Dharti* (Earth) and *Swarga* (Heaven)—the twin parents of the whole Universe. The union of these two produced the Sun. The Sun is considered as the most important agent in the creation of the world. It is the soul of all that 'moved' or that 'not moved'. It has been later identified as Rajapati, Vishwakarma and sometimes with the 'Golden Egg' and 'Unborn Being'. This 'Unborn Being' is also named as *Atma* (Soul) who produced the Universe through intermediary body. Unlike our modern scientists, the ancient Indian astronomers believed in geo-centric universe, i.e. the Earth at the centre and the Sun being as one of its planets.

The Srimad-Bhagavatam[8] presents an earth-centred conception of the cosmos. It tells of innumerable universes; each one is contained in a spherical shell surrounded by layers of elemental matter that mark the boundary between mundane space and the unlimited spiritual world. The region within the shell (Figure 4.2) is called the *Brahmanda*, (or *Brahma* egg). It contains an earth disk or plane—called *Bhu-mandala*—that divides it into an upper, heavenly half and a subterranean half, filled with water. *Bhu-mandala* is divided into a series of geographic features, traditionally called *dvipas*, or "islands," *varshas*, or "regions," and oceans.

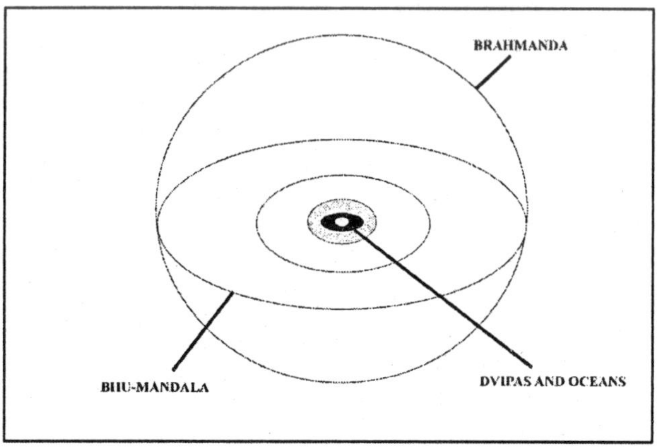

Fig. 4.2: Brahmanda

One can compare *Bhu-mandala* with an astronomical instrument called an astrolabe of ancient Greeks. On the astrolabe, an off-centred circle represents the orbit of the sun—the ecliptic. The Earth is represented in stereographic projection on a flat plate, called the mater. The ecliptic circle and important stars are represented on another plate, called the rete. Different planetary orbits could likewise be represented by different plates, and these would be seen projected onto the Earth plate when one looks down on the instrument. The *Bhagavatam* similarly presents the orbits of the sun, the moon, planets, and important stars on a series of planes parallel to *Bhu-mandala*. The details given in the *Bhagavatam* show that its cosmology is quite sophisticated. Not only does the *Bhagavatam* use the ecliptic model, but it turns out that the disk of *Bhu-mandala* corresponds in some detail to the solar system (Fig. 4.3).

The solar system is nearly flat. The sun, the moon, and the five traditionally known planets—Mercury through Saturn—all orbit nearly in the ecliptic plane. One striking feature of the Bhagavatam's descriptions has to do with size. If we compare *Bhu-mandala* with the Earth, the solar system out to Saturn, and the Milky Way galaxy, *Bhu-mandala* matches the solar system closely, while radically differing in size from Earth and the galaxy. Furthermore, the structures of *Bhu-mandala* correspond with the planetary orbits of the solar system. If we compare the rings of *Bhu-mandala* with the orbits of Mercury, Venus, Mars, Jupiter, and Saturn, we find several close alignments that give weight to the hypothesis that *Bhu-mandala* was designed as a map of the solar system. The dimensions of *Bhu-mandala* in the *Bhagavatam* are consistent with modern data on the size of the sun's orbit and the solar system as a whole.

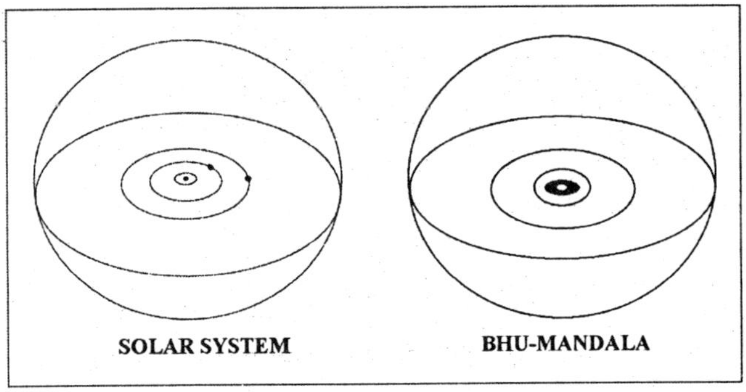

Fig. 4.3: The Ecliptic Model

By combining the basic concepts of Taoism and Confucianism, the ancient Chinese created a traditionally consistent and uniquely harmonious view of the universe. Taoism is a philosophy of duality and balance, and its focus lies in the concept of *yin* and *yang*. These two terms represent opposable forces in the universe. *Yin* is cool, wet, and female, while *Yang* tends to be hot, dry and male. The Confucian version of *Tao* and *Te* were referred to as *Jen* and *Li*. There were primarily three models of the Chinese interpretation of the physical orientation of the celestial orientation. The early Chinese drew up their pictures of the universe with a focus on the relations between Heaven and Earth. Three cosmological models (Fig. 4.4) were advocated by different schools and held true for a long time and were replaced in the seventeenth century by the Western theories.

Gai Tian' was the theory of the dome shaped heaven. It placed what we today call Ursa Major at the centre of the celestial dome, and China at the centre of the Earth. The school of 'Hun Tian' envisioned a spherical heaven shaped much like a hen's egg where the earth is like the yolk. Finally, the theory of 'Xuan Ye' suggested the unprecedented thought of an infinite universe with suspended celestial bodies. The main characteristics of the ancient Chinese models of the universe include: (i) the models did not address what we know as

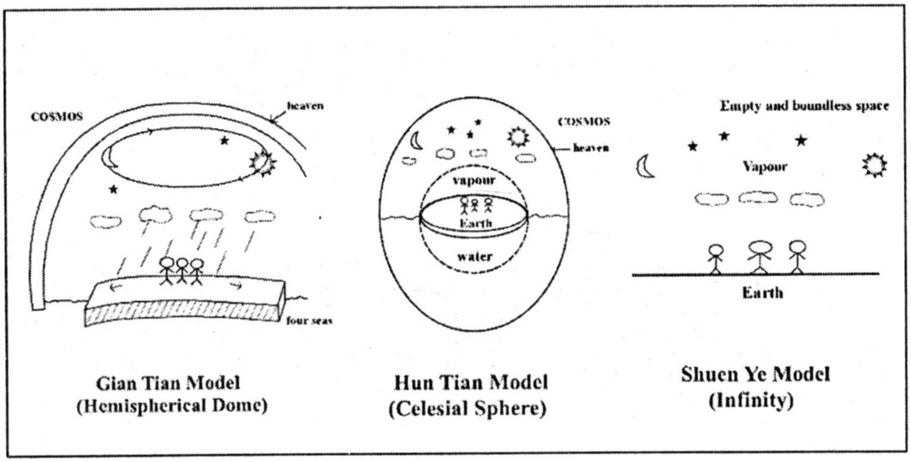

Fig. 4.4: Three Early Chinese Models of the Universe

"universe", but instead "Heaven and Earth"—a "researchable" universe. Precisely, "Heaven and Earth" denotes a space to be observed from the Earth and to be home of all visible heavenly entities. It was believed that beyond this space there is unknown infinite cosmos; (ii) the Earth was flat; and (iii) the heavens were round; the heavenly bodies moving freely. These features are remarkably distinguished from those of the Greek cosmological models in which the universe was finite, Earth was round (spherical) in its centre, and the heavens were layers of solid spheres; the heavenly bodies were attached to the layers respectively. These theories lead to many technological advances in astronomy such as armillary rings and spheres. Like many other cosmologies, there was an incredibly strong emphasis on harmony and balance in the Chinese interpretation of the universe that obviously originated from philosophical ideas. While the Greeks focused on the constellations on the horizon and created a solar calendar, the Chinese observed the circum polar stars, which lead them to devise a lunar calendar instead. The universe theory of all the above civilizations is congruent with the geocentric model.

Astrology and Astronomy

Astrology as a system of beliefs holds that the relative position of celestial bodies and related details can provide information about personality, human affairs and other terrestrial matters. The Astronomy, interwoven with Astrology has, thus, fascinated all mankind from the beginning. Astrology had developed in the Graeco-Roman world during the Hellenistic period, in clear synthesis with Babylonian astronomy. Ptolemy's treatise on astrology, the 'Tetrabiblos', was the most popular astrological work of antiquity. It exercised great influence on the Islamic world and the medieval Latin West. The astronomy of ancient people is related mainly to celestial mechanics. The astronomical observations were made with the help of some simple instruments like gnomon and scaphion. Deriving its name from the famous astronomical observatory at the University of Nalanda the science of Astronomy is called '*Khagola-shastra*' in India. The first references to it, are to be found in Rig Veda (2000 B.C.). Bhaskara II and his works represent a significant contribution to astronomical knowledge in the twelfth century. Written in 1150 his Siddhanta Shiromani consists of two parts: 'Goladhyaya' (sphere) and 'Grahaganita' (mathematical astronomy). Ancient Chinese astronomy was primarily a government activity. It was the astronomer's role to keep track of the solar, lunar, and planetary motions as well as divine what astronomical phenomena may mean for the ruling emperor.

The concept of *Prithvi* (Earth) was the most basic in the study of geography. It has been profusely used in the Vedas and Puranas. The study of the earth or *Prithvi* has been called '*Bhugol*' or Geography. Thus, the use of the term *Bhugol* for the discipline of Geography appropriately suggests that the ancient Indians endorsed the earth being a sphere. This fact was further affirmed by their observations on the periodical occurrence of eclipses and the circular shadow of the earth casted on moon during lunar eclipse; and the rise and setting of Sun, which according to the prevalent belief, never sets but goes on the other side of the sphere earth. Aryabhatta was the first one to have propounded the theory that the earth is a sphere in the fifth century. He was also the earliest to discover that the orbits of the planets around the Sun are ellipses. His book, *The Aryabhattiya,* described numerical and geometric rules for eclipse calculations. Aryabhatta also discussed about the locations of the planets in relation to its movement around the sun, the light of planets and the moon as reflections from the sun, the eclipse of the moon and the sun, day and night, the contours of the earth, and the length of the year exactly as 365 days.[10] His working was later adopted by the Greeks and then the Arabs. Among Greeks, various arguments were put forward by Aristotle in support of the sphericity of the Earth; and they included, for instance, is the fact that during a solar eclipse, the shadow caused by the interposition of the earth between the sun and the moon is always convex and therefore, the earth is spherical; the stars which are visible in a certain locality may not be visible further to the north or south, a phenomenon which could be explained within the understanding of a spherical earth; and from mere observation the earth has the appearance of being curved. When observed from an elevated site, the sky seems to surround the earth, like a dome, to meet the apparently curved horizon. If observed over the seasons, the dome would appear to revolve, with many of the heavenly bodies changing their position in varying degrees, but returning annually to a similar place in the heavens. From the shore, a ship can be seen to be descending, gradually, below the horizon, with the hull disappearing from view first, to be followed by masts and sails. If one had a companion observing from a higher point, the companion would see the ship for a long period before it disappeared from view. Making observations regarding the sphericity of celestial bodies, the ancient Chinese Scholar, Shen Kuo, likened the Moon to a ball of silver, which does not produce light, but simply reflects light if provided from another source (the Sun). He explained that when the Sun's light is slanting, the Moon appears full. He then explained if one were to cover any sort of sphere with white powder, and then viewed from the side it would appear to be a crescent, hence he reasoned that celestial bodies were spherical.[11] Shen is also known for his cosmological hypotheses in explaining the variations of planetary motions, including retrogradation.[12] Shen's hypotheses were similar to the concept of the epicycle in the Graeco-Roman tradition. Shen likened the relationship of the Moon's path to the ecliptic, the path of the Sun, "to the figure of a rope coiled about a tree". Besides, Aryabhatta was the first astronomer to make an attempt at measuring the Earth's circumference since Erastosthenes (200 B.C.). His approximation remained the most accurate for over a thousand years. Aryabhatta's system of astronomy was called the 'Audayaka' system (days are reckoned from '*uday*', dawn at 'lanka', equator). Aryabhatta appears to have believed that the earth rotates about its axis.

This is indicated in the statement, referring to Lanka, which describes the movement of the stars as a relative motion caused by the rotation of the earth: Like a man in a boat moving forward sees the stationary objects as moving backward, just so are the stationary stars seen by the people in Lanka (i.e. on the equator) as moving exactly towards the West. Sri Lanka is here a reference point on the equator, which was taken as the equivalent to the reference meridian for astronomical calculations. The facts put forward by ancient Indians related to the size and dimension of the earth were quite near to accuracy. It was well known to the ancient Indian scholars that the earth is an oblate spheroid slightly flattened at the poles. Also, its equatorial diameter is greater than its polar diameter. The estimates made by Suryasidhanta and Aryabhatta in this regard were very close to those established in the modern period. Among Greeks, Hipparchus developed theories for the Sun and Moon demonstrating 'that they are represented by uniform circular motions' as Ptolemy had explained later. The discovery for which Hipparchus is most famous is the discovery of precession which is due to the slow change in direction of the axis of rotation of the earth. This work came from Hipparchus' attempts to calculate the length of the year with a high degree of accuracy. Two definitions of a year appear in his work, i.e. related to the Sideral and Tropical year. Sideral year is the orbital period of earth around the Sun, taking the stars as the reference frame; it is 20 minutes longer than the Tropical year. A Tropical year, on the other hand, is the time that Sun takes to return to the same position in the cycle or seasons, as seen from the Earth, e.g. the time from Vernal Equinox to Vernal Equinox, or from Summer Solstice to Summer Solstice; it is a period of 12 months measured from any date. Hipparchus applied his knowledge of spherical angles to the problem of denoting locations on the Earth's surface. The ESA's[13] Hipparcos Space Astrometry Mission was named after him. Ptolemy synthesized and extended Hipparchus' system of epicycles and eccentric circles to explain his geocentric theory of solar system. He believed the planets and sun to orbit the Earth in the order Mercury, Venus, Sun, Mars, Jupiter and Saturn. This system became known as the Ptolemaic system. This is described in the book Mathematical Syntaxis (widely called the Almagest), one of the most influential books of classical antiquity. His work has been preserved, like most of Classical Greek science, in Arabic manuscripts. The Ptolemaic explanation of the motions of the planets remained the accepted wisdom until the Polish scholar Copernicus proposed a heliocentric view in 1543. Aryabhatta had also described a geocentric model of the solar system, in which the Sun and Moon are each carried by epicycles which in turn revolve around the Earth. In this model, the motions of the planets are each governed by two epicycles, a smaller *'manda'* (slow) epicycle and a larger *'śīghra'* (fast) epicycle.[14] The order of the planets in terms of distance from earth are taken as: the Moon, Mercury, Venus, the Sun, Mars, Jupiter, Saturn, and the asterisms. Most historians of astronomy consider that this two epicycle model reflects elements of pre-Ptolemaic Greek astronomy. Another element in Aryabhatta's model, the 'śīghrocca', the basic planetary period in relation to the Sun, is seen by some historians as a sign of an underlying heliocentric model.[15] He states that the Moon and planets shine by reflected sunlight.

Besides, Philolaus' astronomical system is also popular. It is famous as the first to move the earth from the centre of the universe and make it an orbiting planet instead. The central fire rather than the sun is at the centre of Philolaus' cosmos. The heavenly bodies are arranged in concentric circles around this central fire. Philolaus' astronomical system Copernicus referred to Philolaus as his precursor in to his *De revolutionibus orbium coelestium* (On the Revolutions of the Heavenly Spheres), the seminal work on the heliocentric theory. The heavenly bodies are arranged in ten concentric circles around this central fire. Beginning from the outside, the fixed stars come first, then the five planets, the sun, the moon, the earth and a mysterious counter-earth. The fixed stars serve as the backdrop against which the other heavenly bodies move. It is the motion of the earth that accounts for the apparent daily motion of the sun across the sky and hence for night and day. In his book, Bhaskaracharya mentions about force of attraction resembling gravity, discovered centuries later by Newton[16]. The study of astronomy in Bhaskara's work is also based on a model of the solar system which is heliocentric and whose movements are determined by gravitation. Heliocentrism had also been talked about ealier in 499 by Aryabhatta, who argued that the planets follow elliptical orbits around the Sun. There is an old Sanskrit *Sloka* (couplet):"*Sarva Dishanaam, Suryaha, Suryaha, Suryaha.*" This couplet means that there are suns in all directions. This couplet which describes the night sky as full of suns indicates that in ancient times Indian astronomers had arrived at the important discovery that the stars visible at night are similar to the Sun visible during day time. In other words, it was recognized that the sun is also a star, though the nearest one. This apart, many Indian astronomers had formulated ideas about gravity and gravitation. Brahmagupta, in the seventh century had said about gravity that "Bodies fall towards the earth as it is in the nature of the earth to attract bodies, just as it is in the nature of water to flow." About a hundred years before Brahmagupta, another astronomer, Varahamihira had claimed for the first time that there should be a force which might be keeping bodies stuck to the earth, and also keeping heavenly bodies in their determined places. Thus the concept of the existence of some tractive force that governs the falling of objects to the earth and their remaining stationary after having once fallen; as also determining the positions which heavenly bodies occupy, was recognised. It was also recognized that this force is a retractive force. The Sanskrit term for gravity is *Gurutvakarshan* which is an amalgam of *Guru-tva-akarshan; Akarshan* means to be attracted. Thus the fact that the character of this force was of attraction was also recognized. The function of attracting heavenly bodies was attributed to the sun. The idea that the sun was looked upon as the power that attracts heavenly bodies is supported by the virile terms like Raghupati and Aditya used in referring to the sun. While the male gender is applied to refer to the sun, the earth (*Prithivi, Bhoomi* etc.) is generally referred to as a female. The literal meaning of the term *Gurutvakarshan* also supports the recognition of the heliocentric theory. Many ancient Indian astronomers have also referred to this concept; Aryabhatta has suggested it in his treatise *Aryabhattiya*; and Bhaskaracharya has made references to it in his Magnum Opus *Siddhanta-Shiromani*. Thus the heliocentric idea is believed to have existed in a rudimentary form in the days of the Rig Veda and was refined further by astronomers of a later age. The heliocentric theory of gravitation was, a thousand years later, articulated by Copernicus and Galileo. Where the studies of all the others depended on observational astronomy, Anaximander's merits lie in the field of speculative astronomy. One may discern three of his astronomical speculations that the celestial bodies make full circles and pass also beneath the earth, that the earth floats free and unsupported in space, and that the celestial bodies lie behind one another. These three propositions make up the core of Anaximander's astronomy, constituting the origin of the Western concept of the universe. Anaximander's vision implied depth in the universe, that is, the idea that the celestial bodies lie behind one another. Anaximander's order of the celestial bodies is clearly that of increasing brightness. Due to the inclination of the axis of the heaven, the celestial bodies do not circle around the earth in the same plane as the earth's-flat-surface, but are tilted. This tilting of the heaven's axis has been one of the biggest riddles of the universe.

The Greeks were the first to describe the circumstances under which eclipses occur and the way light is reflected by the moon. The most intriguing for historians of astronomy was Hipparchus who turned his attention to a wide variety of astronomical questions, including the computation of lunar and solar eclipses. He was the first to develop models for the motion of Sun and Moon and the reliable methods to predict solar eclipses. Even with observations with crude instruments, the astronomers in ancient India as well were able to arrive at near perfect measurement of astronomical movements and predict eclipses. Instead of the prevailing cosmogony where eclipses were caused by pseudo-planetary nodes Rahu and Ketu, they explained eclipses in terms of shadows cast by and falling on earth. The ancient Indian scholars were aware of the causes of occurrences of eclipses (*grahanas*). Some rituals and ceremonies of the Indian society were being performed on the days when eclipses occurred. The Aryans[17] considered an eclipse as unlucky (inauspicious) and a messenger of disaster. It was also believed that if a solar and lunar eclipse occurs in the same month, it becomes more disastrous. Varahamihira considered the effects of eclipse month-wise and emphasized that eclipse in *Pusa* (December) leads to famine and its occurrence in April and May results in good rainfall, while and eclipse in *Phagun* (March) and *Asadh* (June) is unlucky. It was believed that the looking at eclipses is bad, because seeing the eclipse (particularly Solar) is harmful for the naked eyes. Observing total solar eclipses was a major element of forecasting the future health and successes of the Emperor in China, and astrologers were left with the onerous task of trying to anticipate when these events might occur. Solar eclipses were important enough to be recorded in chronicles and on "oracle" bones. Although the eclipse records are often incomplete and the dating of the bones is not reliable, the eclipse observations from the Chou dynasty and Warring States period (1050-221 B.C.) onwards have been reliably dated.

Hipparchus made extensive observations of star positions, and is credited with the production of the first known catalogue of stars in ancient Greece. The 'Almagest' of Ptolemy contains an updated version of star catalogue. Its list of forty-eight constellations is ancestral to the modern system of constellations. The Greeks applied their knowledge of spherical angles to the problem of denoting locations on the Earth's surface. '*Akshansh* and *Deshantar*' are the terms used for latitudes and longitudes respectively in the ancient Indian literature. Puranas have a reference of three imaginary lines of latitudes passing through Equatorial belt, North Pole and South Pole. Accordingly, three major regions have been identified in the Literature, viz. Equatorial (*Nirakshadesha*), Northern Polar (*Meru*) and Southern Polar (*Bhadvanala*). The North Pole has been called as *Zenith* and the South Pole as *Nadir*. The South Pole was truly considered as the antipode of the North Pole, i.e. diametrically opposite to it. However, the world was not believed to exist beyond Equator, as the region here was compared to hell of the earth. The Eastern part, on the other hand, was believed to be 'the land of Gods'. This thinking is in consonance with that of the Europeans in the Early Medieval period, when the Dark Ages prevailed and the East in 'T-in-O' Maps was assumed to be the place of Adam and Eve. The ancient Indian scholars have also drawn Prime Meridian. These imaginary lines, the position of Sun and various stars have helped them to determine local time at various places.

A number of astronomical instruments were in use. With the invention of astrolabe the Greeks were able to measure the geographical latitude and time by observing stars. The ancient Chinese invented the first planetarium. By 2300 B.C., ancient Chinese astrologers and astronomers already had sophisticated observatory buildings. Based on the theories of his predecessors, Zhang Heng systematically developed the celestial sphere theory. An armillary constructed on the basis of his hypotheses bears a remarkable similarity to Ptolemy's earth-centred theory. Su Song created a celestial atlas (in five separate maps), which had the hour circles between the *xiu* (lunar mansions) forming the astronomical meridians, with stars marked in an equidistant cylindrical projection on each side of the equator, and thus, was in accordance to their north polar distances.[18] Su Song's star map represents the oldest existent star maps in printed form. Being the head official for the Bureau of Astronomy, Shen Kuo was an avid scholar of medieval astronomy, and improved the designs of several astronomical instruments. Shen is credited with making improved designs of the gnomon, armillary sphere, and clepsydra clock.

Physical Studies

The earth studies of ancient scholars, particularly of Indians, were very elaborate and dealt with its Cardinal points, Meteorology, Geomorphology, Seismology and Physical Divisions. The Local gazetteers featured a wealth of geographic information in China.[19] Initially, the idea of four main directions had been formulated in Rigveda, viz. *Purva* (East), *Paschim* (West), *Uttar* (North) and *Dakshin* (South). By adding Zenith (*Meru*) and Nadir (*Bhadvanala*) it was raised to six. But, afterwards, ten directions have been frequently mentioned in the Puranic literature. The designation of these directions is significant in the sense that it bears concept of the Gods dominating in each of these directions. The ten directions and the ruling deity of each is mentioned in Table 4.3.

Table 4.3: Cardinal Points and the Ruling Daities as per Puranic Literature

DIRECTION		RULING DEITY	
Purva	(East)	*Indra*	(The God of Rain)
Agneyay	(Southeast)	*Agni*	(The God of Fire)
Dakshina	(South)	*Yama*	(The God of Death)
Nairitya	(Southwest)	*Niriti*	(The God of Disaster)
Paschim	(West)	*Varuna*	(The God of Water)
Vayavya	(North-west)	*Marut/Vayu*	(The God of Air)
Uttar	(North)	*Kubera*	(The God of Wealth)
Isana	(Northeast)	*Isa*	(The God of Power)
Urdhava	(Zenith*)	*Brahma*	(The Creator of Universe)
Adhoh	(Nadir^)	*Sesanaga*	(The Universal Ocean)

*The point directly above the observer.
^ The point directly below the observer.

The studies on Meteorology stretch back millennia. The ancient Indian scholars have identified the vacuum between the earth and the heaven as *Antariksha*, and they were aware of its vast extent or thickness and the occurrence of various weather phenomena here, as rain, winds, clouds, lightening, fog, and frost etc. *Ramayana* furnishes a lot of this kind of information. Bhaskaracharya has conceived the thickness of this *Antariksha* around the earth to be 12 *yojans* (or 96 kms; one *yojan* being equal to about 8 kms.). As far as the knowledge about the seasons is concerned, it is based largely on the studies in India.

Rigveda mentions five seasons. In Valmiki Ramayan, however, six seasons have been identified, as listed in Table 4.4.

Table 4.4: The Seasons as Mentioned in Ancient Indian Literature

SEASON		MONTHS	
Basant	(Spring)	*Chaitra-Baisakh*	(March-April)
Grishma	(Summer)	*Jyestha-Asadh*	(May-June)
Prourit/Varsha	(Rainy)	*Shravan-Bhadrapad*	(July-August)
Sharad	(Autumn)	*Ashvin-Kartik*	(September-October)
Hemant	(Winter)	*Mangsir-Paus*	(November-December)
Shishir	(Severe Winter)	*Magh-Phagun*	(January-February)

The observations of Chinese scholars on weather and climate are also commendable. They are known to have prepared a number of weather reports on the basis of empirically based facts. Preserved in the National Archives some of these reports date back to 13 centuries before Christ. Early speculation and hypothesis pertaining to meteorology had a long tradition in China. This is proved by the accurate description of the process of the water cycle by Wang Chong (27–97).[20] Shen discussed vividly about tornadoes—the first known description of them being in East Asia—and gave reasoning that rainbows were formed by the shadow of the Sun in rain, occurring when the Sun would shine upon it. It is believed that Shen's explanation of the rainbow as a phenomenon of atmospheric refraction is basically in accord with modern scientific principles.[21] Shen hypothesized that rays of sunlight refract before reaching the surface of the earth, hence people on earth observing the Sun are not viewing it in its exact position. Shen also hypothesized that the altitude of the apparent Sun is higher than the actual altitude of the Sun. There are also evidences of the interest of earlier Chinese scholars in Paleoclimatology. Shen's discovery about the petrified bamboos in a region Yanzhou, located in northern China, apparently turned to 'stone' during the Zhiping reign period (1064–1067) proved that the bamboos had never been known to be grown there; the climate there must have shifted geographically over time.[22] Considering that damp and gloomy low places provide suitable conditions for the growth of bamboo, Shen deduced that the climate of Yanzhou must have fit that description in very ancient times.

The seismological studies reflect in the works of all the ancient scholars alike. The Greeks found the cause of earthquakes in water; their hypothesis was rational because it provided an explanation which did not invoke hidden entities. It was an advance upon the traditional Homeric view that they resulted from an angry supernatural god, Poseidon, shaking the earth through his rapid striding. Among the early theories about earthquakes developed by ancient Greeks, they were believed to have caused, for instance, by excess water near the surface crust of the earth bursting into the Earth's hollows (Anaxagoras); or the saturation of the Earth with water (Democritus); or due to massive pieces of the Earth falling into the cavernous hollows due to drying (Anaximenes); and/or by instability of vapour (*pneuma*) caused by the drying of the moist Earth by the Sun's rays (Aristotle).[23] For 'earthquakes' the term *bhukamp* has been used by Indians in Puranas. It was assumed that the deities of Air, Fire and Water cause the earthquakes. Known for excellent record keeping, the Chinese kept accurate records of not only celestial events, but of earthquakes as well. From the earliest times, the Chinese were concerned with the destructive force of earthquakes. The earlier Chinese people believed that seismic events were supernatural in origin: signs from heaven from angry gods designed to punish those below. The ancient Chinese explained them as disturbances with cosmic *yin* and *yang*, along with the heavens' displeasure with acts committed (or the common peoples' grievances ignored) by the current ruling dynasty. However, the learned scholars of Han Dynasty, like Zhang Heng, discounted these superstitions and tried to explain them by scientific means. Heng also designed a device capable of indicating tremors, their distance from the seismic event, and the event's location. Against the grain of earlier theories proposed by his fellow Chinese and contemporary Greeks, Zhang Heng believed that earthquakes were caused by wind and air, writing: The chief cause of earthquake is air, an element naturally swift and shifting from place to place. As long as it is not stirred, but lurks in a vacant space, it reposes innocently, giving no trouble to objects around it. But any cause coming upon it from without rouses it, or compresses it, and drives it into a narrow space ... and when opportunity of escape is cut off, then 'With deep murmur of the Mountain it roars around the barriers', which after long battering it dislodges and tosses on high, growing more fierce the stronger the obstacle with which it has contended.[24]

A significant amount of written works, particularly in China, is devoted to geomorphology. Fan Chengda's geographical treatise, known as the *Gui Hai Yu Heng Chi*, focused primarily on the topography of the land, along with the agricultural, economic and commercial products of each region in China's southern provinces[25]. Du Yu, one the earliest of such writers, believed that the land of hills would eventually be levelled into valleys and valleys would gradually rise to form hills. Xu Xiake travelled throughout the provinces of China to write his enormous geographical and topographical treatise, documenting various details of his travels, such as the locations of small gorges, or mineral beds such as mica-schists. Xu's work was largely systematic, providing accurate details of measurement, and his work read more like a twentieth century field surveyor than an early seventeenth century scholar.[26] Shen Kuo devised a geological hypothesis for land formation (geomorphology), based upon findings of inland marine fossils, knowledge of soil erosion, and the deposition of silt.[27] As stated earlier, he also proposed a hypothesis of gradual climate change, after observing ancient petrified bamboos that were preserved underground in a dry northern habitat that would not support bamboo growth in his time. Shen Kuo's formulation was based upon several observations as evidence. This included his observation of fossil shells in a geological stratum of a mountain hundreds of miles from the ocean. He inferred that the land was reshaped and formed by erosion of the mountains, uplift, and the deposition of silt, after observing strange natural erosions of the Taihang Mountains and the Yandang Mountain near Wenzhou. He hypothesized that, with the inundation of silt, the land of the continent must have been formed over an enormous span of time.

In the ancient period, the knowledge about various parts of the world was limited. It was due to the poor means of communication and transportation. Even then, the attempts were made to divide the world into several regions, on the basis of available information. Such descriptions exist in Puranas. The term *Dwipa* has been used to designate various realms (continents) of the earth. Accordingly, the known world during the Puranic period was divided into seven *Dwipas* or 'regions' (Fig. 4.5).

These divisions exclude the American Continents, Greenland, England and Antarctica, since they were discovered only during the Age of Discovery in the late medieval period. The regions, as depicted in Fig. 4.5, seem to have derived their names from the existing popular trees or grasses here.

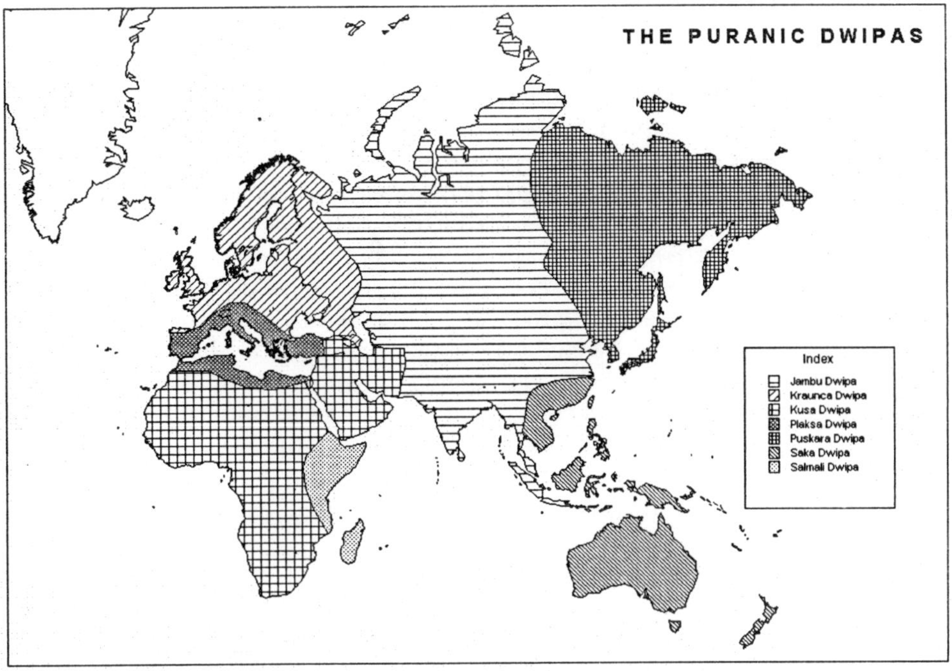

Fig. 4.5: The Puranic Dwipas

Mathematical Sciences

The mathematical sciences are a group of field of study that includes, in addition to mathematics, those academic disciplines that are primarily mathematical in nature as Arithmetic, Algebra, Geometry, Trigonometry, Algorithm and Calculus. The Vedic literature is replete with concepts of zero and other mathematical concepts that were transmitted to other parts of the world and led to further developments laying foundations of the discipline. Among the various fields of Greek interest, mathematics was one, and Pythagoras excelled in Mathematics. Known as "the father of numbers", Pythagoras thought numbers were divine and believed that everything was related to mathematics; and through mathematics everything could be predicted and measured, just as in music, in rhythmic patterns or cycles. In India, mathematics has its roots in Vedic literature. The Vedic Literature is replete with concept of zero and other mathematical concepts that were transmitted to other parts of the world and led to further developments laying foundations of the discipline. In the classical period of Indian mathematics (400 A.D. to 1200 A.D.), various treatises on mathematics were authored by Indian mathematicians. The ancient and medieval Indian mathematical works, all composed in Sanskrit, usually consisted of a section of *sutras* in which a set of rules or problems were stated with great economy in verse in order to aid memorization by a student. This was followed by a second section consisting of a prose commentary that explained the problem. The oldest extant mathematical document produced on the Indian subcontinent is the birch bark Bakhshali Manuscript, discovered in 1881 in the village of Bakhshali, near Peshawar (modern day

Pakistan) and is likely from the seventh century A.D.[28] The brick work of Harappa and Mohenjodaro excavations show that people of ancient India (2500 BCE) possessed knowledge of Geometry. Aryabhatt formulated the rules for finding the area of a 'triangle', which led to the development of Trignometry. Baudhayana was the author of one of the earliest *Sulba-sutras* (texts dealing with geometry and mathematical principles). His writings contain several important mathematical concepts that were later attributed to the Greeks. One such example is the Pythagorean Theorem.[29] There are evidences of the Babylonian cuneiform tablets containing Pythagorean triples indicating that there was sophisticated understanding on the topic in Mesopotamia as well as early as in 1850 B.C.[30] Excavations at Harappa, Mohenjo-daro and other sites of the Indus Valley Civilization have uncovered evidence of the use of "practical mathematics". They used a standardized system of weights and produced weights in regular geometrical shapes, which included hexahedra, barrels, cones, and cylinders, thereby demonstrating knowledge of basic geometry.[31] The inhabitants of Indus civilization also tried to standardize measurement of length to a high degree of accuracy. They designed a ruler—the *Mohenjo-daro ruler*—whose unit of length (approximately 1.32 inches or 3.4 centimetres) was divided into ten equal parts. Bricks manufactured in ancient Mohenjo-daro often had dimensions that were integral multiples of this unit of length.[32]

Indian mathematicians made early contributions to the study of arithmetic, calculus, algorithm, decimal number system and the concept of zero.[33] Arithmetic is the oldest and most elementary branch of mathematics. It is used by almost everyone, for tasks ranging from simple day-to-day counting to advanced science and business calculations. Arithmetic was discovered by Indians in about second century B.C. Bhaskaracharya's book Lilavati is regarded as the first book on modern arithmetic. The Arabs learnt and adopted it from India and spreaded it to Europe.[34] Calculus is the study of change[35], in the same way that geometry is the study of shape and algebra is the study of operations and their application to solving equations. A course in calculus is a gateway to other, more advanced courses in mathematics devoted to the study of functions and limits, broadly called mathematical analysis. *The Siddhanta Shiromani* of Bhaskara II is an astronomical treatise and contains many theories related to the preliminary concepts of infinitesimal calculus and mathematical analysis. Evidence suggests Bhaskara was deeply acquainted with 'differential calculus' and suggested that the differential coefficient vanishes at an extremum value of the function, indicating knowledge of the concept of 'infinitesimals'.[36] Madhava (1340-1425) and the Kerala School mathematicians from 14th to sixteenth century expanded on Bhaskara's work and further advanced the development of calculus in India. Algorithms are used for calculation, data processing, and many other fields. Also the technique of calculation, called algorithm, which is today widely used in designing software programmes (instructions) for computers was derived from Indian mathematics. This technique which was named by westerners after Al Khwarazmi as "Algorismi" gave us the modern term Algorithm. Al Khwarazmi was the chief exponent of this Indo-Arab amalgam in mathematics who evolved this technique of calculation from Indian sources. Just like other mathematical inventions, the decimal number system in use today was first recorded in Indian mathematics.[37] It was then transmitted to the Islamic world, and eventually to Europe.[38] The earliest surviving evidence of decimal place value numerals in India and South East Asia is from the middle of the first millennium C.E. Decimal numerals recording the years 683 C.E. have also been found in stone inscriptions in Indonesia and Cambodia, where Indian cultural influence was substantial.[39] The inclusion of zero in mathematics opened up new dimensions in the field. The concept of zero reached the west through Arabs. Numerals are found in the inscriptions of Ashoka the Great in the third century B.C.E. This knowledge travelled to Europe. In Arab countries even now numerals are known as *Hindse*: i.e. from India. The first, however, to compute with the zero as a number and to know negative numbers, was Brahmagupta. Brahmagupta's treatise *Brahma-sputa-siddhanta* was translated into Arabic. For several centuries this translation remained a standard text of reference in the Arab world. It was from this translation of an Indian text on Mathematics that the Arab mathematicians perfected the decimal system and gave the world its current system of enumeration which we call the Arab numerals, which are originally Indian numerals. The representation of numbers in a 'positional system'[40] is another contribution of ancient Indians (Bhāskara-I) to mathematics.

The field of astronomy, especially mapping the positions of the stars and planets on the celestial sphere, served as an important source of geometric problems in the past. Geometry originated in Egypt where it developed out of necessity; it was was introduced into Greece by Thales who had visited Egypt. Initially a body of practical knowledge concerning lengths, areas, and volumes, in the third century B.C. geometry was put into an axiomatic form by Euclid, whose treatment—Euclidean geometry—set a standard for many centuries to follow. The development of geometry is preserved in a work of Proclus, *A Commentary on the First Book of Euclid's Elements*. The Indian mathematics recognised geometry as *Rekha Ganita* (Line Computation). The *Sulbasutra* of Baudhayana contains geometric solutions. The *Sulbasutra*, which literally mean 'Rule of the Chord' give geometrical methods of constructing altars and temples. The Arab scholar Mohammed Ibn Jubair al Battani studied Indian use of ratios from *Rekha Ganita* and introduced them among the Arab scholars.

Astronomical calculations are complex and involve many variables that go into the derivation of unknown quantities. Algebra is a short-hand method of calculation and by this feature it scores over conventional arithmetic. Europe borrowed the knowledge of Algebra from Arabs, who acquired this from India. Algebra is the Arabic name for *Bijaganitam* (Bija means 'another' or 'second' and Ganitam means mathematics). It was recognized as a parallel system of computation, different from the conventional one which was used since the past and was till then the only one. Whatever the origin of algebra, it is certain that this technique of computation originated in India and was current around 1500 years back. Aryabhatta, an Indian mathematican, who lived in the fifth century A.D. has referred to *Bijaganitam* in his treatise on Mathematics, *Aryabhattiya*.[41] Another Indian mathematician-astronomer, Bhaskaracharya had also authored a treatise, entitled *'Siddhanta-Shiromani'* containing a section on *Bijaganitam* around twelfth century A.D. The term 'algebra' is a corruption of the term *'Al jabr'* which the Arabs gave to *Bijaganitam*. Incidentally the term *Bijaganit* is still use in India to refer to this subject.

Trigonometry is believed to have evolved in the Hellenistic world. Hipparchus made an early contribution to trigonometry producing a table of chords, an early example of a trigonometric table. He also introduced the division of a circle into 360 degrees into Greece. Trigonometry was introduced into ancient India through the translation of Greek works and was further advanced here.[42] A later landmark in Indian mathematics was the development of the series expansions for trigonometric functions (sine, cosine, and arc tangent) by mathematicians of the Kerala School in the fifteenth century C.E. The *Surya Siddhanta* (c. 400), a treatise of Indian astronomy, contains the roots of modern trigonometry. Later, its Arabic and Latin translations were very influential in Europe and the Middle East. Aryabhatta's contribution in this field of mathematics is unparalleled. Besides, the *Siddhanta Shiromani*

Bhaskara-II (written in 1150) demonstrates knowledge of trigonometry, including the sine table and relationships between different trigonometric functions. He also discovered spherical trigonometry, along with other interesting trigonometrical results.

Cartography

The Greek contribution in the field of cartography and map-making is unparalleled. Ptolemy was the most influential among the ancient Greeks and Romans. He designed a graticule and plotted the world using the concept of conical projection. On the Chinese realm, Pei Xiu and Zhang Heng were the earliest ones to promote and use cartographic techniques. They were familiar with the use of scale. Zhang Heng was the first to establish the grid reference system in Chinese cartography[43]. The earliest extant maps found in archeological sites of China date to the fourth century B.C. and were made in the ancient State of Qin.[44] From the first century A.D. onwards, official Chinese historical texts contained a geographical section, which was often an enormous compilation of changes in place-names and local administrative divisions controlled by the ruling dynasty, descriptions of mountain ranges, river systems, taxable products etc. The earliest known reference to the application of a geometric grid and mathematically graduated scale to a map was contained in the writings of the cartographer Pei Xiu.[45] Pei Xiu outlined and analyzed the advancements of cartography, surveying, and mathematics up until his time.[46] The Chinese makers of maps were familiar with the use of scale. Pei Xiu outlined six principles that should be observed when creating a map. The first three principles outlined the use of scale (*fan lu*), direction (*zhunwang*), and road distance (*daoli*), while the last three principles are used to properly calculate distances on uneven terrain as represented on a flat, two dimensional map.[47]

Humanistic Studies

The human aspect was intelligently dealt in various Historical, Regional, Anthropological, Psychological and Linguistic studies by the scholars of antiquity. The Greeks (Herodotus, Polybius and Posidonius) and Chinese (Ban Gu, Sima Qian) were the pioneers in the field of history. Among Greeks Herodotus, is the first known historian of classical antiquity. His *'The Histories'* is centred on the theme of Persia's imperial progress. It is a description of the years before the wars between Greece and Persia and the wars themselves (499-479 B.C.). Although his writing covers only a short period in the history of Greek civilization, his account of the history of the wars is the first example of historical writing in the Western World. Before Herodotus the idea of recording history was not widespread and not attempted in a systematic manner. Polybius was another historian to attempt to present history as a sequence of causes and effects. Polybius provided a unified view of history rather than a chronology. Posidonius, on the other hand, continued the World History of Polybius. His 'Histories' continue the account of the rise and expansion of Roman dominance. It was not concerned with isolated political history of peoples and individuals, but included discussions of all forces and factors (geographical factors, mineral resources, climate, nutrition), which let humans act and be a part of their environment. For example, Posidonius considered the climate of Arabia and the life-giving strength of the sun, tides and climatic theory to explain people's ethnic or national characters.[48] The Chinese historical form of dynasty history, or *Jizhuanti* history of dynasties, was codified by Ban Gu, but historians regard Sima Qian's work as their model, which stands as the "official format" of the history of China. In writing 'Shiji', Sima initiated a new writing style by presenting history in a series of biographies. His work was divided into particular subjects, including annals, chronicles, treatises—on music, ceremonies, calendars, religion, economics, and extended biographies. The modern historian Hsu Mei-ling states that Ban Gu's written work in geography set the trend for the establishment of geographical sections of history texts, and most likely sparked the trend of the gazetteer in ancient China.[49] Before Sima, histories were written as certain events or certain periods of history of states. His idea of a general history affected later historiographers like Zheng Qiao in writing '*Tongshi*' and Sima Guang in writing '*Zizhi Tongjian*.'

Regarding the study of human beings, the explanations offered suggest that human culture developed as a response to necessity and the hardships of our environment.[50] According to Anaximander, for instance, life originated from the moisture that covered the earth before it was dried up by the sun. Democritus suggests that human life was originally like that of animals; it describes the gradual development of human communities for purposes of mutual aid, the origin of language, crafts and agriculture[51]. The study such phenomena as perception, cognition, emotion, personality, behaviour and interpersonal relationships had also been a concern in the past. Philolaus, for instance, presented account of psychic faculties: The head [is the seat] of intellect, the heart of soul (*psychê*) and sensation, the navel of rooting and first growth, the genitals of the sowing of seed and generation. The brain (contains) the origin of man, the heart the origin of animals, the navel the origin of plants, the genitals the origin of all (living things). For all things both flourish and grow from seed. Philolaus thus recognizes four distinct psychic faculties. He made a clear distinction between intellect (*nous*), which is limited to human beings, and perception, which animals possess as well. This hierarchy of four faculties is strikingly associated with the hierarchy of living beings (human, animal, plant) as well as a hierarchy of parts of the body (head, heart, navel, genitals). Philolaus made a significant advance over earlier Presocratics by making a firm distinction between thinking and perception. His systematization of psychic faculties goes beyond anything found in his predecessors and is the recognizable precursor of Aristotle's distinction between the faculties for nutrition, appetite, sensation, movement and thought.[52]

The core text for linguistics in ancient times has been discovered in the '*Aṣṭādhyāyī*' of Pāṇini, the *sine qua non* of learning. Pāṇini's grammar can be considered to be the world's first formal system, well before the nineteenth century innovations of Gottlob Frege and the subsequent development of mathematical logic. To design his grammar, Pāṇini used the method of "auxiliary symbols," in which new affixes are designated to mark syntactic categories and the control of grammatical derivations. This technique was rediscovered by the logician Emil Post and is now a standard method in the design of computer programming languages.

The ancient scholars were concerned with documenting geographical information of regions far outside of their realm as well. The issues related to man-nature relationships, resources and population were important in China. In China, the earliest known geographical writing dates back to the fifth century B.C..[53] Although Chinese had been writing of civilizations of the Middle East, India, and Central Asia since the traveller Zhang Qian (Second century B.C.), later Chinese would provide more concrete and valid information on the topography and geographical aspects of foreign regions as well. The Tang Dynasty Chinese diplomat Wang Xuan-ce wrote the book *Zhang Tian-zhu Guo Tu*

(Illustrated Accounts of Central India) after travelling to Magadha (modern north-eastern India) during seventh century A.D. The book included a wealth of geographical information. Chinese geographers such as Jia Dan described the sea route going into the mouth of the Persian Gulf, and that the medieval Iranians (the people of Persia) had erected 'ornamental pillars' in the sea that acted as lighthouse beacons for ships that might go astray. Arabic writers such as al-Mas'udi and al-Muqaddasi confirmed Jia's reports about lighthouses in the Persian Gulf. The geography of medieval Cambodia (the Khmer Empire) was documented by Zhou Daguan in his book *Zhen-La Feng Tu Ji* (1297 A.D.).

Sciences

The sciences like physics, chemistry, botany, biology, pharmacology, medicine etc. are rooted in the works of ancient Greeks and Indians. Among Greeks, the foundation of physics lies in the writings of the scholars like Empedocles and Posidonius. Empedocles asserted that the matter is created by four 'elements': earth, air, fire and water, by being 'mixed' in different combinations and proportions. The four elements correspond, at macroscopic level, with the traditional quadripartite division of the cosmos into earth, sea, air, and the fiery ether of the heavenly bodies. Posidonius advocated a theory of cosmic "sympathy", the organic interrelation of all appearances in the world, from the sky to the earth, as part of a rational design uniting humanity and all things in the universe. The knowledge of Indian scholars in this regard was more extensive. The concepts of atom (*Anu, Parmanu*) and relativity (*Sapekshavada*) were explicitly stated by them nearly 600 years before Christ in an elaborate manner. From the Vedic times, Indians too classified the material world into four elements, viz. Earth (*Prithvi*), fire (*Agni*), air (*Maya*) and water (*Apa*); but, also added a fifth element, viz. ether or *Akasha* to these. These five elements or *Pancha Mahabhootas* were identified with various human senses of perception; earth with smell, air with feeling, fire with vision, water with taste and ether with sound. Since ancient times Indian philosophers believed that these other elements were physically pulpable and hence comprised miniscule particles of matter. The last miniscule particle of matter which could not be subdivided further was termed 'atom' (*Parmanu*). This Indian concept of the atom was developed independently and prior to the development of the idea in the Graeco-Roman world. It was Kanada, a sixth century B.C. scholar, who first propounded that the *Parmanu* (atom) was an indestructible particle of matter constuting the universe. Each *Parmanu* has a peculiar property which depends, on the substance to which it belongs. It was because of this conception of peculiarity of *Parmanu* (atoms) that this theory propounded by Kanada came to be known *'Vaisheshika-Sutra'* (Peculiarity Aphorisms). In this context Kanada seems to arrive at conclusions which were surpassed only many centuries after him. About his ideas on atom, Kanada observed that an inherent urge made one *Parmanu* combine with another. When two *Parmanu* belonging to one class of substance combined, a *dwinuka* (binary molecule) was the result. This *dwinuka* had properties similar to the two parents *Parmanu*. In the material universe, according to him, *Parmanu* belonging to different classes of substances combine in different combinations giving us a variety of *dwinuka*, which in other words means different types of substances. Apart from such combination of different *Parmanu*, Kanada also put forth the idea of chemical changes occurring because of various factors. He claimed that variation in temperature could bring about such changes. He cited the examples of blackening of a new earthen pot and the ripening of fruit to illustrate the chemical change in substances brought about by the heat.

Another Indian philosopher, Pakudha Katyayana who also lived in the sixth century B.C. and also a contemporary of Gautama Buddha, had also propounded ideas about the atomic constitution of the material world. These philosophers considered the Atom to be indestructible and hence eternal. The Vaisheshika School of philosophers believed that an atom was a mere point in space. Indian theories about the atom are greatly abstract and enmeshed in philosophy as they were based on logic. These Indian ideas about atom and atomic physics could have been transmitted to the West during the contacts created between India and the West by the invasion of Alexander. Thus, it remains a fact that Indian ideas about atom are the oldest Parallel to the development of the concepts of atom and atomic permutations and combinations in physics there also was a similar development of ideas in the area of chemistry. In ancient India, chemistry was called Rasayan Shastra, Rasa-Vidya, Rasatantra and Rasakriya all of which roughly mean 'Science of liquids'. There also existed chemical laboratories and chemicals works, which were called Rasakriya-nagaram and Rasakriya-shala which literally mean 'School where liquids are activated'.

The science of living organisms, or biology reflected only in the works of Greeks in earlier times. The most successful scientific writings in this field were those of Aristotle. He was a careful and meticulous observer who fascinated by the task of classifying animal species and arranging them into hierarchies. He dealt with over five hundred animal species in this way and dissected nearly fifty of them. His mode of classification was reasonable and, in some cases, strikingly modern. His ideas led to the development of a rational theory of evolution, as put forward by Darwin, 2200 years after the time of Aristotle. The scientific study of plant life or Botany began with tribal efforts to identify edible, medicinal and poisonous plants, making it one of the oldest sciences. From this ancient interest in plants, the scope of Botany has increased to include the study of over 550,000 species of living organisms. Again, it was a Greek philosopher, Theophrastus, a pupil of Aristotle, is regarded as the founder of Botany. He covered most aspects of Botany: descriptions of plants, classification, plant distribution, propagation, germination, and cultivation. He distinguished between two main groups of flowering plants—dicotyledons and monocotyledons in modern terms—and between flowering plants and cone-bearing trees (angiosperms and gymnosperms). Theophrastus classified plants into trees, shrubs, under-shrubs, and herbs. He described and discussed more than 500 species and varieties of plants from lands bordering the Atlantic and Mediterranean.

Where the ancient Indians held the great concern for medicine, the ancient Chinese did the same for pharmacology.[54] The '*Charaka Samhita*' of Acharya Charak is among the earliest surviving Sanskrit medical manuals, and the most authoritative. In "*Charak Samhita*" Acharya Charak revealed the facts on human anatomy, embryology, pharmacology, blood circulation and various diseases. In it, he has described the medicinal qualities and functions of 100,000 herbal plants. *Charaka Samhita* was translated into Persian and Arabic in eighth century, and is still used today. It represented a major advance over the superstitious ways of treating medical problems. Instead of appeasing deities and making sacrifices, practitioners were now looking at clinical problems and deciding how to treat them based on the specific disease. They developed concepts of health and disease which they applied in practice. The science of medicine became known as Ayurveda, or science of life. Unlike the germ theory which developed much later, Ayurveda proposed that human disorders arose from an imbalance of three vital substances, or humours, present in every living creature: wind, bile, and phlegm (that is, mucus). Besides, the Instruments described in the *Sushruta Samhita* were further modified in the Western World. Many Indian ideas of medicine were incorporated into the Unani system of medicine of the Arabs. Among Chinese, Shen Kuo documented and systematically described hundreds of different plants, agricultural crops, rare vegetation, animals, and minerals found in China.[55] In the year 1070 Su Song and a team of scholars compiled and edited the *Bencao Tujing* ('Illustrated Pharmacopoeia', original source material from 1058–1061), which was a groundbreaking treatise on pharmaceutical botany, zoology, and mineralogy. It also includes valuable information on metallurgy and the steel and iron industries during eleventh century China. He created a systematic approach to listing various different minerals and their use in medicinal concoctions, such as all the variously known forms of mica that could be used to cure ills through digestion. Su made systematic descriptions of animals and the environmental regions they could be found, such as different species of freshwater, marine, and shore crabs.

Technology

The ancient civilizations had acquired proficiency in various technological fields as, for instance, metallurgy, Aviation, Civil Engineering and Architecture. Particularly, the Ancient Chinese are known for not only their travels and trade, but also for their many inventions and technological advancements. The civil engineering and architecture blossomed to a highly precise science; this found its expression in innumerable monuments of ancient India, as indicated by the discovery of urban settlements of Mohenjo-daro and Harappa. The Indians also excelled in the extraction of metals from ore, and also in the casting of metals. Though Indians could have had borrowed the idea of smelting metals from an outside source, they seem to have had used metals in warfare from around 1500 B.C. when the Aryans are said to have invaded the Indus Valley cities. The next definite reference to the use of metals by Indian soldiers is by the Greeks. The Greek historian Herodotus has observed in the fifth century that "Indians in the Persian army used arrows tipped with iron". Indian steel and iron were reportedly being used by the Romans for manufacturing armour as well as cutlery. But these references apart, it is in India itself that we find actual objects that reflect the advancement of the technique of smelting. Ashoka pillar at Mehrauli, New Delhi and another iron pillar in Karnataka stand proof of India's metallurgical heritage. Another instance of Indian metallurgy is the copper statue of Gautama Buddha found at Sultan Ganj in Bihar. This testifies to the fact that Indian metallurgy had reached a high stage of perfection more than 1500 years ago. Through his innovative discoveries, Acharya Bharadwaj has been recognized as the pioneer of aviation technology. His "Yantra Sarvasva" has described three categories of flying machines: (1) One that flies on earth from one place to another. (2) One that travels from one planet to another. (3) One that travels from one universe to another. His brilliance in aviation technology is further reflected through techniques described by him:

1. *Profound Secret:* The technique to make a flying machine invisible through the application of sunlight and wind force;
2. *Living Secret:* The technique to make an invisible space machine visible through the application of electrical force;
3. *Secret of Eavesdropping:* The technique to listen to a conversation in another plane; and
4. *Visual Secrets:* The technique to see what's happening inside another plane.

Pinhole Camera, Water Clock, Armillary Sphere, Seismograph, Clock Tower, Magnetic Needle Compass, Movable type printing, Odometer and South Pointing Chariot were among several Scientific Instruments and Appliances invented by the ancient and medival people. Shen Kuo experimented with the pinhole camera in the fourth century B.C. Although the Iraqi Muslim scientist Ibn al-Haytham was the first to experiment with camera obscura, Shen Kuo was the first to attribute geometrical and quantitative properties to the camera obscura.[56] The Water clock or Clepsydra (a timekeeping device) was in use in China as long ago as the Shang Dynasty. The inflow clepsydra with an indicator rod on a float had been known in China since the beginning of the Han Dynasty in 202 B.C. and had replaced the outflow type. The Han Chinese noted the problem with the falling pressure head in the reservoir, which slowed the timekeeping of the device as the inflow vessel was filled. Zhang Heng was the first to address this problem, by adding an extra compensating tank between the reservoir and the inflow vessel.[57] The Greek astronomer Eratosthenes invented the first armillary sphere in 255 B.C.[58]. The Chinese armillary sphere was also fully developed by 52 B.C. by Zhang Heng. Zhang Heng's water-powered armillary sphere had profound effects on the Chinese astronomy and mechanical engineering in later generations. His model and its complex use of gears greatly influenced the water-powered instruments of later astronomers. In 132 B.C. , Zhang Heng also presented to the Han court what many historians consider to be his most impressive invention, the first seismometer. It was named *Houfeng didong yi* (instrument for measuring the seasonal winds and the movements of the Earth), and it was able to determine the exact direction of tremors and earthquakes.[59] Shen Kuo's writing on magnetic compasses has proved invaluable for understanding China's earliest use of the compass for seafaring navigation.[60] An obscure commoner and artisan known as Bi Sheng (990–1051 A.D.) invented ceramic movable typeprinting.[61] Bi Sheng's methodical innovation was revolutionary for his time. By fifteenth century, metal movable type printing was developed in Ming Dynasty China and was widely applied. The European printing press, first invented by Johannes Gutenberg (1398–1468), was eventually wholly adopted as the standard in China, yet the tradition of woodblock printing remains popular in East Asian countries still. Zhang Heng is often credited with inventing the first odometer[62] as well. Similar devices were used by the Roman and Han-Chinese empires at about the same period. The South Pointing Chariot was another mechanical device credited to Zhang Heng. It was a non-magnetic compass vehicle in the form of a two-wheeled chariot. Differential gears driven by the chariot's wheels allowed a wooden figurine (in the shape of a Chinese state minister) to constantly point to the south, hence its name.[63]

Travels and Explorations

The travels and explorations fascinated the Europeans and Chinese alike. Hanno the Navigator, for instance, had travelled as far south as Sierra Leone and circumnavigated Africa. In the fourth century B.C. the Greek explorer Pytheas travelled through north-west Europe and circled the British Isles. The explorations of Alexander the Great conquered much of what was then the civilized world and brought Greek ideas and the Greek way of doing things to all the countries he conquered. The discovery of the rest of the world by Chinese travellers is an aspect of the history of geography that is often overlooked in Western writings. Travels beyond Chinese horizons were undertaken by orders of an emperor or by missionaries or traders. The ancient travel books here are known as the

'Travels of *Emperor Mu'*, who ruled from 1001 to 945 B.C. Emperor Mu, it is recorded, had the ambition to travel all-round the world and to leave the marks of the chariot wheels on every land. These books, which may be compared with that of *Odyssey* of Homer (the ancient Greek scholar), also reveal the story of high adventures. There is no doubt that at a very early date, the Chinese travellers had gone far beyond their original cultural hearth area. The discovery of Mediterranean civilization is credited to Chinese traveller Zhang Qian in 128 B.C. Zhang Qian visited directly the kingdom of Dayuan in Ferghana[64], the territories of the Yuezhi in Transoxiana, the Bactrian country of Daxia with it remnants of Graeco-Bactrian rule, and Kangju. He also made reports on neighbouring countries that he did not visit, such as Anxi (Arsacid territories), Tiaozhi (Mesopotamia), Shendu (India) and the Wusun. Zhang Qian also reports about the existence of India (*Shendu*) south-east of Bactria. The name comes from the Sanskrit word "Sindhu", used for the province of Sindh (now a province of Pakistan) by its local people. Zhang Qian also visited directly the area of Sogdiana (Kangju), home to the Sogdian nomads. From his missions he brought back many important products, the most important being alfalfa seeds (for growing horse fodder), strong horses with hard hooves, and knowledge of the extensive existence of new products, peoples, and technologies of the outside world. Zhang Qian's journeys had promoted a great variety of economic and cultural exchanges between the Han Dynasty and the Western Regions. Since silk was the dominant product traded over this route, the trade link became known as the Silk Route.[65] Among the other Chinese travellers one most distinguished was the Buddhist monk, Huan Tsang. In the seventh century after Christ he was able to cross the high windswept plateau of Tibet and the world's highest mountain on the way to India. In 1296 A.D. the Chinese traveller, Chou Ta Kuan, visited Combodia and wrote a detailed account of the strange customs of the Combodians. By thirteenth century A.D. the Chinese merchants were sailing their junks to Java and Malaya and even as far as India. Cheng Ho, a Chinese admiral, between 1405 and 1433 A.D., carried out major work of exploration in this direction. His voyages opened regular trade routes to Java, Sumatra, Malaya, Ceylon and the west coast of India. He also reached the Persian Gulf, the Red Sea, and the east coast of Africa, south of the equator. In the east he went as far as Taiwan.

Thus, with the efforts of the European, Indian and Chinese scholars, a large amount of geographical knowledge was accumulated. New researches were thereby attracted and the research made progress. With these three parallel schools of thought, geography witnessed a great degree of professionalization. With time, however, the Graeco-Roman (European) school started becoming dominant and it developed new methods that come to be regarded as more interesting or significant. With time the West became the centre of continuously increasing scholarly research. In other words, geography witnessed its first paradigmatic phase of development with the Graeco-Roman School.

NOTES

1. Grayling, A.C. (1998).
2. Heinrich Robert Zimmer (1890–1943). was an Indologist and historian of South Asian art, most known for his works, *Myths and Symbols in Indian Art and Civilization* and *Philosophies of India*.
3. Encyclopedia of Philosophy, Volume 3, (1967), Macmillan, Inc.
4. Müller (1899), Chapter 7, "Yoga Philosophy", p. 104.
5. Stanford Encycopedia of Philosophy (Internet).

6. Barnes, Jonathan (1982).
7. Husain, 1988.
8. *Bhagavata Purana* also known as *Śrīmad Bhāgavata Mahā Purāṇa, Śrīmad Bhāgavatam* or *Bhāgavata,* is one of the Hinduism's eighteen great Puranas (*Mahapuranas,* great histories). Composed in Sanskrit and available in almost all Indian languages, it promotes *bhakti* (devotion) to Krishna integrating themes from the Advaita (monism) philosophy and from the Dvaita (dualism) philosophy. The *Bhagavata Purana,* discusses a wide range of topics including Cosmology, Genealogy, Geography, Mythology, legend, Music, Dance, Yoga and Culture.
9. Crespigny (2007). 1050.
10. Hugh Thurston, (1996), pp. 178–189.
11. Katz (2007). 309.
12. Sivin (1995). III, 14-15.
13. The European Space Agency (ESA), established in (1975), is an inter-governmental organization dedicated to the exploration of space, currently with 18 member states. Headquartered in Paris.
14. Howard Eves (1990), p. 237.
15. Boyer, Carl B. (1991).
16. Jewish Encyclopedia.
17. They were a tribal and nomadic peoples living in the far reaches of Euro-Asia in hostile steppe lands barely scratching out a living. Their culture was oriented around warfare, and they were very good at it. Somewhere in the early centuries of the second millenium B.C., they began to migrate southwards in waves of steady conquest across the face of Persia and the lands of India. There, they would take on the name "superior" or "noble" to distinguish themselves from the people they conquered. Their name is derived from the Indo-European root word, "ar," meaning "noble." In Sanskrit, they were the "Aryas" ("Aryans").
18. Sivin (1995). III, 15.
19. Needham, Volume 3, 512.
20. Sivin (1995). III, 17.
21. Needham (1986). Volume 3, 413–414.
22. Sivin (1995), III, 24.
23. Loewe (1988). 509, 513, 515.
24. Neehdam (1986). Volume 4, Part 2, 484.
25. Hsu (1993). 98.
26. Needham, Volume 3, 517.
27. Hsu (1993), 98.
28. Filliozat (2004), pp. 140–143.
29. Hayashi (2005), p. 363.
30. Joseph (2000), p. 229.
31. Katz (1995), pp. 173–174.
32. Seidenberg. 1978.
33. Sanchez, Julio; Canton, Maria P. (2007), p. 37.
34. Encyclopedia Britannica.
35. Chopra, (1982), pp. 52-54.
36. Ifrah (2000), p. 346.
37. Hayashi (2003), p. 119.
38. Filliozat (2004), pp. 143–144.
39. Joseph, Brian (Ed.); Janda, Richard (Ed.) (2008), p. 163.
40. Ifrah (2000), p. 449.
41. Bourbaki (1998), p. 46.
42. Boyer, Carl Benjamin (1949). Chopra (1982) pp. 52–54.
43. Needham, Volume 3, 500.
44. Hsu (1993). 96–97.
45. Crespigny (2007), 1050.
46. Needham (1986), Volume 4, Part 3, 352; Ruitenbeek (1996), 26.

47. Hsu (1993), 102.
48. Finel, Irving (1995), pp. 26–27.
49. Needham, Volume 3, 501.
50. Dayuan lies south-west of the territory of the Xiongnu, some 10,000 *li* (5,000 kilometres) directly west of China. The people are settled on the land, plowing the fields and growing rice and wheat. They also make wine out of grapes. The people live in houses in fortified cities, there being some seventy or more cities of various sizes in the region.
51. Crespigny (2007), p. 1222.
52. Kurt A. Raaflaub & Richard J. A. Talbert (2009), p. 147.
53. Huffman (1993), pp. 307-23.
54. Pharmacology is not synonymous with pharmacy, which is the name used for a profession, though in common usage the two terms are confused at times. Pharmacology deals with how drugs interact within biological systems to affect function. It is the study of drugs, of the body's reaction to drugs, the sources of drugs, their nature, and their properties. In contrast, pharmacy is a medical science concerned with the safe and effective use of medicines.
55. Crespigny (2007). 1050; Loewe (1968), 107.
56. Sivin (1995). III, 22.
57. Wu (1943). 211-212.
58. An armillary sphere is basically a skeletal celestial sphere with a model of the Earth or, later, of the Sun placed in the centre. Armillary spheres were developed by the Greeks in antiquity for use as teaching tools. In larger and more precise forms they were also used as observational instruments, being preferred by Ptolemy. Armillary spheres became popular again in the late middle ages. With the advent of the Copernican model of a Sun centred Universe pairs of spheres contrasting the Copernican and Ptolomaic models became common teaching/demonstration tools. Such small teaching sphere remained popular through the nineteenth century. The Armillary sphere consists of two major components, the sphere and the stand, as seen in the figure above. The central body in the sphere represents the Earth, which was, of course, considered the centre of the Universe.
59. Yan (2007). 128.
60. Wolf, Eric (1994), p. 227.
61. Cartledge, Paul. (1997).
62. An odometer (mileometer, milometer) indicates distance travelled by a car or other vehicle. The device may be electronic, mechanical, or a combination of the two. The word derives from the Greek words "hodós", meaning "path" or "way", and "métron", "measure".
63. Huffman (1993), 307-23.
64. Dayuan lies south-west of the territory of the Xiongnu, some 10,000 *li* (5,000 kilometres) directly west of China. The people are settled on the land, plowing the fields and growing rice and wheat. They also make wine out of grapes. The people live in houses in fortified cities, there being some seventy or more cities of various sizes in the region. The population numbers several hundred thousand (Shiji, 123, Zhang Qian quote, trans. Burton Watson).
65. Crespigny (2007), 1222.

REFERENCES

Bourbaki, Nicolas (1998). *Elements of the History of Mathematics,* Berlin, Heidelberg, and New York: Springer-Verlag, ISBN 3540647678.

Bowman, John S. (2000). *Columbia Chronologies of Asian History and Culture.* New York: Columbia University Press.

Boyer, Carl Benjamin (1949). *The History of the Calculus and its Conceptual Development.* Hafner. Dover edition (1959), ISBN 0-486-60509-4.

Cartledge, Paul, (1997). *The Cambridge Illustrated History of Ancient Greece,* Cambridge University Press.

Chan, Alan Kam-leung and Gregory K. Clancey, Hui-Chieh Loy (2002). *Historical Perspectives on East Asian Science, Technology and Medicine*. Singapore: Singapore University Press. ISBN 9971-69-259-7.

Chopra, Pran Nath (1982). *Religions and communities of India*, Vision Books, ISBN 9780856920813.

Crespigny, Rafe de. (2007). *A Biographical Dictionary of Later Han to the Three Kingdoms (23-220 CE)*. Leiden: Koninklijke Brill. ISBN 90-04-15605-4.

Ebrey, Patricia Buckley, Anne Walthall, and James B. Palais (2006). *East Asia: A Cultural, Social, and Political History*. Boston: Houghton Mifflin Company. ISBN 0-618-13384-4.

Encyclopedia Britannica.

Encyclopedia of Philosophy, Volume 3, (1967), Macmillan, Inc.

Filliozat, Pierre-Sylvain (2004). "Ancient Sanskrit Mathematics: An Oral Tradition and a Written Literature", in Chemla, Karine; Cohen, Robert S.; Renn, Jürgen *et al.*, *History of Science, History of Text (Boston Series in the Philosophy of Science)*, Dordrecht: Springer Netherlands.

Grayling, A.C. (1998). *Philosophy: A Guide through the Subject*, Oxford University Press.

Hayashi, Takao (2003). "Indian Mathematics", in Grattan-Guinness, Ivor, *Companion Encyclopedia of the History and Philosophy of the Mathematical Sciences*, 1, pp. 118-130, Baltimore, MD: The Johns Hopkins University Press.

Hayashi, Takao (2005). "Indian Mathematics", in Flood, Gavin, *The Blackwell Companion to Hinduism*, Oxford: Basil Blackwell.

'*History of China*', Wikipedia, the Free Encyclopedia.

Howard Eves (1990). *An Introduction to the History of Mathematics* (6 ed.). Saunders College Publishing House, New York.

Hugh Thurston (1996). *Early Astronomy*, New York: Springer-Verlag.

Hsu, Mei-ling. "The Qin Maps: A Clue to Later Chinese Cartographic Development," *Imago Mundi* (Volume 45, 1993): 90-100.

Huffman, Carl A. (1993). *Philolaus of Croton: Pythagorean and Presocratic*, Cambridge University Press.

Husain, M. (1988). *Evolution of geographic Thought*, Jaipur: Rawat Publications.

Ifrah, G. (2000): *The Universal History of Numbers*. New York: John Wiley & Sons.

Irving L. Finkel, (1995). "A join to the Map of the World: a notable discovery", British Museum Magazine: *The Journal of the British Museum Friends*, pp. 26-27.

Jewish Encyclopedia.

Joseph, Brian (Ed.); Janda, Richard (Ed.) (2008). *The Handbook of Historical Linguistics*. Blackwell Publishing (published 30 December 2004). p. 163. ISBN 978-14051274.

Joseph, G.G. (2000). *The Crest of the Peacock: The Non-European Roots of Mathematics*, Princeton, NJ: Princeton University Press.

Katz, Victor J. (1995). "Ideas of Calculus in Islam and India", *Mathematics Magazine (Math. Assoc. Amer.)* 68 (3): 163–174.

Katz, Victor J. (2007). *The Mathematics of Egypt, Mesopotamia, China, India, and Islam: A Sourcebook*. Princeton: Princeton University Press. ISBN 0-691-11485-4.

Kurt A. Raaflaub and Richard J. A. Talbert (2009). *Geography and Ethnography: Perceptions of the World in Pre-Modern Societies*, John Wiley & Sons, ISBN 1405191465.

Loewe, Michael. "The Oracles of the Clouds and the Winds", *Bulletin of the School of Oriental and African Studies, University of London* (Volume 51, Number 3, 1988): 500–520.

Müller, Max (1899). *Six Systems of Indian Philosophy; Samkhya and Yoga, Naya and Vaiseshika*. Calcutta: Susil Gupta (India) Ltd. ISBN 0-7661-4296-5.

Needham, Joseph (1986). *Science and Civilization in China: Volume 1, Introductory Orientations*. Taipei: Caves Books Ltd.

Needham, Joseph (1986). *Science and Civilization in China: Volume 3, Mathematics and the Sciences of the Heavens and the Earth*. Taipei: Caves Books Ltd.

Needham, Joseph (1986). *Science and Civilization in China: Volume 4, Physics and Physical Technology, Part 1, Physics*. Taipei: Caves Books Ltd.

Needham, Joseph (1986). *Science and Civilization in China: Volume 4, Physics and Physical Technology, Part 2: Mechanical Engineering*. Taipei: Caves Books Ltd.

Needham, Joseph (1986). *Science and Civilization in China:* Volume 4, *Physics and Physical Technology, Part 3, Civil Engineering and Nautics*. Taipei: Caves Books Ltd.

Needham, Joseph (1986). *Science and Civilization in China: Volume 5, Chemistry and Chemical Technology, Part 1: Paper and Printing*. Taipei: Caves Books Ltd.

Needham, Joseph (1986). *Science and Civilization in China: Volume 5, Chemistry and Chemical Technology, Part 7, Military Technology; the Gunpowder Epic*. Taipei: Caves Books Ltd.

Needham, Joseph (1986). *Science and Civilization in China: Volume 6, Biology and Biological Technology, Part 1, Botany*. Taipei: Caves Books Ltd.

Pingree, David (1992). "Hellenophilia versus the History of Science", *Isis* 83 (4).

Plofker, Kim (2009). *Mathematics in India: 500 BCE–1800 CE*, Princeton, NJ: Princeton University Press. ISBN 0691120676.

Rana, L. (2014). *Ancient Tradition in Geography: The Western & Oriental Perspective*, Classical Publishing Company, Karampura, New Delhi.

Seidenberg, (1978). "The origin of mathematics". *Archive for the history of Exact Sciences*, Vol. 18.

Shu-Chiu LIU (2006). "Historical models and science instruction: A cross-cultural analysis based on students' views", *Asia-Pacific Forum on Science Learning and Teaching*, Volume 7, Issue 2, Article 1 (December, 2006).

Sivin, Nathan (1995). *Science in Ancient China: Researches and Reflections*. Brookfield, Vermont: VARIORUM, Ashgate Publishing.

Stanford Encycopedia of Philosophy (Internet).

Yan, Hong-sen. (2007). *Reconstruction Designs of Lost Ancient Chinese Machinery*. Dordrecht: Springer. ISBN 1402064594.

5

Graeco-Roman School (Paradigm Phase I)

The Graeco-Roman School represents the third phase of development from about 800 B.C. to 500 A.D., i.e. 'Paradigm Phase-1'. As a dominating school, the Greeks and Romans displaced others in quite a short space of time. A paradigm was established leading to concentrated research in geography, an activity described as 'Normal Science'. The Greeks were the first known culture to have actively explored geography as a science. While maps had been made millennia earlier in Sumer, the Greeks were the first to attempt to fully and systematically describe the world. It was the Greeks who developed the procedures to be described initially as 'scientific methods'. Greeks first handled the information gathered by the middle-Easters scientifically. The developments in Greek School were followed by that in Roman. Although, similar and parallel developments had taken place in ancient Indian and Chinese realms as well, but with time Graeco-Roman School began to dominate. The Greek colonies had established and flourished during 8th and sixth century B.C. This period, particularly, is known as the 'Golden period of Greeks'. During this period, the city of Miletus had become the main centre of geographical learning. In fact, geography, as a discipline, has grown out of three simultaneous and related activities, viz. exploration, charting and mapping, and speculation. The exploration led to gathering of facts about earth's surface; charting and mapping aided in representation of this information; and speculation helped in examining, analyzing and drawing conclusions out of the gathered information. First two of the activities, were common among the other civilizations of the time as well, but the ability to speculate the facts was observed only among Greek scholars. Therefore, it is Greeks that the real beginning of our discipline is marked with. Thus, the first amazing period of intellectual ferment that became the part of the written tradition of the Western World was that of Greeks and Romans. They are credited to have begun, the 'mathematical' and 'literary' traditions in geography in true sense. Besides 'geography', a number of other terms, which are part of present geographical vocabulary, were given by the Greek and Roman scholars. Although many of the conceptions as put forward by the ancient Greeks were based on what they had inherited from earlier civilizations, but their own contributions were more far-reaching and valuable in themselves. Greeks distinguished themselves from others because of their intellectual, philosophical and scientific approach. The Romans, in general, carried forward the Greek tradition. Moreover, many of the Roman scholars as Ptolemy, Strabo, Polybius etc., wrote in Greek itself. After about 120 B.C., the Greek Science started loosing its originality. Little of worth was produced after 200 A.D. Due to rise of Roman Empire about first century A.D., the interest in Greek literature declined. Even the Roman Empire had collapsed by 500 A.D. The decline in morals and values, deterioration in public health, environmental problems, terrible inflation, political corruption, unemployment, urban decay, inferior technology etc. were among many reasons for this fall. The final blow came in the form of invasion by the barbarians, the Germanic tribes. All this destroyed the social base of Greek and Roman culture. In spite of all odds, the Roman scholars were able to extend further the Greek literary and mathematical tradition. Strabo, for instance, compiled the regional and historical geography of the world; Ptolemy commendably contributed to practical geography; and Polybius enriched the field of physical geography. A number of Greek centres of learning became popular during ancient period, as Athens, Miletus, Halicarnassus, Corinth, Argos, Thebes, Sparta, Amphipolis, Methone, Potidaea etc.

Ancient Greece, in fact, is the term used to describe the Greek-speaking world in ancient times. It refers not only to the geographical peninsula of modern Greece, but also to areas of Hellenistic[1] culture that were settled in ancient times by Greeks. It covered Cyprus, the Aegean coast of Turkey (then known as Ionia), Sicily and southern Italy, and the scattered Greek settlements on the coasts of what are now Albania, Bulgaria, Egypt, Libya, southern France, southern, east and northeast Spain, Georgia, Romania and Ukraine.[2] Ancient Greece is the period in Greek history lasting for close to a millennium, until the rise of Christianity. It is considered by most historians to be the foundational culture of Western Civilization. The civilization of the ancient Greeks has been immensely influential on the language, politics, educational systems, philosophy, science, art, and architecture of the modern world, fuelling the Renaissance in Western Europe and again resurgent during various neo-Classical revivals in eighteenth and nineteenth century Europe and the Americas. The ancient Greeks saw the poet Homer as the founder of geography. His works the *Iliad* and the *Odyssey*, although, are works of literature but contain a great deal of geographical information. The works show that the Greeks by eighth century B.C. had considerable knowledge of geography of eastern Mediterranean. The poems contain a large number of place names and descriptions. Among others, Thales of Miletus is known to have wondered about the shape of world. He proposed that the world is based on water, and that all things have grown out of it. He also laid down many of the astronomical and mathematical rules that would allow geography to be studied scientifically. His successor Anaximander was the first person known to have attempted to create a scale map of existing world and to have introduced the Gnomon (the instrument functioning like a Sun-dial) to Ancient Greece. Hecateus of Miletus initiated a different form of geography as he wrote a detailed prose account of known world. A similar work, and one that mostly survives today, is Herodotus' 'Histories'. While primarily a work of history, the book contains a wealth of geographic descriptions covering much of the familiar world. Herodotus was the first to have noted the process by which large rivers, such as the Nile, build up deltas, and was also the first to have observed that winds tend to blow from colder regions to warmer ones. The size and shape of Earth were important questions to ancient Greeks. Eratosthenes, for instance, attempted to calculate its circumference. Some of the figures Eratosthenes had used in his calculation were considerably in error and Posidonius set out to get a more accurate measurement. Also Eratosthenes was the one who coined the term 'geography' (from 'geo' meaning 'the earth' and 'graphe' meaning 'the description'). Pythagoras, a sixth century B.C. mathematician and philosopher, was the first to propose a spherical world, arguing that sphere was the most perfect form. Plato embraced this idea and Aristotle presented evidence to verify this. Eudoxus of Cnidus used the idea of a sphere to explain how the sun created differing climatic zones based on latitude. This led the Greeks to believe in a division of world into five regions. Besides, many of the analytic procedures were set forth in the writings of Plato and Aristotle. Plato developed the deductive approach related to logical derivations from literature or theory; whereas Aristotle highlighted the inductive method, dealing with information empirically observed. The theories of Greek scholars clashed with the evidence of explorers. Hanno the Navigator had travelled as far south as Sierra Leone. In fourth century B.C. the Greek explorer Pytheas travelled through north-west Europe and circled the British Isles. He found that the region was considerably more habitable than theory expected. The conquerors also carried out exploration as, for example, Alexander the Great. He set out to learn more about world through his military expeditions and took a large number of geographers and writers with his army who recorded their observations as they moved east.

Roman civilization is often grouped into classical antiquity with ancient Greece. The Roman Empire is the term conventionally used to describe the Ancient Roman society. Ancient Rome was a civilization that grew out of the city-state of Rome, founded in the Italian Peninsula in the nineth century B.C. Gradually, the Roman civilization converted from a monarchy to an oligarchic republic to a vast empire. It came to dominate the Western Europe and the entire area surrounding the Mediterranean Sea through conquest and assimilation after the Greeks in the beginning years of the Christian era. Ancient Rome contributed greatly to the development of law, art, literature, architecture and language in the Western World and its history continues to have a major influence on the world today. The Greek and Roman civilizations not only followed one after the other, both also had strong spatial and cultural linkages with each other. Most of the Greek accomplishments in geography were passed on to the Romans. Roman military commanders and administrators used this information to guide the expansion of their Empire. The Romans also made several important additions to geographical knowledge. Strabo (64 B.C.–20 A.D.), for instance, wrote a 17-volume series called 'Geographia'. Strabo's detailed descriptions give a clear picture of status of geographical knowledge of the time. In his series of books, Strabo described the cultural geography of various people found from Britain to India in the Far East, to Ethiopia in the south and as far as Iceland in the north. A century after Strabo another Roman scholar, Ptolemy, launched a similar undertaking. By this time the Roman Empire had expanded through much of Europe and previously unknown areas such as the British Isles had been explored. The Silk Road was also in operation, and for the first time knowledge of the Far East began to be known. During the second century A.D., Ptolemy made a number of important contributions to geography. Ptolemy' publication 'Geographike hyphegesis' or 'Guide to Geography' compiled and summarized much of the geographic information accumulated at that time. Some of his other significant contributions include the creation of different methods for projecting the Earth's surface on a map, the calculation of coordinate locations for some eight thousand places on Earth, and development of the concepts of geographical latitude and longitude. Ptolemy's 'Geographia' opens with a theoretical discussion about the nature and techniques of geographical inquiry, and then moves to detailed descriptions of the known world. Ptolemy lists a huge number of cities, tribes, and sites and locates them on world map. For the most part ancient Greek geography was an academic field. It was the Romans who made far more extensive practical use of geography and maps. The contribution of ancient Greeks and Romans to various fields of geographical study ranged from physical, human, historical, theoretical, political, regional, mathematical and practical, thus, covering almost the entire gamut of geographical knowledge. Besides, Geography as a discipline got enriched by a number of concepts, viewpoints, and techniques and various terminologies were added to its vocabulary by ancient Greek scholars. Along with coining the term 'geography' for our discipline, a number of other terminologies are also credited to ancient Greeks.

When the Egyptian and Babylonian civilizations had stagnated and declined, there arose in the Mediterranean coast an energetic and inquisitive sea faring group of people, the Greeks. The Greeks assimilated technology and knowledge from almost all cultures they encountered—picking and choosing the best. The main feature of Greek thought, therefore, represents the mathematical and astronomical knowledge of the Assyrians, Babylonians, Egyptians and others. They studied, mapped, chartered and also analyzed it. The Greeks were the pioneers in having and transmitting the knowledge of geographical nature.

Contributions

In the initial phase of its development, geography was enriched by a number of fields by the ancient Greek and Roman scholars. The various areas of geographical enquiry of these scholars particularly included the regional, mathematical and practical, historical, theoretical and physical geography.

Regional Geography

The field of regional geography was largely contributed by Hecateus, Alexander the Great, Pytheas and Strabo. As derived from the work of these scholars, the existing world of the times was believed to have two major regional divisions, namely Europe and Asia. '*Ges Periodos*', the literary work of Hecateus, is actually regional description of these two earliest known continents. Besides, parts of northern Africa had also been discovered. The military and armed explorations of Alexander the Great, the king of Greece, added substantially to the field of regional geography. His conquests pushed Greek knowledge of the earth far to the east as far as and across the Indus River, and southward along the eastern side of Mediterranean to Egypt, where he established his rule. A large part of the Mediterranean region became the part of his kingdom. He founded the city of 'Alexandria' in 332 B.C., which later emerged as one of the great commercial and intellectual centres of the ancient world. Alexander's staff, while travelling, included writers to describe the lands they crossed, and astronomers to make observations. Although, the primary objective of Alexander's conquests was to expand Greek territory, he sent back to the Greek world a wealth of new observations concerning what it was like beyond and they proved highly beneficial to the field of geography. With his accounts the Greeks became familiar with the natural features of Persian Empire, Central Asia, Afghanistan, India and coastal parts of Iran. While Alexander was extending Greek geographic horizons to the east, another Greek explorer was voyaging to the north-west of the Greek world, covering parts of Western and Northern Europe. This was Pytheas, who is known to have made remarkable voyages between 330-300 B.C. Pytheas described about the customs of people of Britain in detail. He observed the change in character of agriculture from south to north in Great Britain. His writings contain a reference of a sea full of ice that could neither be traversed on foot nor in a boat. He has mentioned a place north of Norway, where the length of the longest day was between 17 and 19 hours, and where the Sun remained above horizon during whole of the longest day. He was also the first Greek to tell about ocean-tides and their relation to the various phases of moon. Thus, Pytheas reported things that were so contrary to Greek experience that the scholars of his day discredited him and treated his important observations and information as pure fantasies. However, today we acknowledge him as a great scholar of the ancient period.

For Strabo, the Roman scholar, geography was not merely an aid in the task of government, as it was believed earlier; but it was equally important to an individual too. Because, in his view, geography acquainted man with the habitable land, with oceans, with vegetation, fruits, and with the various socializations of earth's surface etc. that helped in identifying the various related problems and happiness of the man. He called all these elements as 'natural attributes of a place', and established their relationship with man and other places of the surface of the earth. This concept is still the basic idea in all geographical studies and has a firm and popular place in present day geography. These natural attributes of the place distinguish it from other places and regions. These ideas underlie the "Areal Differentiation" viewpoint, as presented later in geographical studies by Richard Hartshorne, a twentieth century American scholar. Strabo considered 'Place' in two different perspectives, viz. in terms of its astronomical (abstract) location and in terms of its terrestrial (on land) position.

Mathematical and Practical Geography

The development of mathematical and practical geography in this period is linked primarily to Thales, Anaximander, Eratosthenes, Hipparchus, Posidonius, Aristotle and Ptolemy. Thales and Anaximander, particularly, are known as the originators of mathematical tradition in geography. Greeks gave the system of latitudes and longitudes and proved that the earth is a sphere. The first of the Greek scholars to be concerned about the measurement and the location of things on the face of the earth was Thales. He was very sharp in understanding mathematical and geographical regularities. There are six geometric propositions credited to him. They are: (1) The circle is divided into two equal parts by the diameter; (2) The angles at either end of the base of an isosceles triangle are equal; (3) When two parallel lines are crossed diagonally by a straight line, the opposite angles are equal; (4) the angle in a semi-circle is a right angle; (5) the sides of similar triangles are proportional; and (6) two triangles are congruent if they have two angles and a side respectively equal. Thales was first to apply various geometrical concepts (circles, diameter, angles etc.) in the geographical studies. Anaximander is known for the introduction into the Greek world of a Babylonian instrument called 'Gnomon' (an instrument having functions like a sundial). It was simply a pole set vertically above a flat surface on which the varying positions of the earth could be measured by the length and direction of the shadow cast by the vertical pole. With the help of Gnomon, it was possible to make a variety of observations regarding noontime, duration of sunshine, time of solstices and equinoxes, the time of sunrise and sunshine and so on.

Eratosthenes and Posidonius, both, further extended the mathematical tradition in geography, but the works of former were more elaborate and substantial. The common contribution of the two included calculation of circumference of the earth for the first time in history. However, the calculations of Eratosthenes had remarkable accuracy. His findings were based on the use of 'Gnomon' and theorems of Thales. Hipparchus, another mathematician and astronomer of this period, attempted to determine exact positions of every point on the earth's surface. He was first to divide circle into 360 degrees, based on Assyrian mathematics. Hipparchus defined a grid of latitudes and longitudes. The equator, he pointed out, was a great circle (one that divides the earth into two equal parts) and meridians, drawn converging at the poles, are also great circles. The parallels become shorter and shorter as they approach the poles. Since the earth makes one complete revolution in 24 hours and there are 360 meridians drawn from equator to poles, each hour the earth turns through 15 degrees of longitudes. Hipparchus was also the first to wrestle with the problem of showing the curved surface of the earth on a flat surface.

Aristotle, whose ideas were later adopted by Eratosthenes and others, was of the view that the Earth was a sphere, placed in the centre of the universe around which the celestial bodies revolve. This misconception prevailed for centuries. However, during the time of Aristotle, the sphericity of the earth had been proved, latitudes and longitudes of many places had been calculated and the practice of plotting world maps on a grid had begun. He was the first philosopher who wrote with definite arguments about the spherical shape of the earth[3]. The arguments given in support of his statement about the sphericity of the earth were:

1. He deduced it from the law of gravitation, or as he expressed it, the tendency of all things towards the centre. By the action of this, when the earth was in the course of formation, and the component elements were coming together, the mass thus formed by acceleration was so constituted that its entire circumference must be equidistant from its centre.
2. He inferred the shape from what is seen to take place in lunar eclipse, i.e. the shadow falls on moon is spherical.

Among all the ancient scholars of Greek and Roman period Ptolemy was most influential. His contribution in the field of mathematical geography is highly commendable and has been acknowledged throughout ages. Ptolemy developed the sound principles of mathematics and applied to the study of practical geography. He recalculated the circumference of the earth. He designed a graticule and plotted the world using the concept of conical projection with remarkable accuracy. He worked out the dimension of the inhabitable world and plotted the places of physical and human interest of the contemporary times. Ptolemy authored two great books, of which one is 'the Syntaxis', popularly known as 'Almagest', which is a great work in classical astronomy. It long remained the standard reference work on the movement of the celestial bodies. The other book is known as *'The Outline of Geography'*. This work is devoted to the exact determination of the position of places by means of latitudes and longitudes. In the opinion of Ptolemy, geography is a science that deals with the art of map-making. This conception dominates the entire first book of Ptolemy. The basic objective of Ptolemy's attempts was to reform the map of the world, on the basis of the astronomical principles. Therefore, he was able to prepare the first complete map of the known world. Ptolemy's concept of Universe agreed with that of Aristotle: "the earth was a sphere that remained stationary in the centre while the celestial bodies moved around it in circular courses". This remained an accepted doctrine until the Age of Discovery. Another important contribution of Ptolemy was "*Guide to Geography*", the largest geographical gazetteer of the times, which contained 8 volumes and the basis on which he revised the world map. In the art of map making, the great contribution made by Ptolemy was his improvement over the previously drawn maps and he did this by adopting projections for the world map, showing the graticule of latitudes and longitudes. In fact, in regard to the mathematical construction of the projection of his maps, Ptolemy was far ahead of his predecessors. The writings and maps of Ptolemy inspired the geographers and explorers of Great Age of Discovery. With the death of Ptolemy, the geographic horizons that had been widened both physically and intellectually by the Greeks closed in again.

Historical Geography

Mainly Herodotus and Strabo contributed this field of learning. Although Herodotus is usually described as the first great historian, a large part of his work is easily identified as geography. In fact, He is credited with the very old idea that all history must be treated geographically and all geography must be treated historically. In fact, the notion of geography as "the hand-maiden of history" came from Herodotus. According to him Geography provides the physical background, the stage setting, in relation to which the historical events take on meaning. He provided some excellent examples of what we would call today 'historical geography', i.e. the re-creation of past geographies and the tracing of geographical change through time. The contributions of Herodotus are based on his personal observations during many years of travel. Strabo, on the other hand, compiled the writings of his predecessors, particularly Homer, Herodotus, Aristotle, Eratosthenes, Posidonius and Hipparchus, in his book '*Geographie*'. This work consisted of 17 volumes. It was the world's first administrative handbook as the purpose was to provide a text of information for the Roman Administrators and Military Commanders. The major part of his work is devoted to detailed description of the various parts of the known world. First two volumes are general; eight books cover Europe; six Asia; and one covers present Africa. This work is mainly historical, but Strabo also described in it—the character, physical properties, surface configuration and natural production of each country known in his times. Strabo considered geography essentially as an encyclopedic description of the known inhabited world.

Theoretical Geography

Two great Greek philosophers, namely Plato and Aristotle, made important contributions to the development of geographical ideas. They respectively were the masters of deductive and inductive reasoning. Plato insisted that the observable things on the earth were only poor copies of ideas, or perfect predicates from which observable things had degenerated or were in process of degenerating. According to him, the existing patterns related to man's activities are the result of his thinking, ideas or imagination. He argued that every system is gradually moving towards chaos, i.e. there is always degeneration of things from their original perfect state. He gave numerous examples in support of this from the world around. He always argued that in seeking explanations one should always proceed from the general theory to particular situation, i.e. to explain things in the context of some general rule or theory. Because he believed that there is a hidden cause behind every occurrence (existence). This kind of thinking is called deductive reasoning. Aristotle, on the other hand, was one of the scholars of Plato's academy. But, according to him the best way to build theory was to observe facts and the best way to test theory was to confront it with observations. Where Plato believed in building theory by reasoning from the general to the particular, Aristotle did it in the reverse order, i.e. by reasoning from the particular to the general. Thus, the latter approach may be called inductive. Aristotle recognized that observations made through senses can never provide explanations. Our senses, he said, can tell us that, e.g. 'fire is hot' but cannot tell us that 'why it is hot'. Aristotle formulated four fundamental principles of scientific explanations, i.e. of answering the question: "what makes this thing the way it is?" They are: (1) one way is to describe its nature, to tell its essential characteristics; (2) a second way is to specify the kind of matter, the substance, of which it is composed; (3) a third way is to tell what caused the process through which the thing became as it is; and (4) a fourth way, which is complementary to the third, is to tell the purpose the thing fulfills. Unlike Plato, Aristotle assumed that things were in process of physical change leading to a final perfect shape. This model for scientific explanations constituted the world's first paradigm for the guidance of scholars. The multifarious works of this great philosopher created a keen desire in his pupils to make direct observations. One of his outstanding and adventurous pupils was Alexander, who made his own heroic efforts for travelling.

Physical Geography

The Greeks made a remarkable development in the field of physical geography[4]. The work of Greeks contain numerous references on mountains, delta-building, tides, change of weather, rain, earthquakes and their causes, volcanoes and the transformation in the topographic features, besides the plant studies. The scholars like Homer, Posidonius, Herodotus, Theophratus, Plato and Aristotle made valuable contributions in this direction.

From the study of Aegean Sea, the Greeks expanded the horizon of knowledge in the study of seas and oceans distinguishing the varying properties of their coastlines, salinity, waves, tides and winds. Posidonius was the most outstanding physical geographer. He was considered as an authority on oceanography, on which he contributed a book, entitled "*The Ocean*". The studies on tides and the measurement of the depth of the sea and related aspects go to his credit.

Interestingly, Herodotus was a historian by profession, but he always placed and studied historic events in geographic setting. He introduced Greek alphabet 'delta' (Δ) to geographical dictionary, on the basis of his observations of the black soil along the Nile River. He linked the shape of silt-deposits on the flood plain to that of 'Delta'. He also corrected the misconceptions regarding Caspian Sea, for which the contemporary identification was that of a 'sea'. He supported evidences for it's being a 'lake'—with no contacts with the outside world.

Physical geography was the field of interest for Aristotle and Plato as well. Aristotle, for instance, explained the phenomenon of expansion of land in the shallow seas and formation of delta. He correctly pointed out the process of alluvial deposition by which so many places of land area being eroded and added into the sea. He affirmed this with the study of Sea of Azov, which was continuously becoming shallower and he feared that one day it would be filled up completely and would convert into a dry land. The delta formation of Nile was also attributed to the enormous silt carried by the river from its upper reaches (Ethiopia). Theophratus was a student of Aristotle. He examined the relationships of plants and climate, and this marked the beginning of the study of plant geography.

Plato discussed some of the barren lands of Attica (Greece) and explained that such waste tracts in the past were full of vegetative covers and fertile soils. Under the impact of the external forces, the forests have been depleted and soils leached resulting into barren topography. Such waste lands he said are like the skeleton of sick man, all the fat and soft earth having been wasted away, and only the bare framework of the land being left. Plato considered man also an active agent who changed the face of the earth.

The Greeks, right from the Homerian period, have recognized four major winds, having different properties, directions and the associated weather type. They are: *Boreas*: the north wind—strong, cool, with clear skies; *Eurus*: the east wind—warm and gentle; *Notus*: the south wind—wet and sometimes violent; and *Zephyrus*: the west wind—balmy (fragrant) but with gale force. Moreover, the Greeks also divided the world into: Torrid, Temperate and Frigid zones. They established a relationship between temperature and ecumene regions of the world.

Inheritance

The major contribution of Greeks and Romans was a body of philosophical and cosmological ideas, besides a number of terminologies that still constitute the basic vocabulary of geography. The ancient Greeks were the originators of literary and mathematical traditions in geography. Academically, these ancient scholars formed strong ties of geography with history and this inter-disciplinary bond could not be broken for several centuries to come, and has been intensely represented in the writings of geographers till the modern times. The roots of the concept of 'Areal differentiation' as adopted by Hartshorne in the modern period lie in the writings of Strabo, the first century B.C. Roman scholar. Ptolemy's geocentric view of the universe, i.e. "the earth was a sphere that remained stationary in the centre, while the celestial bodies moved round it in circular courses", remained an accepted doctrine until the Age of Discovery. One of the major contributions of the Greek scholars was towards suggesting the methods of theory building and hypothesis testing. Plato favoured theory building by intuition and reasoned from general to particular (the deductive approach). On contrary to it Aristotle suggested the method of reasoning from particular to general (the inductive approach). These methods of deductive and inductive reasoning were introduced to geography as early as 4th and 5th centuries B.C. The Greeks are also credited for the introduction of Gnomon, one of the first scientific instruments ever made. The Gnomon is an instrument that functions as a sundial and can be used for finding the declination of the Sun through the year, among other things.

There were both positive and negative gains from the Graeco-Roman School of thought. As far as the negative side is concerned, one of the negative views that prevailed during this period was related to 'Geo-centric Concept of the Universe', which considered the Earth to be lying at the centre of the Universe and all heavenly bodies revolving around it. This was contrary to the present day belief in the 'Helio-centric Concept of the Universe', which proves the Sun at the centre and the planets revolving around it. Moreover, no attempts were made to translate the works of the Greek or Roman writers into other languages in use. Therefore, the interest in philosophical and scientific writings declined and Geography had to face a period of recession in the centuries to come.

However, the positive gains of the period were more numerous. Geography as a discipline got enriched by a number of concepts, viewpoints, and techniques and various terminologies were added to its vocabulary particularly by the ancient Greek scholars. Besides coining the term 'geography' for our discipline, a number of other terminologies are also credited to the ancient Greeks. The important ones include, for instance, 'cartography', 'chorography', 'longitude', 'latitude', 'torrid', 'frigid', 'temperate' etc. All these terms are still in use in the modern period. Thus, the contribution of ancient Greeks and Romans to various fields of geographical study ranging from physical, human, historical, theoretical, political, regional, mathematical and practical covered almost entirely the gamut of geographical knowledge.

NOTES

1. The term 'Hellenistic' refers to the spreading of Greek culture over the non-Greek peoples that were conquered by Alexander the Great. The Hellenistic civilization was a fusion of Greek and Oriental cultures that eventually gave Christianity the opportunity to flourish. The main cultural centres expanded from mainland Greece, to Pergamon, Rhodes, Antirch and Alexandria. Modern historians see the death of Alexander the Great in 323 B.C. as the beginning of the Hellenistic period.
2. Rana, L. (2008), pp. 3-15.
3. Husain, (1988), 32.
4. *Ibid.*

REFERENCES

Husain, M. (1988). *Evolution of Geographical Thought,* Second Revised Edition, Jaipur: Rawat Publications.

Rana, L. (2008). *Geographical Thought: A Systematic Record of Evolution,* Concept Publishing Company, New Delhi.

6
Dark Age in Europe (Crisis Phase I)

The first phase of Crisis with Revolution in geography was witnessed during 500–1100 A.D., popularly known as Dark Age in Europe. The Dark Age[1] is a historical periodization traditionally referring to the Middle Ages. The term denotes an intermediate and a long period of decline between Classical Antiquity and the Modern era. However, the term does not refer to the entire medieval period; it is generally restricted to the Early Middle Age.[2] The Early Middle Ages was labeled so during the eighteenth century Age of Enlightenment, a characterization based on the relative scarcity of literary and cultural output from this time. The Dark Age is most commonly known in relation to the European Early Middle Ages. It marks the intellectual darkness characteristic of the time between the fall of Rome and the Renaissance. It asserts demographic, cultural and economic deterioration that occurred in Western Europe following the decline of the Roman Empire. The term employs traditional 'light-*versus*-darkness' imagery to contrast the era's 'darkness' with earlier and later periods of 'light' where there was relatively a lack of records in the former case in comparison to the latter having an abundance of records.[3] The period seems 'dark' from the point of the scarcity of historical record, and artistic and cultural output. The early medieval phase was seen as a period of social regress dominated by religion. By 500 A.D. the Roman Empire in the West was destroyed and this marked the onset of 'Dark Age in Europe', which continued up to 1100 A.D.

History

The death of Theodosius-I[4] in 395 was followed by the division of the empire into two. The Western Roman Empire was disunited and turned into a mosaic of warring Germanic kingdoms in the fifth century. Particularly, after third century onwards its territories were settled by succeeding waves of barbarian[5] tribal confederations. The barbarian invasions sweeping in beyond Imperial borders brought the end of the Roman Empire. Collapse of Rome was the major event of the time. Starting in the second century itself, the Roman civilization began to decline, witnessing a fall in urbanization, seaborne commerce and population. Early in the third century Germanic peoples reached the Black Sea, creating formidable confederations. Prominent among these barbarian tribal confederations, were the Goths, Vandals, Huns, Avars and Magyars. They rejected the classical culture of Rome. The arrival of the Huns in 372–375 ended the history of these kingdoms.[6] The Roman Empire was overrun and completely displaced by the waves of invaders that flooded Western Europe.

Roman culture north of the Po River was almost entirely displaced by the migrations. The new people greatly altered established society, including law, culture, religion, and patterns of property ownership. Early Medieval Period marked the start of the Middle Ages of European history. The period saw a continuation of trends evident since late classical antiquity, including population decline, especially in urban centres, a decline of trade, a small rise in global warming and increased migration. The Early Middle Ages also marked the beginning of the cultural distinctions between Western and Eastern Europe north of the Mediterranean. Where the Western Roman Empire was at the verge of collapse, the Eastern Roman Empire, or Byzantine Empire, continued to survive.

Major Developments

During the Medieval period many changes took place all over the world. The political map was altered with the decline of old kingdoms and empires, and the rise of new ones. These changes were very marked in Western Europe. The political unity of the Roman Empire dwindled in Europe. Christianity influenced the social, economic, literary and philosophical developments during this period. The Christian Era was marked by the loss of ancient scientific concepts about the world and their replacement by unscientific, uncritical cosmogonies based largely on scriptures. During this period, travelling and explorations, owing to political instability was hazardous. However, the missionary travels were the only source of knowledge for the regional account of different nations. Most of the classical correct concepts were forgotten, and the old errors appeared again about the world map and habitable portions of the earth. Moreover, the involvement of Christians and Muslims in crusade wars further aggravated the problem. The crusades were the Roman Catholic Holy Wars to recapture Jerusalem and the Holy Land from the Muslims.

The Romans were in practice of two-field agriculture, with a crop grown in one field and the other left fallow and ploughed under to eliminate weeds. Systematic agriculture largely disappeared and yields declined. The period saw a recession in agricultural and economic activity. There was a consequent collapse in trade and manufacture for export. Major industries that depended on trade, such as large-scale pottery manufacture, also vanished. Administrative, educational and military infrastructure quickly disappeared, and the loss of the established *cursus honourum* (course of offices) led to the collapse of the schools and to a rise of illiteracy. The volume of trade reached its lowest level. Waves of epidemics wiped out large rural populations. This continued a pattern that had been underway since the third century. Much learning under the Roman Empire was in Greek, and with the re-emergence of the wall between East and West, little eastern learning continued in the West. Due to the demographic displacement that accompanied the end of the Western Roman Empire, by this point most Western Europeans were descendants of non-literate barbarians rather than literate Romans. In this sense, education was not lost so much as it had yet to be acquired.[7]

De-urbanization reduced the scope of education, and by the sixth century teaching and learning moved to monastic and cathedral schools, with the study of biblical texts at the centre of education. However, in the Eastern Roman Empire (Byzantium), learning (in the sense of formal education involving literature) was maintained at a higher level than in the West. The classical education system, which would persist for hundreds of years, emphasized grammar, Latin, Greek, and rhetoric. In the ancient world, Greek was the primary language of science. Advanced scientific research and teaching was mainly carried on in the Hellenistic side of the Roman Empire, and in Greek. Later Roman attempts to translate Greek writings into Latin had limited success. As the knowledge of Greek declined, the Latin West found itself cut off from some of its Greek philosophical and scientific roots. For a time, Latin-speakers who wanted to learn about science had access to only a couple of books that summarized Greek handbooks as those by Boethius (470–524).[8]

For almost a thousand years, Rome was the most politically important, richest and largest city in Europe. Around 100 A.D., it had a population of about 450,000, and declined to a mere 20,000 during the Early Middle Ages, reducing the sprawling city to groups of inhabited buildings interspersed among large areas of ruins and vegetation.[9] At the end of the eighth century, the former Western Roman Empire was decentralized and overwhelmingly rural.

The Eastern Roman Empire in Constantinople was left as the legal successor to the classical Roman Empire. After Greek replaced Latin as the official language of the Empire, historians refer to the empire as "Byzantine". Westerners would gradually begin to refer to it as "Greek" rather than "Roman". The inhabitants, however, always called themselves *Romaioi*, or Romans. The Eastern Roman Empire aimed to retain control of the trade routes between Europe and the Orient, which made the Empire the richest polity in Europe. Making use of their sophisticated warfare and superior diplomacy, the Byzantines managed to fend off assaults by the migrating barbarians. By the early eighth century, notwithstanding the shrinking territory of the empire, Constantinople remained the largest and the wealthiest city of the entire world.

From the seventh century onwards, the Byzantine history was greatly affected by the rise of Islam and the Caliphates[10]. Muslim Arabs first invaded historically Roman territory under Abū Bakr, first Caliph of the Rashidun Caliphate, who entered Roman Syria and Roman Mesopotamia. Under Umar, the second Caliph, the Muslims decisively conquered Syria and Mesopotamia, as well as Roman Palestine, Roman Egypt, parts of Asia Minor and Roman North Africa. In the mid seventh century A.D., following the Muslim conquest of Persia, Islam penetrated into the Caucasus region situated between the Black Sea and the Caspian Sea. This expansion of Islam continued under Umar's successors and then the Umayyad Caliphate, which conquered the rest of Mediterranean North Africa and most of the Iberian Peninsula. Over the next centuries Muslim forces were able to take further European territory, including Cyprus, Malta, southern France, Crete, and Sicily and parts of southern Italy.

Much of Roman civilization north of the Po River had been wiped out in the years after the end of the Western Roman Empire, between the 5th and 8th centuries. But, the conditions in Western Europe began to improve after 700. New political and social infrastructure began to develop. Much of this was initially Germanic and pagan. The interaction between the culture of the newcomers, their warband loyalties, the remnants of classical culture, and Christian influences, produced a new model for society.[11] The turmoil of the so-called Barbarian invasions in the beginning of the period gradually gave way to more stabilized societies and states as the origins of contemporary Eastern Europe began to take shape during the High Middle Ages.

In the course of the eleventh century, Islam's scientific knowledge began to reach Western Europe, *via* Islamic Spain. The works of Euclid and Archimedes, lost in the West, were translated from Arabic to Latin in Spain. The modern Hindu-Arabic numeral system, including a notation for zero, had been developed by Hindu mathematicians in the 5th and 6th centuries. Muslim mathematicians learned of it in the seventh century and added a notation for decimal fractions in the ninth and tenth centuries. Around 1000, Gerbert of Aurillac (later Pope Sylvester II) made an abacus with counters engraved with Arabic numerals. A treatise by Al-Khwārizmī on how to perform calculations with these numerals was translated into Latin in Spain in the twelfth century.

The Catholic Church, the only centralized institution to survive the fall of the Western Roman Empire intact, was the sole unifying cultural influence in the West, preserving Latin learning, maintaining the art of writing, and preserving a centralized administration through its network of bishops ordained in succession. The Early Middle Ages are characterized by the urban control of bishops and the territorial control exercised by dukes and counts. The rise of urban communes marked the beginning of the High Middle Ages. Western Europe, as in 1000 CE, remained less developed compared to the Islamic world, with its vast network of caravan trade, or China, at this time the world's most populous empire under the Song Dynasty. Constantinople had a population of about 300,000, but Rome had a mere 35,000 and Paris 20,000. By contrast, Córdoba, in Islamic Spain, at this time the world's largest city contained 450,000 inhabitants. In the culture of Europe, several features surfaced soon after 1000 that mark the end of the Early Middle Ages: the rise of the medieval communes, the reawakening of city life, and the appearance of the burgher class, the founding of the first universities, the rediscovery of Roman law, and the beginnings of vernacular literature.

Impact

The Early Medieval Phase saw geographic horizons closing in due to theocratic[12] slant in human thinking and writings. The wide-ranging Greek geographical concepts were largely forgotten, except among the Arabic scholars. Observations piled up in Christian monasteries, but the intellectual climate was not favourable for the formation of new interpretations. The approach to the science, including geography, became theocratic. The image of the world was moulded to fit the Biblical references. This kind of thinking had an adverse impact on the art of map-making, which got distorted. The tradition of making fairly accurate maps, as started by the Greeks and Romans, got lost. Instead, there emerged maps, which were purely fancy and unscientific. The earlier concepts of sphericity and the nearly correct distances of places, latitudes and longitudes were ignored.

The approach to studies of any kind became theocratic. Consequently, during this period, Geography made a little progress as the organized body of knowledge. The quest for scientific learning stopped; instead, there was regression of the already acquired scientific knowledge. At best, the scholars of this period, made accurate but sterile copies of the works of ancients, rejecting anything that did not conform to the dogmas of the Church. Such an intellectual environment stifled any development of critical scientific analysis. Concepts of the world, which had been developed in Greek and Roman times, were reshaped to conform to the teachings of the Church. The image of the world was molded to fit the Biblical references, which were considered more valid. The earth was considered as a flat disc with Jerusalem at the centre (Fig. 6.1).

This kind of thinking had an adverse impact on the art of map-making, resulting in great distortions. The tradition of making fairly accurate maps, as started by the Greeks and Romans got lost. Instead, there emerged maps, which were purely fancy and unscientific. The ideal map of this period was 'Orbis-Terrarum' or: 'T-in-O' Map (Fig. 6.1). 'Orbis-Terrarum' means 'wheel of the world'. From the early medieval times to the Renaissance, this legend was applied to any map of the world; 'world' of course meaning 'the World as it was known at the Time'. This originally included the Mediterranean Sea and its surrounding territories. In the Early Middle Ages, the religious mapmakers wished to illustrate geography of faith rather than mere terrestrial territory. Their 'Orbis-Terrarum' maps literally used as 'O' with a 'T' in it–a circular ocean surrounding the earth, which itself surrounding a T-shaped body of water (the stem of T representing the Mediterranean Sea and its top representing the Aegean and Black Seas on the one hand and the Nile River and Red Sea on the other). The three divisions–Europe, Asia and Africa–were accepted as standard. The centre of the inhabited world, just above the centre of the T, was Jerusalem. At the Far East, beyond the limit of the inhabited world, was paradise (the house of *Adam* and *Eve*). Moreover, in all these maps were inserted the mythical places, beasts and dragons. This type of cartography was in practice for a long time. These maps were made very beautiful and extremely decorative, but did not have any relation with actual facts, as they were far from reality with no scientific base.

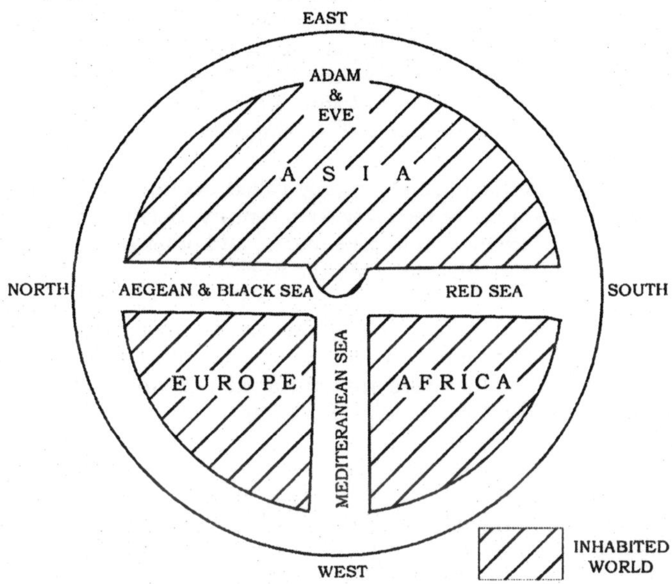

Fig. 6.1: 'T-in-O' Map of Early Medieval Period

So far the sprawl of the habitable world is concerned many wrong ideas were put forward. There was vagueness in the east-west and north-south extension of the world. The sphericity and the nearly correct distances of places, latitudes and longitudes were ignored. As for the travels and journeys, they continued but geography was hardly benefited. Because the purpose of these travels was not scientific or speculative. These were mainly religious or missionary movements made by pilgrims to holy shrines. Moreover, the descriptions made were also purely aesthetic, i.e. beautiful without logic.

Onset of High Medieval Phase

Against this background, there emerged the intellectual curiosity, integrity and catholicity of the Arab scholars that led to the great achievements of Islamic civilization. Thus, when Europe was plunged into Dark Ages, the Arab School of Thought was rising to prominence in medieval times. In contrast to earlier times, new zeal of learning spread through the world of Islam. The Arab civilization, in fact, became the main link between the ancient civilization and the modern world. The rise of Arab school marks the end of the recession in development of geographic literature and return to more scholarly research.

NOTES

1. Jordan, Chester William (2004).
2. Thompson, Bard, (1996).
3. *Ibid.*
4. Kennedy, Hugh (1995), pp. 249–272.
5. The word **barbarian** generally refers to an uncivilized, uncultured person, either in a general reference to a member of a nation or ethnicity perceived as having an inferior level of civilization, or in an individual reference to a brutal, cruel, insensitive person whose behaviour is unacceptable in a civilized society. The word "Barbarian" comes into English from Medieval Latin *barbarinus,* from Latin *barbaria,* from Latin *barbarus,* from the ancient Greek word *barbaros* which meant a non-Greek, someone whose (first) language was not Greek.
6. Edward Gibbon (1776).
7. Anicius Manlius Severinus Boëthius, commonly called Boëthius (477–524 A.D.), was a Roman senator, consul, and philosopher of the early sixth century. Boëthius composed his *Consolation of Philosophy,* a philosophical treatise on fortune, death, and other issues, which became one of the most popular and influential works of the Middle Ages.
8. The term 'theocracy' is commonly used to describe a form of government in which a religion or metaphysical faith plays the dominant role. In other *words,* theocracy is a constitution governed by God; accordingly, theocratic means a belief in this constitution, in which God is regarded as the sole sovereign; the laws of the realm are considered as divine commands rather than human ordinances and the priests are considered as the officers of the invisible ruler.
9. Glenn, R. (1997).
10. Norman Cantor, (1994), pp. 102; 147-48.
11. Norman Cantor, (1994). *Ibid.* p. 52.
12. Theodosius-I (Latin: *Flavius Theodosius Augustus;* 347–395), also known as Theodosius the Great, was Roman Emperor from A.D. 379 to A.D. 395, as the last emperor to rule over both the eastern and the western halves of the Roman Empire.

REFERENCES

David Knowles, (1964). *The Evolution of Medieval Thought,* Random House.

Edward Gibbon (1776). *The History of the Decline and Fall of the Roman Empire.* I. W. Strahan and T. Cadell.

Glenn, R. (1997). "The population of ancient Rome", *Antiquity,* December 1.

John Marenbon (1983) 1988. *Early Medieval Philosophy (480-1150): An Introduction,* London: Routledge.

Kennedy, Hugh (1995). "The Muslims in Europe". In McKitterick, Rosamund, *The New Cambridge Medieval History: c.500-c.700.* Cambridge University Press.

Marcia L. Colish (1997). *Medieval Foundations of the Western Intellectual Tradition: 400-1400.* (New Haven, CT: Yale University Press).

Norman Cantor, (1994). *The Civilization of the Middle Ages,* New York: Harper Collins.

Richard Southern (1953). *The Making of the Middle Ages,* Yale University Press.

Thompson, Bard (1996). *Humanists and Reformers: A History of the Renaissance and Reformation.* Grand Rapids, MI: Erdmans. p. 13.

7

Medieval Thought and Rise of Arab School (Paradigm Phase 2)

The medieval phase represents two major philosophical trends in geographic development, viz. European (the Christian) and Middle-Eastern (the Islamic). As discussed earlier, the conception of a middle age in Europe corresponds to a period in the history of the Latin West[1], i.e. the area of Roman Catholism. The medieval philosophy is conventionally understood as the philosophy of Western Europe between the decline of Roman Empire and the Renaissance (500 to 1450 A.D.). If one takes medieval philosophy to include the Patristic period[2], then the area gets expanded to include Greek-speaking eastern Europe, as well as North Africa and parts of Asia Minor. However, the Paradigm Phase-2 in geography is associated with the Rise of Arab School of Thought during High Medieval Phase (800–1450 A.D.).

European

The Platonic and neo-Platonic influences dominated philosophical thinking in the West during the first part of Middle Ages.[3] Here, the Medieval philosophy was shaped by two pieces of the cultural legacy that survived the collapse of Roman civilization, viz. Latin language and Christianity (the Church). The Latin language provided medieval thinkers with access to some important ancient resources, which led early medieval thinkers to an acquaintance with classical ideas. Saint Augustine (354–430)[4] was the most significant of these sources. His thought, and in particular his philosophical approach to Christianity and his Christianized Neoplatonist philosophical outlook, profoundly affected virtually every area of medieval philosophy. Christianity had grown in importance in the late Roman Empire and the Church remained, until the twelfth century, virtually the only institution capable of supporting intellectual culture. It sustained formal education in schools associated with its monasteries and cathedrals, and provided for the preservation of ancient texts in its libraries and scriptoriums. Moreover, in addition to providing the institutional basis for medieval philosophy, Christianity was an important stimulus to philosophical activity. Its ideas and doctrines constituted a rich source of philosophical subject matter. Medieval philosophy, therefore, took root in an intellectual world sustained by the Church and permeated with Christianity's texts and ideas. Augustine laid the foundation for medieval Christian philosophical theology. He argued that Christianity can be strengthened and enriched by philosophy; Christian philosophers should begin by believing on the authority of the Bible and the church. The Augustinian method of belief seeking understanding was taken for granted by the vast majority of philosophers in the Middle Ages. Further, Augustine's writings provided a wealth of rich and compelling examples of philosophical reflection on topics ranging from the nature of evil and sin to the nature of the Trinity. Boethius (480–524)[5] stood with Augustine in this respect as an important model for later thinkers. He composed several short theological treatises bringing together the Aristotelian logic with the doctrines of Christian creed. Inspired by the philosophical analysis and argumentation prominent in these writings, medieval philosophers enthusiastically took up, developed and extended the enterprise of philosophical theology.

The full flowering of the philosophical tradition that grows from these beginnings occurs in the period from 1100 to 1400. Two developments, in turn, are of particularly importance for understanding the rapid growth and flourishing of intellectual culture in these centuries. The First is the influx into the West of a large and previously unknown body of philosophical material newly translated into Latin from Greek and Arabic sources; and the Second is the emergence and growth of the great medieval universities. As far the Recovery of Texts is concerned, the Medieval philosophers before 1100 A.D. had access to only a few texts of ancient Greek philosophy: those comprising 'the old logic' (Aristotle's *Categories* and *De interpretatione* and Porphyry's *Isagōgē*) and a small part of Plato's *Timaeus*. The scholarly generation thereafter witnessed the appearance in the Latin West of the remainder of Aristotle's logical works. Over the next hundred years, most of Aristotle's natural philosophy (most importantly the *Physics* and *On the Soul*) and the *Metaphysics* and *Ethics* became available for the first time. The new texts became the subject of increasingly sophisticated and penetrating scholarly research; they were incorporated into the heart of the university curriculum. The philosophers in the 13th and 14th centuries had never before encountered a philosophical system of such breadth and sophistication. By the middle of thirteenth century, the universities at Paris and Oxford were the leading centres of European philosophical activity. Virtually all the great philosophers from 1250 to 1350 studied and taught in the schools at one or both of these centres. As a result of the migration and centralization of philosophical activity to the universities, the philosophical education was gradually expanded and standardized and the philosophers themselves became highly trained academic specialists. These features of later medieval philosophy make it genuinely scholastic, that is, a product of the academic environment of the schools. As obvious, the logic, natural philosophy, metaphysics and ethics occupied the centre of the curriculum leading to the basic university degrees.

Besides Augustine and Boethius, Anselm of Canterbury (1033–1109)[6] and Peter Abelard (1079–1142)[7] were among the other important patristic authors of philosophical significance. Augustine's authority has been felt broadly not just in philosophy but also in theology, popular religion, and political thought.[8] After Augustine, the first thinker of philosophical note was Boethius. His importance rests on his translations and commentary activity. He came up with the lofty goal to translate Plato and Aristotle into Latin. In addition to his translations, Boethius wrote a number of logical treatises of his own. Together all these logical writings, both the translations and the others, constitute what later came to be called the "Old Logic" (= *logica vetus*).[9] Basically, everything the Middle Ages knew about logic up to the middle of the twelfth century was contained in these books. As a result, Boethius is one of the main sources for the transmission of ancient Greek philosophy to the Latin West during the first half of the Middle Ages. He also proved to be influential in the twelfth century and afterwards for the metaphysical views contained in a series of short studies known collectively as the Theological Tractates. After Boethius, the classical Graeco-Roman world entered a period of relative stagnation (Dark Age), a period that lasted until after the year 1000. Then, shortly after the turn of the millennium, things began to revive. The Germanic "barbarian" tribes that had so disturbed the late Roman Empire had long since settled down. Trade began to revive, travel became relatively safe again, at least compared to what it had been, new cities began to emerge, and along with them new social arrangements began to develop. Education was part of this general revival, and with it philosophy. The major medieval philosophers before the year 1000 are fewer in number. But after 1000 their numbers grew exponentially. Simultaneously, philosophy became increasingly technical and "academic". Anselm of Canterbury represents a major transitional figure. Anselm is no doubt best known as the originator of the famous "ontological argument" for the existence of God. His writings illustrate the increasing role of "dialectic" in philosophy and theology. By the early twelfth century, the revival of education that had begun shortly after the millennium was in full swing. During the first half of the century, the most important philosopher by far was undoubtedly Peter Abelard. Abelard represents the full flower of "early medieval philosophy," Abelard's writings further amplify the tendency, already seen in Anselm, to increase the use of reasoning and argumentation in theology. Prior to Abelard, philosophy in the Middle Ages had not been an exclusively academic affair. Soon, however, this all changes, and Philosophy becomes an increasingly specialized discipline, pursued by and for those whose livelihood is found only in educational institutions. Philosophy and theology become more clearly distinguished from one another; both become more systematic, rigorous and precise. By the end of twelfth century, almost all of Aristotle's works available today had been translated into Latin and, together with the commentaries and other newly translated texts, gradually began to circulate. These new logical writings, as distinct from the "Old Logic" (= *Logica vetus*) stemming from Boethius, became known collectively as the "New Logic" (= *Logica nova*). After them, the Physics, Metaphysics and other Aristotelian writings gradually became known. This relatively sudden injection of so much new and unfamiliar material into Western Europe was nothing less than revolutionary. It was no longer possible for philosophers and theologians to regard their task as simply one of deepening and elaborating traditional views that had come mainly from the Church Fathers and other familiar and approved authorities. It was now a matter of dealing with an entirely unfamiliar framework, with new ideas, accompanied by powerful arguments for them.

As part of the revival that began after the turn of the millennium, new forms of education began to emerge in Western Europe. In general, one may distinguish four main types of educational practices in the Middle Ages, viz. Monastic schools, Individual "Masters", Cathedral schools and Universities. Monastic schools were schools that had been regularly associated with monasteries ever since the sixth century. Beginning in the mid-eleventh century, individual scholars would occasionally set up a "school" of their own and gather students around them. Such schools were sometimes itinerant and depended entirely on the appeal of the teaching "master." Perhaps the closest analogue to this arrangement would be the modern "martial arts" schools one often finds in present-day cities. The practice declined after 1150. The Cathedral schools were schools associated with the official church of a bishop, and played a role similar to that of the monastic schools for monasteries. Cathedral schools flourished c. 1050–1150. Frequently, universities grew out of cathedral schools. Thus, the cathedral school at Paris developed by the early-thirteenth century into the University of Paris. The important cathedral school drew students from all over Europe. Such a school became known as a '*studium generale*'. Some of these '*studia generalia*' survived and became known as "universities." Universities were divided into "faculties." The four most common ones were the faculties of arts, law, medicine, and theology. Most universities had arts faculties, in addition to one or more of the others. The arts faculty was for the basic training of students, before they proceeded to one of the "higher" faculties. In the medieval university, philosophy was cultivated first and foremost in the arts faculty. Medieval philosophy invented the philosophy of religion. A whole host of traditional problems in the philosophy of religion first took on in the Middle Ages the forms in which we still often discuss them today. As for logic, the later Middle Ages was one of the great periods in the history of logic.[10] From the time of Abelard through at least the middle of the fourteenth century, the peculiarly medieval contributions to logic were developed and cultivated to a very high degree. It was no longer a matter of interpreting Aristotle, or commenting on the works of the "Old Logic" or the "New Logic"; wholly new genres of logical writing sprang up, and entirely new logical and semantic notions were developed. In metaphysics, the Middle Ages has a well deserved reputation for philosophical excellence. The problem of universals, for example, was one of the topics that were discussed at this time with a level of precision and rigor it would be hard to find matched before or since. In natural philosophy and philosophy of science, medieval philosophy was of course very strongly influenced by Aristotle. Particularly from the fourteenth century on, the increasing use of mathematical reasoning in natural philosophy eventually paved the way for the rise of early modern science.

Thus, at the most basic level, medieval philosophers shared a common view of the world in terms of Metaphysics, Psychology, Ethics, Logic and Natural Philosophy. Aristotle's *Physics* in particular received enormous attention, and medieval philosophers developed sophisticated tools of logical, conceptual and mathematical analysis to deal with problems raised by Aristotle's discussions of motion, change, continuity and infinity. From the late twelfth century, medieval philosophy is profoundly affected by the Islamic philosophers. Scientific treatises by Islamic thinkers provided the material and impetus for significant developments in astronomy, mathematics and other fields.

Islamic

There was rise of Islam in Arab in seventh century and the new zest of learning in medieval period spread through the world of Islam. The vast expanse of the Muslim Empire facilitated coming together of intellectual and scientific traditions of various civilizations, particularly Greek and Indian. The Arabian conquest of Palestine, Syria, Persia and Bactria, where Greek culture still flourished, gave to Muslims a rich fruit of Greek learning. Their contacts with the Indians helped to diffuse the system of numerals from India to the West, and this system was called '*Hindsa*' in Arabic language. Starting in the eighth Century the renewed study of Greeks and Indians revived much of the forgotten learning. But the classical spirit of inquiry was not revived. The renaissance remained primarily a religious movement, more concerned with preserving the knowledge of the past than expanding it through scientific investigation. Nevertheless, the Arabs did keep the ancient learning alive, and it was from their hand that the Europe received it, as the Dark Ages drew to a close. The Arab scholars and travellers presented a new outlook of the world, which later led to the onset of the Age of Discovery. The Arab civilization, in fact, became the main link between the ancient civilizations and the modern world. Arab Empire started disintegrating after A.D. 1400 and the Arab period culminated in the Age of Discoveries and Explorations.

After the fall of the Roman Empire in fifth century, the average European's knowledge of the world around them was limited to their local area and to maps provided by the religious authorities. The Islamic empire began to expand beyond the Arabian Peninsula after the death of the prophet and founder of Islam, Mohammad, in 632 A.D. In eighth century, all of northern Africa, the Iberian Peninsula (Spain and Portugal), India and Indonesia became Islamic lands. Around 762, Baghdad became the intellectual capital of the empire. Over time, Baghdad accumulated a wealth of knowledge and many key geographical works from the Greeks and Romans. Ptolemy's *Almagest*, which was a reference to the location and movement of heavenly bodies along with his *Geography*, a description of the world and a gazetteer of places, were two of the first books translated, thus keeping their information in existence. With their extensive libraries, the Islamic view of the world between 800 and 1400 was much more accurate and elaborate than that of Christian's. The Muslims were natural explorers, as pilgrimage is part of their religion. Pilgrimage led to further exploration beyond the Arabian Peninsula. By eleventh century, Islamic traders had explored the eastern coast of Africa to 20 degrees south of the Equator (near contemporary Mozambique). Thus, Islamic geography was primarily a continuation of the Greek and Roman scholarship which had been lost in Christian Europe. There were some additions to the collective knowledge by their geographers, like Al-Idrisi, Ibn-Batuta, Ibn-Khaldun etc. By translating important Greek and Roman texts and by contributing to the knowledge of the world, Islamic scholars helped provide the information which allowed the discovery and exploration of the New World in the 15th and sixteenth centuries. With the rise of Arab School, during High Medieval period (800-1400 A.D.), geography witnessed another paradigmatic phase of development.

Arab School

The Arab civilization became the most advanced during the medieval period. In the early years of the seventh century, Prophet Muhammad[11] founded Islam. This new religion, within a very short period of time, not only united the warring tribes but also led to the establishment of a big empire and building of one of the finest civilization of time. Nearly all Arabia had accepted the new religion and became a new unified state before the death of Prophet Mohammad in A.D. 632. After his death, his successors, known as 'Caliphs' or 'Khalifas', held the religious and political authorities. From Arabia, Islam spread fast to many other parts of the world. Within about a hundred years, the Khalifas had conquered Iran, Syria, Egypt, Central Asia, North Africa and Spain. The Arabs had also come to India. Wherever the Arabs went, their religion went with them. This astonishingly rapid conquest was due to several reasons. It was partly missionary zeal, but economic and social causes favoured the expansion. The Arabs, who were the inhabitants of a poor country, wanted richer lands. They were excellent fighters. The people of the neighbouring countries, tired of oppressive rule and fascinated by simplicity of the teachings of Islam, welcomed the invaders. The Arab empire was the largest that the world had so far seen. In the pre-Islamic period of their history, the Arabs were illiterate and unfamiliar with the art of writing. United by Islam, the Arabs started on their quest for knowledge. They sought knowledge wherever they could find it. In fact, the Arab Civilization was far superior to that of Europe in these centuries.

In 762 A.D. the Muslims founded the new city of Baghdad and for more than a century Baghdad was the centre of intellectual world. With the patronage of Caliph Harun-al-Rashid an academy entitled 'Baitul-Hikma' was established. At this acadmy, scholars from all over the world were invited to teach and assist the Arab academicians and to help them in the translation of the Greek, Latin, Persian and Sanskrit works into Arabic. The successive rulers who ascended the throne also gave a great support to the scholars. The Arabs invited the Indian scholars at Baghdad to learn Indian Mathematics and numerals. The Arabs studied the works of Aryabhatt and Surya-Sidhant[12]—the Sanskrit treatise containing the principles of trigonometry. The Arabic translation of the Sidhanta was a landmark in the history of Arab astronomy[13]. Thus the Arabs collected geographical ideas from the Greeks, Romans, Iranians and Indians. On the basis of their observations, explorations and studies they developed their own concepts and theories with great vigour and speed and produced innumerable books on various aspects of geography and allied sciences; and carried their knowledge to far-off lands from the shores of Atlantic to the Pacific and Indian Ocean. The establishment of vast empire facilitated the coming together of intellectual and scientific traditions of various civilizations, from the Western to Oriental. The Arabs made all knowledge their own and developed it further. Their knowledge of science was most advanced in their time. They borrowed knowledge of medicine from Greece and the system of numerals (*Hindsa*) from India. Their system of medicine is still called 'Yunani' that means 'Greek'. Besides, the Arabs also developed a decorative style of writing called 'Calligraphy'. In the Muslim universities, which spreaded from Persia to Spain, the scholars studied the Greek heritage again. The Arab traders travelled widely and brought back new information that the scholars compared with Greek ideas or incorporated into Ptolemy's maps. Besides Baghdad, the main centre of the Muslim intellectual world, the other major centres of learning that emerged during this period included: Jerusalem in Israel; Cairo in Egypt; Toledo in Spain; Istanbul in Turkey; Damascus in Syria; and Shiraj in Iran.

Many factors helped to stimulate this interest in learning, as: the large extent of Arabic Empire which stretched over so many different areas from the Atlantic to the borders of China; the excellent system of roads, a legacy of Rome; the desert routes which encouraged movement and facilitated the pilgrimages to Mecca demanded by the faith; Trade, which naturally arose out of the diversity within the Empire and was further stimulated by the high rank of traders in Muslim society; the conquest of Syria, Persia and Bactria—where the Greek culture still flourished and gave to the Muslims a rich fruit of Greek learning; and the cultural and economic ties with the Indian Subcontinent. In the West, on the other hand, from the beginnings roughly about the year 1000 A.D., greater stability came to the lands of Europe. Here too, the 'High Middle Ages' describe the expansionalist culture and intellectual revival from the late eleventh century to the fourteenth century.

Contribution

Many Arab scholars contributed to the development of geographic thought during High Medieval Phase. They included, for instance, Al Battani, Al Masudi, Al Beruni, Ibn Sina, Al Idrisi, Ibn Batuta and Ibn Khaldun (Table 7.1).

Their outstanding contributions go to various fields of geographical study as Mathematical, Practical, Regional, Historical, Physical and Human.

Table 7.1: Medieval Scholars

Al Balakhi (850-934)
Al Battani (858-929)
Al Masudi (888-957)
Al Beruni (973-1050)
Ibn Sina (980-1037)
Al Idrisi (1099-1165)
Ibn Batuta (1304-1377)
Ibn Khaldun (1332-1406)

Mathematical and Practical Geography

In Arab School, the Greek science was the starting point. The works of Aristotle and Ptolemy were elaborately studied. It is not surprising, therefore, that mathematical geography received the maximum attention. The Arabs more or less adopted the Greek tradition. The factors like semi-desertic conditions, open atmosphere and clear skies, and vast distances to scan that facilitated the astronomical observations also helped in the promotion of studies in mathematical and practical geography. Sea-Charts and Maps based on cylindrical and orthomorphic projections were made by the Arabs. The early Arabs considered the earth as the centre of the Universe, around which revolved the seven planets. These planets, in order of their distance from the earth were: Moon, Mercury, Venus, Sun, Mars, Jupiter and Saturn[14]. It was imagined that each of these planets occupied a particular 'sky' or rather using the expression of Ptolemy a separate 'sphere' and that each sky was under the authority of the 'planet' to which it had been assigned. Ibn Sina was one of the foremost philosophers of the golden age of Islamic tradition. His philosophical encyclopaedia, *Kitab al-Shifa* ("Book of Healing") brought Aristotelian and Platonian philosophy together with Islamic theology in dividing the field of knowledge into two major streams, viz. theoretical knowledge and practical knowledge. There were several attempts to determine the size and shape of the earth. Al Battani, for instance, made important astronomical discoveries, observations and measurements. He wrote a number of books on astronomy and trigonometry. His most famous book was his astronomical treatise with tables, which was translated into Latin in the twelfth century and became popular as *"De scienta stellerum-De numeris stellerum et motibus."* His treatise on astronomy was extremely influential in Europe till the Renaissance, with translations available in several languages. Al Battani's well-known discovery is the remarkably accurate determination of the solar year as being 365 days, 5 hours, 46 minutes and 24 seconds, which is very close to the latest estimates. He calculated the length of equator at 27000 miles, whereas Ptolemy in 157 A.D. had established it at 24000 miles. He also determined with striking accuracy the obliquity of the ecliptic, the length of the seasons and the true and mean orbit of the sun. He proved, in sharp contrast to Ptolemy, the variation of the apparent angular diameter of the sun and the possibility of annular eclipses. He rectified several orbits of the moon and the planets and propounded a new and very ingenious theory to determine the conditions of visibility of the new moon. He also provided very neat solutions by means of orthographic projection for some problems of spherical trigonometry. Al Biruni, a Persian mathematician and astronomer, calculated the latitude of Khwarazm (Uzbekistan), his native place, using the maximum altitude of the sun. He refined the ancient estimate of the Earth's radius from approximately 6,314 km (measured by Eratosthenes in 240 A.D.) to 6,339.6 km. This feat was not surpassed in the Western World until the sixteenth century. The Arab geographers borrowed from the Greeks, the division of the globe of the earth into five zones; each zone represented particular temperature conditions. The five zones are: (i) the Torrid Zone—situated between the two tropics, (ii) the two frigid zones—placed near the poles, and (iii) the two temperate zones—that occupied the intermediate positions. For the determination of latitudes, the Arabs like Eratosthenes and other Greeks made use of the sun's shadow when it happened to be on the meridian. The shadow was marked by a column (gnomon). The Arabs improved over the works of Greeks.

Besides, another greatest contribution of the Arabs goes to the advancement of mathematical knowledge, particularly Trignometry. In mathematics, Al Battani, for instance, was the first to replace the use of Greek chords by sines, with a clear understanding of their superiority. He also developed the concept of cotangent and furnished their table in degrees. Introduction of the Hindu invention of the zero in arithmetic (*'Hindsa'*) was another major development in Mathematics. Before this, the Roman numerals were in use and their range was limited. The zero liberated arithmetic and people could now count in millions, billions and trillions. In mathematics, the Arabs learned the Indian numerals and spread their use far and wide, so that in the West they are to this day called "Arabic numerals". They improved upon the existing knowledge of Algebra and trigonometry. However, surprisingly enough, the attempts of Arab scholars, distorted by their love of ornamentation, did not bring about any marked improvements in cartography and map-making. Moreover, they still believed in geocentric view of the planetary system.

Regional Geography

A wealth of regional descriptions exists in the works of Arab scholars, particularly Al Idrisi and Ibn Batuta. Both physical and human aspects are covered in detail. According to Arabs the western limit of the habitable world is at the end of the Mediterranean sea, the eastern at the Sila (Japan), the northern in the land of Yajuj Majuj (Siberia) and the southern to the south of the equator. Ibn-Batuta was of Negro origin, but was educated in Muslim religion. He was the most remarkable traveller, whose voyages extended east as far as Northern China and south along the coast of Africa well beyond the Equator, covering Egypt, Syria, Persia, Arabia, Zanzibar Isles, India, Maldives, Ceylon, Sumatra and China. One of the most remarkable travellers of all time, Ibn Batuta visited China sixty years after Marco Polo and in fact travelled 75,000 miles, much more than Marco Polo. Ibn Batuta's contribution in geography is unquestionably as great as that of any geographer yet the accounts of his travels are not easily accessible. His writings, with the book *'Travelogue'* finished in 1357, are an important source of knowledge on large parts of the known world of medieval times. Ibn Batuta was the only medieval traveller who is known to have visited the lands of every Muslim ruler of his time. He is considered a most reliable source for the geography of his period and an authority on the cultural and social history of Islam. The writings of Ibn Batuta described the ecumene areas of the world. The travels of Ibn Batuta took him to many parts of Arabia never visited earlier. His travels extended as much as 10° south of the Equator and more. This venture brought empirical proof that Aristotle, the ancient Greek scholar, had been wrong in asserting that the Torrid Zone was too hot and unfavourable for human habitation. Batuta, on the contrary, confirmed that numerous tribes occupied it. The major focus of the writings of Al Idrisi, on the other hand, was on Arabia and India. He has described the country of India as an agriculturally rich country, and mentioned the presence of about 150 towns and cities at that time. Al Idrisi also provided the detailed descriptions of these regions. Al Idrisi was a great traveller. His major contribution lies in medicinal plants as presented in his several books, especially *Kitab al-Jami-li-Sifat Ashtat al-Nabatat*. In addition to the above, he made original contributions to geography, especially as related to economic, physical factors and cultural aspects. He made a planishere in silver for King Roger II, and described the world in *Al-Kitab al-Rujari* (Roger's Book), also entitled *Nuzhat al-Mushtaq fi Ikhtiraq al-Afaq* (The delight of him who desires to journey through the climates). This is practically a geographical encyclopaedia of the time, containing information not only on Asia and Africa, but also Western countries. Al-Idrisi, later on, also compiled another geographical encyclopaedia, larger than the former; it was entitled as *Rawd-Unnas wa-Nuzhat al-Nafs* (Pleasure of men and delight of souls) also known as *Kitab al-Mamalik wa al-Masalik*. His work was soon translated into Latin and, especially, his books on geography remained popular both in the East and the West for several centuries.[15]

Historical Geography

Al-Masudi and Ibn Khaldun were the great Moslem historical geographers. Al-Masudi, the Arab discoverer, is known as the 'Herodotus of the Arabs'. He was the first Arab to combine history and scientific geography in a large-scale work. He travelled extensively in India, the Middle East, and Africa. Al-Masudi wrote a 30-volume history of the world and recounted the experiences of his travels form Europe to India. By presenting a critical account of historical events, he initiated a change in the art of historical writing, introducing the elements of analysis, reflection and criticism, which was later on further improved by Ibn Khaldun. In particular, in his *Al-Tanbih* he makes a systematic study of history against a perspective of geography, sociology, anthropology and ecology. Masu'di had a deep insight into the causes of rise and fall of nations. Ibn Khaldun was the last great Moslem scholar. He is seen as one of the forerunners of modern historiography. He is best known for his book *"The Muqaddimah"* (The introduction to History). Ibn Khaldun represents the climax of history of Islamic economic thought. The Muqaddimah of Ibn Khaldun records an early Muslim view of 'universal history'.

Physical Geography and Climatology

The physical geography finds a special place in the writings of the scholars like Al Biruni, Ibn Sina, Al Idrisi, Al Masudi and Al Balakhi. Their important observations are related to the study of: (i) the processes shaping landforms; (ii) alluvial deposits; and (iii) phenomenon of tides. Al Biruni in his book entitled *'Kitab-al-Hind'* describes the fluvial processes and alluvial deposits of Himalayan region. In this book he recognized the significance of the rounded stones he found in the alluvial deposits south of the Himalayas. The stones became rounded, he pointed out, as they rolled along the torrential mountain streams. Furthermore, he recognized that the alluvial material dropped close to the mountain was relatively coarse in texture and that it became finer in texture farther away from the mountain. Ibn Sina wrote about the processes shaping landforms, viz. weathering and denudation. According to his findings, the streams play an important role in this and the high mountains are more exposed to the process of weathering than any other landform. He keenly observed the work of agents of denudation and weathering in the mountains and ascertained that the mountain streams erode the slopes. He formulated the idea that mountains were being constantly worn down by rivers and that the highest peaks occurred where the rocks were especially resistant to erosion. Mountains are raised up, he pointed out, and are immediately exposed to this process by wearing down, a process that goes on slowly and steadily. Ibn Sina also noted the presence of fossils in the rocks of high mountains. Various Arab scholars and navigators observed the phenomena of tides, and proved that they are caused because of the gravitational pull of the sun and the moon. Al Masudi, navigating the Caspian Sea, recorded the spring and neap tides. Besides, Al Idrisi corrected the erroneous ideas about the Indian Ocean and the Caspian Sea. The Indian Ocean was supposed to be enclosed, whereas the Caspian Sea was shown as an arm of the World Ocean in 'T-in-O' maps of Early Medieval period. He also tried to redefine more accurately the courses of many of the rivers including that of Denube and Niger rivers. However, the most notable under 'physical' were the contributions made to climatology. The important contributions in the field of Climatic Geography have come in the form of coining of the term *mausim* (Monsoon); preparation of the first Climatic Atlas; and division of the world into Climatic Regions. The term 'monsoon' has an Arabic origin, from *mausim*. Arabs were the first to give the idea of periodic nature of monsoons. The studies of Al Masudi, particularly, deal with the Indian Monsoons. The first Climatic Atlas of the world, entitled 'Kitabul Ashkal' was prepared by Al Balakhi in 921 A.D. He gathered the climatic data and information from the Arab travellers. Besides, the Arabs attempted to divide the world into 14 Climatic Regions and observed that climate varies not only by latitude, but also by the positions east and west. They also presented the view that the Southern hemisphere has most of the water and that most of the land area is in the Northern hemisphere and this has reflection in the climate as well. Undoubtedly, this idea had been conceived from the Greeks, basically.

Human Geography

The studies carried under Human Geography by Arab scholars are dominated by the writings of Ibn Khaldun and deal mainly with the influence of physical factors on life forms, including human beings. Ibn Khaldun, in his monumental work, 'Muqaddimah', for instance, has described and discussed human society in various aspects. The work has been divided into 6 sections, viz.: (1) the civilization, geography and anthropology; (2) the nomadic culture and its contrast with the sedentary culture, sociological and historical causes and consequences of the conflicts arising continually from that of fundamental opposition; (3) Dynasties and Kingdoms; (4) Life in villages and cities; (5) Professions or the Means of Livelihood; and (6) Classification of Sciences. The work of Ibn-Khaldun begins with a discussion of man's physical environment and its influence on the living styles of people. He discussed various stages of social organization, identifying the desert nomads as the most primitive and the purest. He suggested that the sedentary city dweller is dependent on luxuries and becomes normally soft. He discussed the forms of government, describing a sequence of stage that marks the rise of dynasty to power, followed by its decline through corruption to its fall. Thus, he explained, in a way, the rise of Arabian power, and also foresaw its fall. Khaldun maintained that the northern hemisphere is more densely populated than the southern and also that the population is meagre along the equator; but away from equator, there are greater concentrations of population up to 64th parallel. Further away there is once again very little or no population at all. The intense heat of the equatorial belt, he considered as the discouraging factor for the concentration of population. A harmonious blending of heat and cold in the temperate region, according to him, is conducive to human growth and settlements. Away from the temperate regions, the excessive cold of the Polar areas is again deterrent to human growth. About the origin of settlements, he stated that 'those who settle are attracted by fertility of land and by the seas with which they can defend themselves against invaders'. In course of time the population increases, and so too the pressure on land, and the habitation becomes a permanent settlement. The settlement grows in the course of time and becomes city—which shows a definite social and socio-economic structure, in ways of division, dearth and abundance, supply and demand. The roots of all cities have always been these small settlements at sometimes or the other. Moreover, he tried to explain various human activities with the help of environment. He explains why the Arabs are nomads and why the Negroes are of easy disposition.

However, the prime focus of the writings of Ibn-Khaldun was on the human society, culture and its development. Modern geographers are particularly impressed with: (i) his cultural interpretation of the physical environment; and (ii) his analysis of the role of the city in regional economy. Ibn-Khaldun tried to correlate man and environmental relationships in a scientific way. According to him, the people of warmer climates are noted for their passionate nature. Those of the colder climate incline to stolidity and lack of vivacity. Those in the temperate climates excel in wisdom and are neither excessively passionate nor markedly stolid. He also explained that the Negros are black because they live in the warm and humid climatic zones while the people of temperate and cold regions are whitish in colour. Similarly, people try to build their houses and settlements on the southern slopes and close to the springs and water sources. In fact, the roots of the concept of climatic determinism lie in the writings of Ibn-Khaldun, who believed that the climate has a tremendous influence on the nature of human beings, their colour and settlements. According to him, in warm climates the people are more passionate; in colder climates people are stolid and are of harsh nature; whereas in temperate climates they are full of wisdom as they occupy the middle position. Similarly, the settlement patterns are also influenced by the climates, according to Khaldun.

The Arab Empire started disintegrating after 1400 A.D. and the Arab School also came to a decline. However, the important contributions of the Arabs may be summarized as: Laying roots for the philosophy of environmental determinism (climatic); Generating a voluminous literature on the ecumene and non-ecumene areas of the known world of their times; and Presenting a new outlook of the world that later led to the onset of the Age of Discovery. The scholarly works of Arabs provided a new and more realistic picture of the world. But, on the whole, the literary developments among the Arabs were hampered by the over-refined state of language and love of storytelling in which the geographical nature of facts was over-shadowed. Nevertheless, the Arabs did keep the Greek learning alive, and it was from their hand the Europe received it, as the Dark Age drew to a close.

The Renaissance Movement had begun in the times of Arabs and got momentum in the Age of Discovery. Starting in the eighth century the renewed study of Greek revived much of the forgotten learning during the Arab period. But the classical spirit was not revived, i.e. no new concepts were established or formulated by the Arab scholars in a recognizable manner. The renaissance remained largely a religious movement, more concerned with preserving the knowledge of the past than expanding it through scientific investigation.

NOTES

1. **"Greek East"** and **"Latin West"** are terms used to distinguish between the two parts of the Graeco-Roman world, specifically the eastern regions where Greek was the *lingua franca,* and the western parts where Latin filled this role. During the Roman Empire a divide had persisted between Latin-and Greek-speaking areas; this divide was encouraged by administrative changes in the empire's structure between the third and 5th centuries, which led ultimately to the establishment of separate Eastern and Western halves of the Empire. After the fall of the Western Empire, the terms "Greek East" and "Latin West" are applied to areas that were formerly part of the Eastern or Western Empires, and also to areas that fell under the Greek or Latin cultural sphere but which had never been part of the Roman Empire. In this sense, particular attention is given to differences in Christianity in the two parts, specifically between Western Christianity and Eastern Christianity.
2. 'Patrology' or 'Patristics' is the study of the so called "Fathers (*patres*) of the Church." The **Patristic Period** is the period in the history of the Christian Church from around 100 A.D. until the eighth century. The period is characterized the theological writings of the Church Fathers such as Augustine.
3. Gilson (1955), p. 144.
4. Augustine of Hippo (354–430), also known as Saint Augustine or Saint Austin, was an early Christian theologian and philosopher whose writings influenced the development of Western Christianity and Western Philosophy. He is viewed as one of the most important Church Fathers in Western Christianity for his writings in the Patristic Era. Among his most important works are *The City of God* and *Confessions.*
5. Anicius Manlius Severinus Boëthius, commonly called Boëthius (480–524 A.D.), was a Roman senator, consuland philosopher of the early sixth century. Boëthius is best known today for *The Consolation of Philosophy*, which became one of the most popular and influential works of the Middle Ages.
6. Saint Anselm of Canterbury (1033–1109) composed dialogues and treatises with a rational and philosophical approach, sometimes causing him to be credited as the founder of Scholasticism. Despite his lack of recognition in this field in his own time, Anselm is now famed as the originator of the ontological argument for the existence of God and of the satisfaction theory of atonement. He was proclaimed a Doctor of the Church by a bull of Pope Clement XI in 1720.
7. Peter Abelard (1079–1142) was the pre-eminent philosopher and theologian of the twelfth century. Prior to the recovery of Aristotle, he brought the native Latin tradition in philosophy to its highest pitch. His genius was evident in all he did. He is, arguably, the greatest logician of the Middle Ages and is equally famous as the first great nominalist philosopher. He championed the use of reason in matters of faith (he was the first to use 'theology' in its modern sense), and his systematic treatment of religious doctrines are as remarkable for their philosophical penetration and subtlety as they are for their audacity. Abelard seemed larger than life to his contemporaries: his quick wit, sharp tongue, perfect memory, and boundless arrogance made him unbeatable in debate—he was said by supporter and detractor alike never to have lost an argument—and the force of his personality impressed itself vividly on all with whom he came into contact.
8. Spade (1994). pp. 57–58.
9. Chadwick (1981) and Gibson (1981).
10. Bocheński (1961. pp. 10–18).
11. Muhammad (571–632) established the religion of *Islam* and the Muslim community. According to traditional Muslim biographers, Muhammad was born in Mecca and died in Medina, both in the Hejaz region of present day Saudi Arabia. The name 'Muhammad' etymologically means "the praised one" in Arabic. Within Islam, Muhammad is known as "The Prophet" and "The Messenger".
12. *The Surya Siddhanta* was an astronomical treatise written in India, which had rules laid down to determine the true motions of the luminaries, which conform to their actual positions in the sky. It is not known who wrote the *Surya Siddhanta* or when it was first compiled, but the most

common dates are usually around the fourth century B.C. Later Indian mathematicians and astronomers such as Aryabhatta made references to this text, while later Arabic and Latin translations were very influential in the Middle East and Europe.

13. Husain, (1988), p. 83.
14. *Ibid.*
15. *Ibid.* p. 84.

REFERENCES

Chadwick, Henry (1981). *Boethius: The Consolations of Music, Logic, Theology and Philosophy*. Oxford: Clarendon Press.

Copleston, Frederick Copleston (1950). *A History of Philosophy*. Vol. II: *Medieval Philosophy: From Augustine to Duns Scotus*. Westminster, Md.: The Newman Press.

Copleston, Frederick Copleston (1953). *A History of Philosophy*. Vol. III: *Late Medieval and Renaissance Philosophy*. Westminster, Md.: The Newman Press.

Gibson, Margaret, ed. (1981). *Boethius: His Life, Thought and Influence*. Oxford: Basil Blackwell.

Gilson, E. (1955). *History of Christian Philosophy in Middle Ages*, New York.

Gracia, Jorge J.E. and Timothy Noone (2003). *A Companion to Philosophy in the Middle Ages*. Blackwell Companions to Philosophy. Oxford: Blackwell Publications.

Henry, Desmond Paul (1967). "Medieval Philosophy." *The Encyclopedia of Philosophy*. New York: Macmillan. Vol. 5, pp. 253–58.

Husain, M. (1988). *Evolution of Geographical Thought*, Second Revised Edition, Rawat Publications, Jaipur.

Knowles, David (1964). *The Evolution of Medieval Thought*. New York: Vintage Books.

Luscombe, D.E. (1997). *Medieval Thought: History of Western Philosophy*, Vol. 2. Oxford: Oxford University Press.

MacDonald, Scott, and Kretzmann, Norman (1998). "Medieval Philosophy." *Routledge Encyclopedia of Philosophy*. London: Routledge. Vol. 6, pp. 269–77.

Rana, L. (2008). *Geographical Thought: A systematic Record of Evolution*, Concept Publishing Company, New Delhi.

Robinson, Fred. C. (1984). "Medieval: the Middle Ages." *Speculum* 59, pp. 745–56. (Presidential address to the annual meeting of the Medieval Academy of America).

Schoedinger, Andrew B., ed. (1996). *Readings in Medieval Philosophy*. New York: Oxford University Press.

Spade, Paul Vincent (1994). "Medieval Philosophy." In Anthony Kenny, ed. *The Oxford Illustrated History of Philosophy*. Oxford: Oxford University Press. Chap. 2, pp. 55–105.

Weinberg, Julius R. (1964). *A Short History of Medieval Philosophy*. Princeton, N. J.: Princeton University Press.

8

Renaissance: Age of Discovery (Crisis Phase II)

The transitional phase from Medieval to Modern period (1450-1750 A.D.) is labelled as the Age of Discovery, the Renaissance period. Precisely it was a period from the early fifteenth century continuing into the eighteenth century. Largely inspired by the writings of Arab scholars, geographic horizons were pushed back during this period. The discovery of new lands, improvements in cartography and correction of old maps, and numerous scientific and technological inventions characterized this age. Later, various geographical societies were constituted to finance explorations and promote research in the discipline. With the improvements in the art of navigation and the use of magnetic compass the travels became very popular. The explorers discovered the world that was unexplored earlier. Geography was enriched by an immense collection of facts and information about the world. Ptolemy's maps were restudied and the missing details were filled in and the discrepancies removed. For the first time, there appeared the complete and accurate world maps and globes. Mercator's projection was devised (1569) and the new maps were prepared on it. As a consequence of several developments there came a scientific revolution not only in materialistic terms but also in the ideas and thinking of man. These developments were related to, for instance, the establishment of the heliocentric concept of universe, a number of astronomical observations by Nicholus Copernicus, the presentation of Laws of Gravitation by Isaac Newton and the invention of Printing Machine in Europe. This marked the Crisis Phase-2 in Geography. The dependence of geography increased on other disciplines. The ground became ripe so that geography could move from consolidation to splitting and specialization. This brought in two fundamental innovations to shake the world of scholarship. First, the 'academic freedom' of the discipline; and Second, the 'separation' of the academic world into distinct fields, each devoted to the study of a specific group of related themes. There emerged lot of scope for scientific investigations and, therefore, generalizations began in geography.

The Age of Discovery was the age of colonization and exploration, during which the European ships travelled around the world in search of new trading routes and partners to feed burgeoning capitalism in Europe. In the process, Europeans encountered peoples and mapped lands previously unknown. The Age of Exploration was rooted in new technologies and ideas growing out of the Renaissance and these included advances in cartography, navigation, and shipbuilding, besides the discovery of new lands. The post-Arab period was not only the Age of Discovery (1400-1750 A.D.), but also a period of transition from Old to New. The important developments of this period include:

Exploration of New Lands; Scientific and Technological Developments; and Improvements in Cartography & Map making.

Exploration of New Lands

Although the travels had become very popular in Arab period itself, a number of other developments gave further impetus to the explorations; and they included: (i) colonization, (ii) use of magnetic compass, and (iii) improvement in the art of navigation. Great explorers emerged in the scene (Table 8.1). Numerous journeys of geographical exploration were commissioned by a variety of nation states in Europe. Most of these voyages were financed because of the potential commercial returns from resource exploitation. The voyages also provided an opportunity for scientific investigation and discovery. These voyages also added many significant contributions to geographic knowledge. Most of the explorers of this period as listed below were of Italian, Portuguese and Dutch origin, the others being Spanish, Russian and English.

Table 8.1: Explorers in Late Medieval Age

NAME	TIME	NATIONALITY	AREA EXPLORED
John Cabbot	(1450-1499)	Italian	Newfoundland
Christopher Columbus	(1451-1506)	Italian	America
Amerigo Vespucci	(1454-1512)	Italian	America
Juan Ponce de Leon	(1460-1521)	Spanish	Florida, USA
Pedro A. Cabral	(1467-1520)	Portuguese	(Brazil) South America
Vasco-da-Gama	(1469-1524)	Portuguese	Indian Ocean
Fernand Magellan	(1480-1521)	Portuguese	Circumnavigation of Earth
Yermak Timofeyevuch	(1532-1585)	Russian	Siberia
Willem Barents	(1550-1597)	Dutch	Northern Shores of Europe
Willem Janszoon	(1570-1630)	Dutch	Coast of Australia
Abel Tasman	(1603-1659)	Dutch	Tasmania & New Zealand
James Cook	(1728-1779)	English	Pacific Ocean

The Italian discoverers, namely John Cabbot, Christopher Columbus and Amerigo Vespucci explored the lands across the Atlantic Ocean. John Cabbot, an Italian-born-English,[1] was the first European to search for the North-west Passage. He discovered the island of Newfoundland in 1497. Christopher Columbus sailed across the Atlantic Ocean, hoping to find a route to India, in order to trade for spices. He crossed the Atlantic Ocean and reached the Americas instead under the flag of *Castile*. History places great significance on his landing in America on October 12, 1492; the entire period of the history of the Americas before this date is usually known as Pre-Columbian. The anniversary of this event, *Columbus Day*, is still being celebrated in many parts of the continent. Columbus is commonly credited as the first European to see the Americas because of the profound impact his contact wrought on history. His voyage marked the beginning of European exploration and colonization of the Americas. Still another Italian was Amerigo Vespucci, who voyaged to and wrote about the Americas. He explored the east coast of South America between 1499 and 1502. His exploratory journeys along the eastern coastline of South America convinced him that a new continent had been discovered, a bold contention in his day when European explorers crossing the Atlantic Ocean, including Christopher Columbus, thought they were travelling to East Asia. Vespucci's voyages became widely known in Europe. In 1507, this new continent was named as "America" after Vespucci's first name, Amerigo.

Pedro Álvares Cabral, Vasco da Gama and Ferdinand Magellan were Portuguese navigators. Cabral is generally regarded as first European discoverer of Brazil (1500). Vasco da Gama discovered an ocean route from Portugal to the East. He was one of the most successful in the European Age of Discovery, and the first person to sail directly from Europe to India through Cape of Good Hope in Africa. Ferdinand Magellan led the first successful attempt to circumnavigate the Earth (Fig. 8.1). But he could not complete his final voyage as he was killed. However, his voyage made him one of the first individuals to cross all the meridians of the globe. He became the first person to lead an expedition sailing westward from Europe to Asia and to cross the Pacific Ocean.

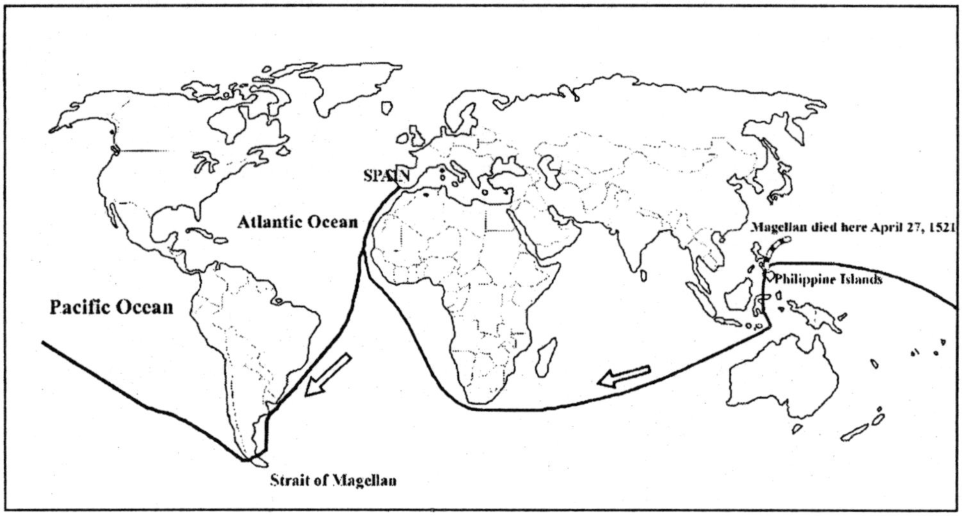

Fig. 8.1: Magellan's Voyage Around the World

The Dutch navigators explored the parts of Europe and Oceania. Among these, Willem Barents was a leader of early expeditions to the far north. Near the end of the sixteenth century, several expeditions tried to find a northeast passage to Asia. No one pursued this quest with such persistence and skill as this Dutch navigator, for whom the Barents Sea is named. Seeking the illusive Northeast Passage, Barents made three voyages in 1594, 1595, and 1596-97 along the frozen and dangerous northern shores of Eurasia. Another Dutch navigator and colonial governor, Willem Janszoon, was the first European known to have seen the coast of Australia. His name sometimes appears as Willem Jansz. He is first recorded as entering into the service of the *Oude compagnie*, one of the predecessors of the Dutch East India Company (VOC), as a mate aboard the *Hollandia*, part of the second fleet dispatched by the Dutch to the Dutch East Indies in 1598. He again sailed from the Netherlands for the East Indies in December 1603 as captain of the *Duyfken*, part of a fleet of twelve ships. He proceeded to chart some 320 km of the coastline, which he thought to be a southerly extension of New Guinea. Finding the land swampy and the people inhospitable,

Willem Janszoon headed home in 1606. He called the land he had discovered "Nieu Zelandt" after the Dutch province of Zealand but this name was not adopted, and was later used by Abel Tasman to name 'New Zealand'. Abel Janszoon Tasman, also Dutch by birth, is best known for his voyages of 1642 and 1644, in the service of the VOC (Dutch East India Company). His was the first European expedition to reach the islands of Van Diemen's Land (now Tasmania) and New Zealand and the first to sight the Fiji islands in 1643. He also mapped substantial portions of Australia.

Among other explorers were Juan Ponce de León (Spanish), Yermak Timofeyevich (Russian) and Captain James Cook (English). Juan Ponce de León[2] was the first European to set foot in Florida. He also established the oldest European settlement in Puerto Rico and discovered Gulf Stream, a current in the Atlantic Ocean. He sailed from Puerto Rico on March 3, 1513, and reached the east coast of Florida (St. Augustine) in April 1513. Ponce de Leon named the land "Pascua de Florida" (feast of flowers). He also claimed this land for Spain. Yermak Timofeyevich was the Cossack leader and explorer of Siberia. His exploration of Siberia marked the beginning of the expansion of Russia towards this region and its colonization. Yermak's exploration of Siberia is reflected in the so called Siberian Chronicles[3]. His life and death have been the sixteenth century subjects for numerous Russian songs, books and paintings. Captain James Cook made three voyages to the Pacific Ocean, accurately charting many areas and recording several islands and coastlines on European maps for the first time. His most notable accomplishments were the British discovery and claiming of the east coast of Australia; the European discovery of the Hawaiian Islands; and the first circumnavigation and mapping of Newfoundland and New Zealand. James Cook's eleven years sailing around the Pacific Ocean contributed much to European knowledge of the area. Several islands such as Easter Island and the Sandwich Islands (Hawaii) were encountered for the first time by Europeans, and his more accurate navigational charting of large areas of the Pacific was a major achievement.

With the discovery of new lands across the Atlantic, Pacific and Indian Oceans, the age of exploration is generally said to have ended in the early eighteenth century. However, the Arctic and Antarctic seas were not explored until the nineteenth century. It also took much longer for Europeans to reach the interior of continents such as Africa and North America than it did the coasts. Columbus, Vasco da Gama, Magellan, Cook and almost all the expeditionary leaders relied on Ptolemy's calculations about latitudes and circumference of the earth. As a result of these voyages huge knowledge was accumulated. There was an exorbitant literature that had to be tested, to be made accessible to others.

Scientific and Technological Developments

Mainly the invention of printing machine and the works of the people like Copernicus, Kapler, Galileo and Newton brought the scientific and technological developments in geography during the Age of Discovery. The printing press was invented at a time when it was really desired. Actually, the older form of printing machine had already come into existence before the Age of Discovery. It was movable type, which allowed individual characters to be arranged to form words; it was invented by Bi Sheng during 1041 to 1048 in China. However, the use of movable type to mass produce printed works was popularized by a German printer, Johannes Gutenberg in 1450. While there are several local claims for the invention of the printing press in other parts of Europe, including Laurens Janszoon Coster in the Netherlands and Panfilo Castaldi in Italy, Gutenberg is credited by most scholars with its initial invention. The invention of Printing Machine in Europe aided the scientific developments by publishing the accounts of explorations, voyages, discoveries and the geography of newly discovered lands.

Up to the fifteenth century most of the scholars took earth as the centre of the universe and tried to study the Universe with a geo-centric approach. But during the Age of Exploration, the heliocentric idea was strengthened. The Polish Scholar, Nicholus Copernicus (1473-1543), carried out numerous observations about the stars, planets and moons for 32 years during 1497-1529 A.D. His experiments provided the first modern formulation of a sun-centred theory of the solar system in his epochal book, *De revolutionibus orbium coelestium*. In 1618, Kepler proved that planetary motions were elliptical rather than circular. Johannes Kepler (1571-1630), a German mathematician and astronomer, was one of the most important scientists in the field of astronomy. He was the founder of "celestial mechanics", having been the first to explain planetary motion. He gave fundamental laws of planetary motion and postulated that the Earth and planets travel about the sun in elliptical orbits. In 1623, Galileo presented the proof that Copernicus was right about the heliocentric Universe. Galileo Galilei (1564–1642) was an Italian physicist, astronomer, and philosopher who is closely associated with the scientific revolution. Galileo's career coincided with that of Johannes Kepler. Galileo's achievements include improvements to the telescope, a variety of astronomical observations, the first and second laws of motion, and effective support for Copernicanism. He has been referred to as the "father of modern astronomy", as the "father of modern physics", and as "father of science". In 1686, Newton presented the "Laws of Gravitation". Isaac Newton (1642–1727) was an English physicist, mathematician, astronomer, alchemist, inventor and natural philosopher that is generally regarded as one of the most influential scientists in history. Newton wrote the *Philosophiae Naturalis Principia Mathematica* wherein he described universal gravitation and the laws of motion, laying the groundwork for classical mechanics. By deriving Kepler's Laws of Planetary Motion from this system, he was the first to show that the motion of bodies on Earth and of celestial bodies is governed by the same set of natural laws. The unifying and deterministic power of his laws was integral to the scientific revolution and the advancement of heliocentrism.

Improvements in Cartography and Map Making

The field of geography that was most negatively affected during Dark Age was the most benefited one during Age of Discovery. It was Cartography. Ptolemy's maps were unearthed and restudied. The missing details were filled up and the discrepancies were removed. Ptolemy's calculations about the latitudes, longitudes and circumference of the earth were corrected. The shape of the earth, i.e. elliptical, was finally ascertained. Missing details being filled up, the world map became almost complete. On this basis the first accurate 'globe' of the world was made. Martin Behaim (1459-1507), a Portuguese navigator and geographer, created the spherical globe depicting the Earth in its true three-dimensional form in 1492. Behaim's invention was a significant advance over two-dimensional maps because it created a more realistic depiction of the Earth's shape and surface configuration. Mercator's Projection was devised (1569 A.D.) and the new maps were prepared on it. Gerardus Mercator (1512-1594) was a Flemish[4] mapmaker and geographer who is best known for the map projection that bears his name. Mercator's projection is based on the concept of 'loxodrome' or 'rhumb line', the line of constant bearing. It makes this projection particularly useful to navigators.

With all the above developments, there came a scientific revolution not only in materialistic terms but also in the ideas and thinking of man. All the disciplines, along with geography, were benefited. Many of the earlier erroneous ideas were eliminated and fantastic stories of strange creatures gradually gave way to sober and reliable accounts or descriptions. The quality of geographic literature, as such, improved.

The Impact

The impact of the developments during Age of Discovery on geography was tremendous. A number of Geographical Societies were constituted in different parts of the world to promote further research in the discipline. Due to vast amount of accumulated knowledge there started branching of geography into different specialized fields of study. Earlier geography was 'all-in-one'. Many old theories or concepts were corrected and re-established; and many new theories about the origin of the earth, continents and man-nature relationships were advocated. The German scholar, Bernard Varenius (1622-1650) made the most fruitful use of all available information and knowledge about the earth during the Age of Discovery; and gave the first scientific literature in geography in the form of his *Geographia Generalis*. The positive gains of this period may be summarized as the Discovery of new lands and completion of world-map; the revival of interest in geography from a scientific perspective; 'Academic Freedom' of geography, i.e. identification as an independent field of study; Accumulation of huge amount of data and knowledge of geographical nature and its publication in the form of *Geographia Generalis*; Advancements in Cartographic technique of data representation and map-making; Beginning of scientific or speculative investigations and generalizations in geography; Emergence of various modern schools of thought in geography (as German, French, American and British); Geography acquiring a new definition and methodology; and Split of geography in various specialized fields of study, e.g. physical, human etc.

In spite of large-scale and worldwide explorations and voyages during the Age of Discovery, certain parts of the world still remained unexplored and untouched. They were the Polar areas, or more specifically the parts of Arctic and Antarctic regions. The scientific explorations of these regions began only in the twentieth century. Further, due to vast knowledge, data and information as accumulated during the Age of Discovery, the field of geography could not remain unified, i.e. having one common consensus on the definition, meaning, content, scope and approach of geography as a scientific discipline. Consequently, the discipline entered into a long period of controversies, as physical versus human, idiographic versus nomothetic, or systematic versus regional. Then, with the passage of time there was more and more use of theoretically deductive approaches. The empirically inductive methods became less popular and there emerged a generation of armchair geographers. However, looking on the positive side of the development, the foundations of Modern period of geographical thought were already laid before the Medieval period came to a close after the Age of Discovery. This was in the form of the writings of Bernard Varenius, the scholar of Pre-Classical phase of German School of Thought.

NOTES

1. Born in Gaeta and later a citizen of Venice, Italy, his Italian name of Giovanni Caboto became shortened to the English style John Cabot after he had lived in Bristol. In Italian, 'Caboto' means "a coastal seaman" and was quite a common name given to sailors and navigators. He was the son of a spice merchant (Guilo Caboto). The Caboto family moved to Venice in 1461 where this young Giovanni worked with his father in a spice shop for many years. John Cabot was attracted to the treasures of the Orient from a relatively early age. Whilst still a young man he moved to Venice, and it was there that he began learning of the value of Eastern spices and materials as it became his trade.
2. Born in Santervas, Spain, in 1460, Ponce de Leon was a soldier fighting Muslims in southern Spain in the early 1490s. He sailed on Christopher Columbus' second expedition to the Americas in 1493, but did not return to Spain with Columbus. Instead, he stayed in Santo Domingo (now called the Dominican Republic). He was appointed governor of the Dominican province of Higuey. He later heard of gold in the neighbouring island of Borinquen (now called Puerto Rico) and brutally conquered the island, claiming it for Spain. He was then appointed governor of this island. Ponce de Leon was then given the right to find and take the island of Bimini (in the Bahamas); he was searching for riches and the fountain of youth (a legendary spring that gave people eternal life and health).
3. A chronicle is the historical account of events. The Siberian Chronicles are the Russian chronicles of the late sixteenth-eighteenth centuries on the history of Siberia. They include the Yesipov Chronicle, Kungur Chronicle, Remezov Chronicle, Stroganov Chronicle, and others. These chronicles represent a valuable source on the early history of the Russian Siberia.
4. Flemishis the territorial name for the Dutch language spoken in historic Flanders, a region mostly comprising the northern part of Belgium, but also including a southern part of the Netherlands and a small area of northern France. The language is officially called Dutch by the governments of Belgium and the Netherlands, but the people living in the historic Flanders region still often use the term Flemish because of its historical and socio-cultural connotations. There are about 5,500,000 speakers in Belgium and the language varies only slightly from Dutch.

REFERENCE

Rana, L. (2008). *Geographical Thought: A Systematic Record of Evolution,* Concept Publishing Company, New Delhi.

9
Pre-Modern Tradition

The Ancient and Medieval developments characterize the Pre-modern period in geography. The Ancient schools of thought represented the Western, Indian and Chinese traditions, whereas the Arabian civilization generated the medieval tradition. Throughout Pre-modern Phase of its development, geography was considered as a descriptive field of study. The Pre-paradigmatic and Professionalization Stages followed by two major paradigmatic shifts and two Crises phases were witnessed during this long period of development. The two major paradigmatic shifts were related to the emergence of Graeco-Roman School in ancient times and Arab School in medieval times. They were succeeded by two Crises phases, viz. Dark Age and Age of Discovery, respectively towards the beginning and end of the medieval period. In general, the religion was the primary source of knowledge in pre-modern times in the European, Oriental and Muslim worlds alike. This long phase of evolution recorded a gradual transition of the geographical tradition from cosmological to scientific. The pre-modern philosophy largely revolved around the questions pertaining to the fundamental causes and principles of the Universe and their understanding. The reasoning and argumentation were established as the crucial philosophical methods.

The scholarly observations of the Western and Eastern civilizations during ancient period mark the Pre-Paradigm phase of geographic development. Towards West the earliest developments took place in western Asia, Egypt and Europe. Various cultural groups were involved besides Greeks and Romans. The Graeco-Roman people had a deep sharing of culture, knowledge, science and technology with many of the preceding civilizations that emerged successively in Mesopotamian region, viz. the Sumerians, Babylonians, Assyrians, Egyptians and Phoenicians. These people invented the first language script, the cuneiform; and made large advances in the techniques of agriculture and science. The origins of philosophy and reasoning are traced back to the early Mesopotamian wisdom. An extremely large body of texts has survived on cuneiform tablets here. The pre–Greek civilizations bequeathed Greeks, particularly, the sexagesimal system of numerals, the earliest known form of writing (cuneiform) and a system of alphabet, besides a number of valuable mathematical and astronomical contributions. The main feature of Greek thought, as obvious, represents the mathematical and astronomical knowledge of the Mesopotamians.

The philosophical developments in ancient Graeco-Roman world are associated with the Ionians, and are represented by four periods, viz. the period of Naturalism, Metaphysical, Ethical and the Religious. Beginning with the Milesian School and culminating in the Neo-Platonic School, there emerged numerous schools of thought in between. The period of Naturalism, for instance, was dominated by the Milesian, Pythagorean, Ephesian, Eleatic, Pluralists' and Atomists' streams. The Metaphysical period includes the philosophies especially those of Sophists, Socrates, Plato and Aristotle. The Ethical Period saw the displacement of metaphysics by ethics and the philosophy followed its ethical development through three streams of thought, viz. Stoicism, Epicureanism and Skepticism. During the Religious or the last period of development, the Greek philosophy sought refuge in religion, with an intermingling of Greek, Egyptian, Chaldean, and Jewish religions; and the phase was represented by the Judaic-Alexandrian, Neo-Pythagorean and Neo-Platonic thought. The Milesians, in general, focused on cosmological concerns, particularly the underlying substance of the universe. Pythagoreanism, revolving around the philosophy of Pythagoras, cultivated the mathematical sciences. The Ephesian school believed everything being in a continuous process of passage from one state to another. The Eleatic School, on the contrary, was based on the doctrine of permanence, changelessness; the basic tenet of its philosophy revealing the fundamental approach to the problem of Being. Being was thought as a space-filling mass and Non-Being as the empty space; the school held that only that perception is true which shows us an unchanging Being. Putting to themselves the problem of 'becoming' (Ephesian) and of 'being' (Eleatic), the Pluralists attempted uniting the two factions and trying to overcome their opposition. They stated that the world as we know it is composed of four basic material particles: earth, air, fire, and water, which are themselves unchanging and indestructible; calling them "the roots of everything" that make all the structures in the world. According to the different proportions in which these four elements are combined with each other the difference of the structure is produced. It is in the aggregation and segregation of elements thus arising, that the atomists found the real process which corresponds to what is popularly termed growth, increase or decrease. Another philosophy that tried to resolve the paradox of unchanging being and a changing world was that of Atomists, who believed that the matter in Universe is composed of an infinite number molecules (atoms) differing in size and form but endowed with perpetual motion which is derived from their essence. As the atoms are endowed with motion, the School admits a second primordial element, the void. The differences in gravity cause the atoms to whirl into motion, thus giving origin to the formation of things. Accordingly, from the primitive void have come the stars and the earth and all beings. Motion of individual distinct atoms has produced an orderly cosmos. During Metaphysical period, the Sophists introduced the study of man for the first time and held that knowledge is essentially empirical and relative to man. Socrates, on the other hand, concentrated on the universal aspects of man, as expressed in his doctrines of 'Concept' and 'Ethics'. They respectively dealt with the Universality and moral intellectualism. To arrive at understanding of these, Socrates used the inductive method of dialogue, popular as the Socratic Method. The greatest of the Socratic schools, after his death, was the Academy of Plato, which stayed closer to the original intent of the teachings of Socrates. Plato's metaphysics divides the world into two distinct aspects: the intelligible world of 'forms' and the perceptual world we see around us. The perceptual world consists of imperfect copies of the intelligible forms or ideas. Nearly every major work of Plato is, in some way, devoted to or dependent on this distinction. Aristotle, on the contrary, placed much more value on knowledge gained from the senses, and correspondingly earned the label of empiricist. Aristotle's method is largely inductive, while Plato's is essentially deductive. In the meanwhile there happened endless wars between the successors of Alexander and the Roman Empire resulting in political and social disruption of the region. This led to the displacement of metaphysics by ethics. During Ethical period, the solutions to the existing problems of life were found in three possible means: to appeal to reason, to appeal to pleasures and to deceive oneself in the belief that every appeal is useless, respectively constituting the three streams of thought which philosophy followed in its ethical development, viz. Stoicism, Epicureanism and Skepticism. During the Religious period of Greek philosophy, the Judaic-Alexandrian School attempted to coordinate Oriental religion with Greek speculation. The Neo-Pythagorean School attempted to construct a world religion upon Pythagorean doctrines, the teaching of Pythagoras aiming at ethico-religious reform. Neo-Platonism, in turn, conceived God as the source and goal of everything. The School tried to resolve this question by overcoming dualism with monism. While, on the one hand, the natural, metaphysical, ethical and religious philosophies inspired the thinking of Western scholars in ancient period, the interest in Natural Sciences also continued to survive. Along with philosophy, the theoretical interest of the time also concentrated on the physical sciences. The knowledge of new lands, new phenomena, and new facts made possible through the expeditions of Alexander also fostered the rise of science. The mathematics, physics, astronomy, zoology and botony were the principal sciences developed during this epoch. Geography, thus, began to be cultivated largely in its astronomo-mathematical aspect during Graeco-Roman period in ancient times in the West.

Towards Orient, the philosophy represents the broad traditions that were popular in India and China. The Indian philosophy had a common underlying theme of religion as expressed through its major stream of thought, viz. Samkhya, Nyaya, Vaisesika, Yoga, Purva Mimamsa and Vedanta. Samkhya, the oldest of all, is a dualistic philosophy that postulates everything in reality stems from '*purusha*', i.e. 'self, *atma* or soul' and '*prakriti*', i.e. 'matter, creative agency or energy'. Nyaya is the closest Indian equivalent to analytic philosophy. Just as Aristotelian logic, it is known for its methodology, based on a system of logic that has subsequently been adopted by the majority of the Indian schools. The Vaisesika School postulates an atomic pluralism, as proposed by the ancient Greek School of Atomists. The main objective of the Purva Mimamsa School was to establish the authority of Vedas. Lastly, the Vedanta or Later Mimamsa School concentrates on the philosophical teachings of the Upanishads. Ancient Indians possessed great knowledge in all areas of the science, technology and human development. In fact, several inventions and discoveries, that are believed to have originated from the Western world have, actually been carried out by ancient Indians. However, the religious flavour is very strong in descriptions of the ancient Indian scholars. Although, the subsequent invaders ruthlessly destroyed the architectural achievements of this country and imposed their despotic rule, they also transmitted Indian sciences to the outside world, leading to translations into Arabic and other languages. Like Europeans and Indians, the ancient Chinese scholars and travellers also possessed a rich heritage of geographical knowledge. China's history has been largely characterized by repeated divisions and reunifications amid alternating periods of peace and war, and violent imperial dynastic change. Cultural and political influences from many parts of Asia, carried by successive waves of immigration, expansion, and assimilation, merged to create the Chinese culture. For centuries, Imperial China was one of the world's most technologically advanced civilizations. As in Egypt, Mesopotamia, and along the Indus River, Chinese civilization began within a major river valley, the Yellow River Valley. China, with a huge geographical expanse, contained an almost infinite number of ethnic groups and languages. Most Chinese dynasties emerged in the historical heartland of China, the China Proper. These earliest dynasties were based on agriculture, augmented by hunting and animal husbandry. They were also in use of a writing system and bronze metallurgy as revealed in inscriptions found on oracle bones.

They initially embraced the philosophy of Confucianism and later the Buddhism. The major contributions of these dynasties include, for instance, the concept of a centralized government, the unification of the legal code, written language, measurement, and currency. The Chinese script is the only logographic script still used in the world. Besides, they also promoted extensive trade with distant foreign countries, through Silk Road. The Chinese merchants explored all of the Indian Ocean. The Shang was the first confirmed historical dynasty. The Han Dynasty was the first one to embrace the philosophy of Confucianism, which became the ideological underpinning of all regimes until the end of imperial China. Under the Han Dynasty, China made great advances in many areas of the arts and sciences. The first opening of trading connections between China and the West, the 'Silk Road' was also enabled during this regime. The Chinese Thought has been mainly represented by Confucianism, Taoism and Buddhism. These three Teaching greatly influenced the Chinese philosophy. Taoism explained the "path" to contentment; Confucius taught people about the importance of being loyal, unselfish, and courteous; and Buddhism focused on achieving absolute peace through an 'Eightfold Path' including rightness of views, intention, speech, action, livelihood, effort, mindfulness and concentration. Among the commonalties of Chinese philosophies are the epistemological optimism, the tendency to not to view man as part of nature, disbelief in the personified supernatural power questioning the nature and existence of God, the belief that the purpose of philosophy is primarily to serve as an ethical and practical guide and the political focus. The Chinese contribution goes to the fields of Astronomy and Astrology, Human and Regional Geography, Physical Geography, Mathematical and Practical Geography, and Accounts of Travels and Explorations. All the above developments in its Pre-Paradigm phase set the ground for geography to emerge as a discipline of scholarship with time, and enter the next phase of development, i.e. Professionalization.

The Professionalization in the discipline had started becoming evident in the attempts of the European, Indian and Chinese scholars during ancient period itself. In spite of physical isolation, fascinating parallels were exhibited in the concepts and methods of study among them. The accomplishments of this phase are linked to the works of numerous philosophers, scholars, mathematicians, administrators and explorers. Geography, as a discipline of scholarship, has filtered through several fields of learning related to the study of Universe, Astronomy and Astrology, Earth Studies, Mathematical Sciences, History and Historiography, Cartography, Humanistic Studies, Sciences, Technology and Travels and Explorations. Among all these, the fields of Philosophy, Astronomy, Mathematics and Science have been commonly contributed by Graeco-Romans, Indian and Chinese; whereas history and cartography have especially interested the Greeks and Chinese, as also the travels and exploration.

The Universe and its origin remained a point of speculation among all the ancient civilizations. The ancient scholars dealt with many problems pertaining to Cosmology, Cosmogony and Cosmography. The early Greek philosophers, for instance, sought to explain the cosmos and all its phenomena by appealing to regularities within the cosmic system itself, without reference to extra-natural causes. The prominent views on the subject were those of Anaxagoras, Anaximenes, Anaximander, Philolos, Empedocles and Heraclitus. The Cosmological theory of Anaxagoras, for instance, suggests the origin of the world caused through the continuous process of motion and separation from a unified mixture. Likewise, Anaximenes also gave an account of the world coming out of the previously existing matter. Philolaus' ideas revolved around a hypothetical astronomical object, the Central Fire, and a sphere of the fixed stars and planets all moving around it. Empedocles, on the other hand, posited two cosmic forces which work upon the elements in both creative and destructive ways, personified as 'Love'—a force of attraction and combination, and 'Strife'—a force of repulsion and separation. The world, he saw, in a constant state of organic evolution due to the combined action of both the forces. The Universe, called Brahamand, is conceived as very immense and wide in the ancient Indian literature. In consonance with the Greek thinking, Acharya Kapil, the founder of Samkhya philosophy, considered 'Prakriti', the primal matter, as the mother of cosmic creation. Four types of theories have been advanced about the origin of Universe. Of these, the theory of Artistic Origin compares the Universe to a house, constructed by a number of Gods by weaving various materials into a pattern, and shaping it by blasting and smelting; the theory of Mechanical Origin suggests the dismemberment of a primeval body and the creation of Universe a part of its sacrifice ceremony; the scholars having faith in the theory of Philosophical Origin believed in the origin of Universe out of a vacuum; and lastly, the theory of Instrumental Origin talks about the occurrence of two parent bodies, viz. the the Earth (*Dharti*) and the Heaven (*Swarga*), from which the Universe was created, the union of these two producing the Sun. With the fusion of the basic concepts of Taoism and Confucianism, the ancient Chinese also created a traditionally harmonious view of the universe. They advocated primarily three models of celestial orientation: 'Gai Tian', the theory of the dome shaped heaven that placed what we today call Ursa Major at the centre of the celestial dome and China at the centre of the Earth; the 'School of Hun Tian' envisioning a spherical heaven shaped much like a hens egg where the earth is like the yolk; and finally, the theory of Xuan Ye suggesting the unprecedented thought of an infinite universe with suspended celestial bodies. These theories lead to many technological advances in astronomy such as armillary rings and spheres.

The first references to astronomy in ancient world are interwoven with astrology. The astronomical observations were made with the help of some simple instruments as the gnomon, scaphion etc. The Greeks made lasting contributions to astronomy in synthesis with Babylonian thought. Of them, the most intriguing is Hipparchus' work which covers a wide variety of astronomical questions. Relying mainly on this work, Ptolemy compiled the astronomical knowledge of ancient Greek and Babylonian world in his Almagest, one of the most influential books of classical antiquity. It contains a myriad of information ranging from earth conceptions to sun, moon, and star movement and also an updated version of a star catalogue. The Ptolemaic explanation of the motions of the planets remained the accepted wisdom until the Polish scholar Copernicus proposed a heliocentric view in 1543. Besides, Philolaus' astronomical system is also popular that proposed the heavenly bodies arranged in concentric circles around a central fire, the first presentation to move the earth from the centre of the universe and make it an orbiting planet instead. Where the studies of all the others depended on observational astronomy, Anaximander's merits lie in the field of speculative astronomy. His vision implied depth in the universe, i.e. the idea that celestial bodies lying behind one another in increasing order of brightness. Due to the inclination of the axis of heaven, the celestial bodies do not circle around the earth in the same plane as the earth's-flat-surface, but are tilted. Indian astronomers also propounded the theory that the earth is a sphere. Through 'Aryabhattiya', the work of Aryabhatta, the Indians were the earliest to discover that the orbits of the planets around the Sun are ellipses, to describe numerical and geometric rules for eclipse calculations, to discuss about the locations of the planets in relation to its movement around the sun, to refer to the light of planets as reflections from the sun, and to explain eclipses in terms of shadows cast by and falling on earth. His working was later adopted by the Greeks and then the Arabs. Aryabhatta too described a geocentric model of the solar system, in which the Sun and Moon are each carried by epicycles which in turn revolve around the Earth. Subsequent Indian astronomers improved on his calculations, but his methods provided the core. Another Indian astronomer, Brahmagupta, calculated the circumference of the earth more or less accurately as known today. Besides, Bhaskara's work is based on a model of the solar system which is heliocentric and whose movements are determined by gravitation. Ancient Chinese astronomy was primarily a government activity. It was the astronomer's role to keep track of planetary motions and to suggest what astronomical phenomena may mean for the ruling emperor. The ancient Chinese invented the first planetarium and had sophisticated observatory buildings. Observing total solar eclipses was a major element of forecasting the future health and successes of the Emperor, and astrologers were left with the onerous task of trying to anticipate when these events might occur. Based on the theories of his predecessors, Zhang Heng systematically developed the celestial sphere theory. The celestial model Zhang Heng constructed was almost a physical model of Ptolemy's earth-centred theory. Shen Kuo, another scholar of medieval astronomy, improved the designs of several astronomical instruments as the gnomon, armillary sphere and clepsydra clock. Shen is also known for his cosmological hypotheses in explaining the variations of planetary motions, including retrogradation.

The Earth Studies of earlier scholars, particularly of Indians, were very elaborate and dealt with its origin, cardinal points, meteorology, geomorphology, seismology and physical divisions. The concept of *Prithvi*, the Earth, was most basic in the study of geography, and has been profusely used in the Vedas and Puranas. The term *'Bhugol'* as used for Geography clearly suggests that the ancient Indians endorsed the earth being a sphere. As far as the origin of the earth is concerned, many of the facts as put forward by the ancient Indian scholars were more or less accurately known. The ancient Indian scholars believed in the solidification of earth from gaseous matter, its crust being made of hard rocks, clayey material and sandy material. They mention the earth to be apparently floating on water like a sailing boat on the river. They were also aware of the fact that there is more land surface in the Northern Hemisphere. The ancient Indian scholars also used the concept of imaginary lines of *'Akshansh'* (latitudes) and *'Deshantar'* (longitudes). On the basis of these, they attempted to divide the Earth into three global realms, viz. Equatorial, Northern Polar and Southern Polar with the identification of North Pole and South Pole as the antipode of each other. They have also drawn Prime Meridian. These imaginary lines, the position of Sun and various stars have helped them to determine local time at various places. However, the world was not believed to exist beyond Equator, as the region here was compared to hell of the earth. The Eastern part, on the other hand, was believed to be 'the land of Gods'. This thinking is in consonance with that of the Europeans in the Early Medieval period, when the Dark Ages prevailed and the East in 'T-in-O' Maps was assumed to be the place of *Adam* and *Eve*. The Puranic literature formulated the idea of ten directions. Besides the main directions of East, West, North and South, there were included Zenith and Nadir, and added four more, viz. Southeast, Southwest, North-west and Northeast. The designation of these directions is significant in the sense that it bears the concept of the ruling deities dominating in each of these directions. The studies on Meteorology stretch back millennia. The ancient Indian scholars were aware of the vacuum between the earth and the heaven as *Antariksha*, its vast extent or thickness and the occurrence of various weather phenomena here. As far as the knowledge about the seasons is concerned, six seasons have been identified in *Valmiki Ramayan*, viz. Spring, Summer, Rainy, Autumn, Winter and Severe Winter. Early speculation and hypothesis pertaining to meteorology also has a long tradition in China. The observations of Chinese scholars on weather and climate are commendable. They are known to have prepared a number of weather reports on the basis of empirically based facts. Besides, the philosopher, Wang Chong, accurately described the process of the water cycle and Shen wrote vivid descriptions of tornadoes. There are evidences of the interest of earlier Chinese scholars in the field of Paleoclimatology as well. Contributing to Seismological studies, the ancient Greeks, Indian and Chinese all provided the rational explanations scientifically alike. Thales and Democritus, for instance, found in water the cause of earthquakes. The deities of Air, Fire and Water cause the earthquakes or *Bhukamp*, as per Indian version. The Chinese kept accurate records of not only celestial events, but of earthquakes as well. They also discounted the superstitions related to the cause of earthquakes. In the ancient period, the knowledge about various parts of the world was limited due to the poor means of communication. Even then, the attempts were made to divide the world into several regions on the basis of available information. The term *Dwipa* has been used to designate various realms (continents) of the earth in India. Accordingly, the known world during the Puranic period was divided into seven *Dwipas* or 'regions', viz. *Jambu, Krauncha, Kusha Plaksha, Pushkara, Shaka* and *Shalmali*. These divisions exclude the American Continents, Greenland, England and Antarctica, since they were discovered only during the Age of Discovery in the late medieval period. The regions appeared to have derived their names from the existing vegetation here.

Along with Mathematics, the group of mathematical sciences also included Arithmetic, Algebra, Geometry, Trigonometry, Algorithm and Calculus. Among the various fields of Greek scholarship, mathematics was one and Pythagoras excelled in it. He thought numbers were divine and believed that everything was related to mathematics; and through mathematics everything could be predicted and measured, just as in music, in rhythmic patterns or cycles. In India, mathematics has its roots in Vedic literature. The oldest extant mathematical document produced on the Indian subcontinent dates back to seventh century A.D. The brick work of Harappa and Mohenjodaro excavations show that people of ancient India (2500 BCE) possessed knowledge of Geometry. Aryabhatt formulated the rules for finding the area of a 'triangle', which led to the origin of Trignometry. There are evidences of the Babylonian cuneiform tablets containing Pythagorean triples indicating that there was sophisticated understanding on the topic in Mesopotamia as well. Excavations at various sites of the Indus Valley have uncovered evidence of the use of "practical mathematics". The people here used a standardized system of ruler and weights; and mass produced weights in regular geometrical shapes, which included hexahedra, barrels, cones, and cylinders, thereby demonstrating knowledge of basic geometry. Indian mathematicians made early contributions to the study of arithmetic, calculus, algorithm, decimal number system and the concept of zero. Bhaskaracharya's book *Lilavati* is regarded as the first book on modern arithmetic. The *Siddhanta Shiromani* of Bhaskara II is an astronomical treatise and contains many theories related to the preliminary concepts of infinitesimal calculus and mathematical analysis. The Kerala School mathematicians from 14th to sixteenth century expanded on Bhaskara's work and further advanced the development of calculus in India. Also the technique of algorithm, which is today widely used in designing software programmes for computers was derived from Indian mathematics. Just like other mathematical inventions, the decimal number system in use today was first recorded in Indian mathematics. It was then transmitted to the Islamic world and eventually to Europe. The earliest surviving evidence of decimal place value numerals in India and South-east Asia is from the middle of the first millennium CE. Decimal numerals, recorded as early as 683 A.D., have also been found in stone inscriptions in Indonesia and Cambodia, where Indian cultural influence was substantial. The inclusion of zero in mathematics opened up new dimensions in the field. The concept of zero reached the west through Arabs. Numerals are found in the inscriptions of Ashoka the Great in the third century BCE. This knowledge travelled to Europe. In Arab countries even now numerals are known as *Hindse,* i.e. from India. It was from an Indian text on Mathematics, again, that the Arab mathematicians perfected the decimal system and gave the world its current system of enumeration which we call the Arab numerals, and which are originally Indian numerals. The first to compute with the zero as a number and to know negative numbers, was Brahmagupta. Brahmagupta's treatise *Brahma-sputa-siddhanta* was translated into Arabic. For several centuries this translation remained a standard text of reference in the Arab world. The representation of numbers in a 'positional system' is another contribution of ancient Indians to mathematics. However, Geometry originated in Egypt where it developed out of necessity; it was introduced into Greece by Thales who had visited Egypt. In the third century B.C. geometry was put into an axiomatic form by Euclid, whose treatment—Euclidean geometry—set a standard for many centuries to follow. The Indian mathematics recognised geometry as *Rekha Ganita* (Line Computation). The *Sulbasutra* of Baudhayana contains geometric solutions. The *Sulbasutra,* which literally mean 'Rule of the Chord' give geometrical methods of constructing altars and temples. The Arab scholar Al Battani studied Indian use of ratios from *Rekha Ganita* and introduced them among the Arab scholars. The term 'algebra' is a corruption of the term '*Al jabr*' which the Arabs gave to *Bijaganitam.* Incidentally the term *Bijaganit* is still in use in India to refer to this subject. Aryabhatta, the fifth century Indian mathematician has reference to it in his treatise on Mathematics, *Aryabhattiya.* Trigonometry is believed to have evolved in the Hellenistic world. Hipparchus made an early contribution to trigonometry producing a table of chords, an early example of a trigonometric table. He also introduced the division of a circle into 360 degrees into Greece. Trigonometry was introduced into ancient India through the translation of Greek works and was further advanced here. The *Surya Siddhanta* (c. 400), a treatise of Indian astronomy, contains the roots of modern trigonometry. Later, its Arabic and Latin translations were very influential in Europe and the Middle East.

As far as Cartography is concerned, Ptolemy, the Greek, was most influential among all. His contribution in the field of cartography and mapmaking is unparalleled. He designed a graticule and plotted the world using the concept of conical projection. On the Chinese realm, Pei Xiu and Zhang Heng were the earliest ones to promote and use cartographic techniques. Zhang Heng was the first to establish the grid reference system. The earliest known reference to the application of a geometric grid and mathematically graduated scale to a map was contained in the writings of the cartographer Pei Xiu who outlined and analyzed the advancements of cartography, surveying, and mathematics up until his time. The Chinese were familiar with the use of scale. Pei Xiu outlined six principles that should be observed when creating a map, the first three principles outlined the use of scale, direction, and road distance, and the last three used to properly calculate distances on uneven terrain as represented on a flat, two dimensional map.

The human aspect was intelligently dealt in various Historical, Regional, Anthropological, Psychological and Linguistic studies by the scholars of antiquity. The Greeks and Chinese were the pioneers in the field of history. Among Greeks, Herodotus is the first known historian of classical antiquity. Although his writing covers only a short period in the history of Greek civilization, his account of the history of the wars is the first example of historical writing in the Western World. Before Herodotus the idea of recording history was not widespread and not attempted in a systematic manner. Polybius provided a unified view of history rather than a chronology. Posidonius, on the other hand, was not concerned with isolated political history of peoples and individuals, but included discussions of all forces and factors (geographical factors, mineral resources, climate, nutrition), which let humans act and be a part of their environment. The Chinese historical form represents dynastic history, or *Jizhuanti* history of dynasties. The Chinese historians regard Sima Qian's work, 'Shiji', as their model, which stands as the "official format" of the history of China. In this, Sima initiated a new writing style by presenting history in a series of biographies. But, Ban Gu's written work in geography set the trend for the establishment of geographical sections of history texts, and sparked the trend of the gazetteer in ancient China. Before Sima, histories were written as certain events or certain periods of history of states. Regarding the study of human beings, the explanations offered suggest that human culture developed as a response to necessity and the hardships of our environment. According to Anaximander, for instance, life originated from the moisture that covered the earth before it was dried up by the sun. Democritus suggests that human life was originally like that of animals; it describes the gradual development of human communities for purposes of mutual aid, the origin of language, crafts and agriculture. The study such phenomena as perception, cognition, emotion, personality, behaviour and interpersonal relationships had also been a concern in the past. Philolaus, for instance, presented account of psychic faculties: The head [is the seat] of intellect, the heart of soul (*psychê*) and sensation, the navel of rooting and first growth, the genitals of the sowing of seed and generation. Philolaus thus recognizes four distinct psychic faculties. He makes a clear distinction between intellect (*nous*), which is limited to human beings, and perception, which animals possess as well. The core text for linguistics in ancient times has been discovered in the '*Aṣṭādhyāyī*' of Pāṇini, the *sine qua non* of learning. Pāṇini's grammar is considered as the world's first formal system, well before the nineteenth century innovations and the subsequent development of mathematical logic. The ancient scholars were concerned with documenting geographical information of regions far outside of their realm as well. The issues related to man-nature relationships, resources and population were important in China. In China, the earliest known geographical writing dates back to the fifth century B.C. Although Chinese had been writing of civilizations of the Middle East, India, and Central Asia since the traveller Zhang Qian (Second century B.C.), later Chinese would provide more concrete and valid information on the topography and geographical aspects of foreign regions as well.

The sciences like physics, chemistry, botany, biology, pharmacology, medicine etc. are rooted in the works of ancient Greeks and Indians. Among Greeks, the foundation of **physics** lies in the writings of the scholars like Empedocles and Posidonius. Empedocles asserted that the matter is created by four 'elements': earth, air, fire and water, by being 'mixed' in different combinations and proportions. The four elements correspond, at macroscopic level, with the traditional quadripartite division of the cosmos into earth, sea, air, and the fiery aether of the heavenly bodies. The knowledge of Indian scholars in this regard was more extensive. The concepts of atom (*Anu, Parmanu*) and relativity (*Sapekshavada*) were explicitly stated by them nearly 600 years before Christ in an elaborate manner. From the Vedic times, Indians too classified the material world into four elements, viz. Earth (*Prithvi*), fire (*Agni*), air (*Maya*) and water (*Apa*); but, also added a fifth element, viz. ether or *Akasha* to these. These five elements or *Pancha Mahabhootas* were identified with various human senses of perception; earth with smell, air with feeling, fire with vision, water with taste and ether with sound. Indian philosophers also believed that these elements were physically palpable and hence comprised miniscule particles of matter. The last miniscule particle of matter which could not be subdivided further was termed *Parmanu*. Another Indian philosopher, Pakudha Katyayana, had also propounded ideas about the atomic constitution of the material world. These philosophers considered the Atom to be indestructible and hence eternal. Parallel to the development of the concepts of atom and atomic permutations and combinations in physics there also was a similar development of ideas in the area of biology, botany and chemistry. The science of living organisms, biology, reflected only in the works of Greeks in earlier times; the most successful scientific writings in this field were those of Aristotle. His ideas led to the development of a rational theory of evolution, as put forward by Darwin 2200 years later. Again, it was a Greek philosopher, Theophrastus, a pupil of Aristotle, who is regarded as the founder of botany. He covered most aspects of botany: descriptions of plants, classification, plant distribution, propagation, germination, and cultivation. Theophrastus classified plants into trees, shrubs, under-shrubs, and herbs. He described and discussed more than 500 species and varieties of plants from lands bordering the Atlantic and Mediterranean. Parallel to the development of the concepts of atom and atomic permutations and combinations in physics there also was a similar development of ideas in the area of Chemistry in earlier times. For pharmacology, the ancient Chinese Scholars, as Shen Kuo and Su Song, held great concern. As far as Medicine is concerned, the *Charaka Samhita* of Acharya Charak is among the earliest surviving medical manuals, and the most authoritative. The work reveals the facts on human anatomy, embryology, pharmacology, blood circulation and various diseases, and a description on the medicinal qualities and functions of about one lakh herbal plants. It represented a major advance over the superstitious ways of treating medical problems. Instead of appeasing deities and making sacrifices, practitioners were now looking at clinical problems and deciding how to treat them based on the specific disease.

The ancient civilizations had acquired proficiency in various technological fields as, for instance, metallurgy, aviation, civil engineering and architecture. Particularly, the Ancient Chinese are known for not only their travels and trade, but also for their many inventions and technological advancements. In the area of smelting metals (Metallurgy), the Indians excelled in the extraction of metals from ore, and also in the casting of metals. There are many such examples that bear testimony to the excellence in smelting metals achieved in India in ancient times. Indian steel and iron were reportedly being used by the Romans for manufacturing armour as well as cutlery. Apart from this, Ashoka pillar at Mehrauli, New Delhi, another iron pillar in Karnataka, and the copper statue of Gautama Buddha found in Bihar stand proof of India's metallurgical heritage. The discovery of urban settlements of Mohenjo-daro and Harappa indicate existence of civil engineering and architecture, which blossomed to a highly precise science of civil engineering and architecture and found expression in innumerable monuments of ancient India. Through his innovative discoveries, Acharya Bharadwaj has been recognized as the pioneer of aviation technology. Pinhole Camera, Water clock, armillary sphere, seismograph, Clock Tower, Magnetic needle compass, Movable type printing, Odometer and South Pointing Chariot were among several inventions in the field of Scientific Instruments and Appliances. Shen Kuo was the first to experiment with the pinhole camera. The Water clock or Clepsydra, a timekeeping device, was in use in China as long ago as the Shang Dynasty. The Greek astronomer Eratosthenes invented the first armillary sphere, an astronomical instrument representing the celestial sphere. Su Song's armillary sphere was the first to be provided with a mechanical clock drive. However, Zhang Heng is the first person known to have applied hydraulic motive power, i.e. by employing a waterwheel and clepsydra, to rotate an armillary sphere. Zhang Heng's water-powered armillary sphere had profound effects on Chinese astronomy and mechanical engineering in later generations. Zhang Heng also presented to the Han court what many historians consider to be his most impressive invention, the first seismometer. Shen Kuo's writing on magnetic compasses has proved invaluable for understanding China's earliest use of the compass for seafaring navigation. An obscure commoner and artisan known as Bi Sheng invented ceramic movable type-printing. Bi Sheng's methodical innovation was revolutionary for his time. Zhang Heng is often credited with inventing the first odometer. Similar devices were used by the Roman and Han-Chinese empires at about the same period. The South Pointing Chariot was another mechanical device credited to Zhang Heng.

The travels and explorations fascinated the Europeans and Chinese alike. Hanno the Navigator, Pytheas and Alexander the Great among Greeks, explored Africa, north-west Europe, British Isles and much of what was then the civilized world. Travels beyond Chinese horizons were undertaken by orders of an emperor or by missionaries or traders which reveal the story of high adventures. The Chinese travellers had gone far beyond their original cultural hearth area. The discovery of Mediterranean civilization, for instance, is credited to Chinese traveller, Chang Chien. Among the other Chinese travellers most distinguished ones included the Buddhist monk, Huan Tsang who was able to cross the high windswept plateau of Tibet and the world's highest mountain on the way to India; Chou Ta Kuan who visited Combodia and wrote a detailed account of the strange customs of the Combodians; and Cheng Ho, a Chinese admiral who sailed towards Java, Malaya and even as far as India. These journeys had also promoted a great variety of economic and cultural exchanges among regions, and opened regular trade routes between the West and the East. Thus, with the efforts of the European, Indian and Chinese scholars, a large amount of geographical knowledge was accumulated. New researches were thereby attracted and the research made progress. With these three parallel schools of thought, geography witnessed a great degree of professionalization. With time, however, the West became the centre of continuously increasing scholarly research; and the Graeco-Roman (European) school started becoming dominant as it developed new methods that come to be regarded as more interesting or significant. In other words, geography witnessed its first paradigmatic phase of development.

The Graeco-Roman School represents Paradigm Phase-1. Displacing others in a short time, the Greeks and Romans led to concentrate research in geography, an activity described as Normal Science. The Greeks were the first to scientifically handle the gathered information and to develop the procedures to be described as 'scientific methods'. Greeks distinguished themselves from others because of their intellectual, philosophical and scientific approach. This phase of geography saw the emergence of various fields of learning ranging from physical, human, historical, theoretical, political, regional, mathematical and practical covering almost entirely the gamut of geographical knowledge. The ancient Greeks were the originators of literary and mathematical traditions in geography. The Greeks are also credited for the introduction of Gnomon, one of the first scientific instruments ever made. The Romans, in general, carried forward the Greek tradition. The existing world of the times was believed to have two major regional divisions, namely Europe and Asia. Besides, parts of northern Africa had also been discovered. The conquests of Alexander the Great pushed Greek knowledge of the earth far to the east as far as and across the Indus River, and southward along the eastern side of Mediterranean to Egypt, where he established his rule. With his accounts the Greeks became familiar with the geography of Persian Empire, Central Asia, Afghanistan, India and coastal parts of Iran. While Alexander extended Greek geographic horizons to the east, Pytheas voyaged north-west of the Greek world, covering parts of Western and Northern Europe. He, particularly, described about people and agriculture in Great Britain; and also made observations related to the vast expanse of the Arctic Ocean, the length of the longest day north of Norway; and ocean-tides with their relation to the various phases of moon. Contributing to regional studies, Strabo, the Roman scholar, tried to establish the relationship of man with what he called 'the natural attributes of a place'. He considered 'Place' in two different perspectives, viz. in terms of its astronomical (abstract) location and in terms of its terrestrial (on land) position. The development of mathematical and practical geography in this period is, although, linked to a number of scholars, Thales and Anaximander, particularly, are known as the originators of mathematical tradition in geography. Thales was first to apply various geometrical concepts in geographical studies. Anaximander is known for the introduction of a Babylonian instrument, Gnomon, into the Greek world, making possible a variety of observations regarding noontime, duration of sunshine, time of solstices and equinoxes, the time of sunrise and sunshine and so on. Further extending the mathematical tradition in geography, Eratosthenes and Posidonius calculated of circumference of earth for the first time in history. Based on Assyrian mathematics, Hipparchus defined a grid of latitudes and longitudes; divided circle into 360 degrees; attempted to determine exact positions of every point on earth's surface; and wrestled with the problem of showing the curved surface of earth on a flat surface for the first time. Aristotle was the first philosopher who wrote with definite arguments about the spherical shape of earth. However, among all Ptolemy was the most influential. He developed the sound principles of mathematics and applied to the study of practical geography; recalculated the circumference of earth; designed a graticule and plotted the world using the concept of conical projection. He also worked out the dimension of the inhabitable world and plotted the places of physical and human interest. Ptolemy's attempts not only helped to prepare the first complete map of the known world, but also reformed the map of world. The writings and maps of Ptolemy inspired the explorers of Great Age of Discovery. Ptolemy's geocentric view of the universe remained an accepted doctrine until the Age of Discovery. Academically, these ancient Greek scholars formed strong ties of geography with history and this inter-disciplinary bond could not be broken for several centuries to come, and has been intensely represented in the writings of geographers till the modern times. The roots of the concept of 'Areal differentiation' as adopted by Hartshorne in the modern period lie in the writings of Strabo, the first century B.C. Roman scholar. Mainly Herodotus and Strabo contributed to Historical Geography. Although Herodotus is credited with the very old idea that all history must be treated geographically and all geography must be treated historically. Strabo on the other hand, compiled the writings of his predecessors, and wrote a voluminous book entitled 'Geographie'. Strabo considered geography essentially as an encyclopaedic description of the known inhabited world. Two great Greek philosophers, namely Plato and Aristotle, made important contributions to the development of geographical ideas suggesting the methods of theory building and hypothesis testing. They respectively were the masters of deductive and inductive reasoning. Where Plato believed in building theory by reasoning from the general to the particular, Aristotle did it in the reverse order. These methods of deductive and inductive reasoning were introduced to geography as early as 4th and 5th centuries B.C. The Greeks made a remarkable development in the field of physical geography. Oceanography, for instance, was authoritatively studies by Posidonius. In spite of being a historian, Herodotus always placed and studied historic events in geographic setting. His observations of silt-deposits on the flood plain along the Nile River led him link it to the shape to Greek alphabet 'Delta', thus introducing 'delta' to geographical dictionary. On the basis of the study of Sea of Azov, Aristotle explained the phenomenon of expansion of land in the shallow seas and formation of delta. He correctly pointed out the process of alluvial deposition by which so many places of land area being eroded and added into the sea. Theophratus examined the relationships of plants and climate, and this marked the beginning of the study of plant geography. Plato discussed the problems and causes of the barren or waste lands and explained that such tracts in the past were full of vegetative covers and fertile soils. Under the impact of the external forces, the forests have been depleted and soils leached resulting into barren topography. Besides, the Greeks, have also recognized four major winds, having different properties, directions and the associated weather type. They are: Boreas: the north wind—strong, cool, with clear skies; Eurus: the east wind—warm and gentle; Notus: the south wind—wet and sometimes violent; and Zephyrus: the west wind—balmy (fragrant) but with gale force. Moreover, the Greeks also divided the world into: Torrid, Temperate and Frigid zones. They established a relationship between temperature and ecumene regions of the world. However, the major contribution of Greeks and Romans was a body of philosophical and cosmological ideas, besides a number of terminologies that still constitute the basic vocabulary of geography. Geography as a discipline got enriched by a number of concepts, viewpoints and techniques. Various terminologies were added to its vocabulary by the ancient Greek scholars. Besides coining the term 'geography', a number of other terminologies are also credited to the ancient Greeks. The important ones include, for instance, 'cartography', 'chorography', 'longitude', 'latitude', 'torrid', 'frigid', 'temperate' etc. All these terms are still in use today.

The first phase of Crisis with Revolution in geography, popularly known as Dark Age in Europe, was witnessed in the Early Medieval Period. This Phase saw geographic horizons closing in due to theocratic slant in human thinking and writings. The approach to the science, including geography, became theocratic. This kind of thinking had an adverse impact on the art of map-making, which got distorted. There emerged maps, which were purely fancy and unscientific. The earlier concepts of sphericity and the nearly correct distances of places, latitudes and longitudes were ignored. Christianity influenced the social, economic, literary and philosophical developments during this. The Christian era was marked by the loss of ancient scientific concepts about the world and their replacements by unscientific, uncritical cosmogonies based largely on scriptures. The missionary travels were the only source of knowledge for the regional account of different places during this period. Consequently, Geography made a little progress as the organized body of knowledge. The quest for scientific learning stopped; instead, there was regression of the already acquired scientific knowledge. Existing concepts of the world were reshaped to conform to the teachings of Church. The image of the world was molded to fit the Biblical references, which were considered more valid. The earth was considered as a flat disc with Jerusalem at the centre. The ideal map of this period was 'Orbis-Terrarum' or 'T-in-O' Map which originally included the Mediterranean Sea and its surrounding territories. The three divisions, namely, Europe, Asia and Africa were accepted as standard. Along with the inhabited world, there were inserted the mythical places, beasts and dragons in all these maps. These maps were made very beautiful and extremely decorative, but did not have any relation with actual facts, as they were far from reality with no scientific base. Also, there was vagueness in the direction. The sphericity and the nearly correct distances of places, latitudes and longitudes were ignored. The travels undertaken were mainly missionary movements made by pilgrims to holy shrines. The descriptions made were also purely aesthetic and without logic. Thus, most of the classical correct concepts were forgotten, and the old errors appeared again about the world maps and habitable portions of the earth. Moreover, the involvement of Christians and Muslims (Arabs) in crusade wars further aggravated the problem. When Europe was plunged into darkness, the Arab School of Thought was rising to prominence in medieval times. In contrast to earlier times, new zeal of learning spread through the world of Islam. The Arab civilization, in fact, became the main link between the ancient civilization and the modern world. The rise of Arab school marks the end of the recession in development of geographic literature and return to more scholarly research. This lead to geography to enter in a new paradigm phase of development.

Medieval philosophers are the historical successors of the philosophers of antiquity, leading the discipline to its Paradigm Phase-2. Medieval Thought in Geographic Development represents two major philosophical trends, viz. European and Middle-Eastern. However, the Paradigm Phase-2 is associated with the Rise of Arab School of Thought during High Medieval Phase. The trend in the West represents the philosophy of Western Europe during the Fall of Rome and the Renaissance. Medieval philosophy in the West, in turn, was shaped by two pieces of the cultural legacy of late antiquity that survived the collapse of Roman civilization. The First of these is the 'Latin language', which provided medieval thinkers with access to some important ancient resources that gave early medieval thinkers an acquaintance with classical ideas. The Second was 'Christianity', stimulating the philosophical activity. Its ideas and doctrines constituted a rich source of philosophical subject matter. Medieval philosophy, therefore, took root in an intellectual world sustained by the Church and permeated with Christianity's texts and ideas in the West. This, in turn, lead to, First the influx into the West of a large and previously unknown body of philosophical material newly translated into Latin from Greek sources, and Second, the emergence and growth of the great medieval universities. There was rise of Islam in Arab in seventh century and the new zest of learning spreaded through the world of Islam. The vast expanse of the Muslim Empire facilitated the coming together of intellectual and scientific traditions of various civilizations, particularly Greek and Indian. The Arabian conquest of Palestine, Syria, Persia and Bactria, where Greek culture still flourished, gave to the Muslims the rich fruit of Greek learning. Baghdad was the major centre of intellectual world. Islamic geography was primarily a continuation of the Greek and Roman scholarship which had been lost in Christian Europe. The Arab scholars and travellers presented a new outlook of the world, which later led to the onset of the Age of

Discovery. With the rise of Arab School, geography witnessed another paradigmatic shift in development. Many Arab scholars contributed to it, their outstanding contributions go to various fields of learning as Mathematical and Practical Geography, Regional and Historical Geography, Physical Geography and Climatology and Human Geography. As far as the Mathematical and Practical Geography is concerned, the Greek science was the starting point in Arab School. It is not surprising, therefore, that mathematical geography received the maximum attention. Besides, the factors like semi-desertic conditions, open atmosphere, clear skies and vast distances to scan that facilitated the astronomical observations also helped in the promotion of studies in mathematical and practical geography. The early Arabs considered the earth as the centre of the Universe, around which revolved the planets. There were several attempts to ascertain the size of the earth. Sea-Charts and maps based on cylindrical projection were made. Borrowing from and improving over the works of Greeks, the Arabs divided the globe of earth into Torrid, Frigid and Temperate zones. Besides, another greatest contribution of the Arabs to the advancement of mathematical knowledge was the introduction of the Hindu invention of the zero in arithmetic (*Hindsa*). Before this, the Roman numerals were in use and their range was limited. A wealth of regional descriptions exists in the works of Arab scholars, covering both the physical and human aspects. The writings of Ibn Batuta described the ecumene areas of the world. Accordingly, the limit of the habitable world was identified up to the end of the Mediterranean sea in the west, Japan in the east, Siberia in the north and about 10 degrees beyond equator in the south. The travels of Ibn Batuta brought empirical proof that Aristotle, the ancient Greek scholar, had been wrong in asserting that the Torrid Zone was too hot and unfavourable for human habitation. Batuta, on the contrary, confirmed that numerous tribes occupied it. The physical geography finds a special place in the writings of the Arab scholars. Their important observations are related to the study of the processes shaping landforms, alluvial deposits and phenomenon of tides. Al Biruni in his book entitled *'Kitab-al-Hind'* describes the fluvial processes and alluvial deposits of Himalayan region. Ibn Sina keenly observed the processes shaping landforms, viz. weathering and denudation. He formulated the idea that mountains were being constantly worn down by rivers and that the highest peaks occurred where the rocks were especially resistant to erosion. He also noted the presence of fossils in the rocks of high mountains. Various Arab scholars and navigators observed the phenomena of tides, and proved that they are caused because of the gravitational pull of the sun and the moon. Al Masudi, for instance, recorded the spring and neap tides while navigating the Caspian Sea. Besides, Al Idrisi corrected the erroneous ideas about the Indian Ocean and the Caspian Sea. The Indian Ocean was supposed to be enclosed, whereas the Caspian Sea was shown as an arm of the World Ocean in 'T-in-O' maps of Early Medieval period. However, the most notable under 'physical' were the contributions made to climatology. The important contributions in the field of Climatic Geography have come in the form of coining of the term *'mausim'* (Monsoon), preparation of the first Climatic Atlas, and division of the world into Climatic Regions. The studies carried under Human Geography by Arab scholars are dominated by the writings of Ibn Khaldun and deal mainly with the influence of physical factors on life forms. The prime focus of the writings of Ibn-Khaldun was on the human society, culture and its development. Modern geographers are particularly impressed with his cultural interpretation of the physical environment and his analysis of the role of the city in regional economy. Ibn-Khaldun tried to correlate man and environmental relationships in a scientific way. Ultimately, this Paradigm Phase is known for laying roots for the philosophy of environmental (climatic) determinism, generating a voluminous literature on the ecumene and non-ecumene areas of the known world of their times and presenting a new outlook of the world that later led to the onset of the Age of Discovery, another crisis phase in Geography.

The Transitional Phase from Medieval to Modern period is labelled as the Age of Discovery or the Renaissance period. Largely inspired by the writings of Arab scholars, the great Age of Discovery began in the late fifteenth century and continuing into the eighteenth century during which the geographic horizons were pushed back. It was the age of colonization and exploration during which the European ships travelled around the world to search for new trading routes and partners to feed its burgeoning capitalism. In the process, Europeans encountered peoples and mapped lands previously unknown. The discovery of new lands, improvements in cartography and correction of old maps, and various scientific and technological inventions characterized this age. With the improvements in the art of navigation and the use of magnetic compass the travels became very popular. Great explorers emerged in the scene and discovered the world that was unexplored earlier. As a result of the developments like the establishment of the heliocentric concept of the universe, numerous astronomical observations by Nicholus Copernicus, the presentation of Laws of Gravitation by Isaac Newton, and the invention of Printing Machine in Europe, there came a scientific revolution not only in materialistic terms but also in the ideas and thinking of man. Geography was enriched by an immense collection of facts and information about the world. Ptolemy's maps were restudied and the missing details were filled in and the discrepancies removed. There emerged lot of scope for scientific investigations and generalizations began in geography. Besides the discovery of new lands The Age of Exploration was rooted in new technologies and ideas growing out of the Renaissance and these included advances in cartography, navigation, and shipbuilding. The post-Arab period was not only the Age of Discovery, but also a period of transition from Old to New. The important developments of this period include the Exploration of New Lands, Scientific and Technological Developments and Improvements in Cartography and Map making. Although the travels had become very popular in Arab period itself, a number of other developments gave further impetus to the explorations; and they included, e.g. colonization, use of magnetic compass and improvement in the art of navigation. Great explorers emerged in the scene. Numerous journeys of geographical exploration were commissioned by a variety of nation states in Europe. Most of these voyages were financed because of the potential commercial returns from resource exploitation. The voyages also provided an opportunity for scientific investigation and discovery. These voyages also added many significant contributions to geographic knowledge. Among these explorers most were of Italian, Portuguese and Dutch origin, the others being Spanish, Russian and English. The Italian discoverers explored the lands across the Atlantic Ocean. John Cabbot was the first European to search for the North-west Passage; he discovered the island of Newfoundland. Christopher Columbus sailed across the Atlantic Ocean, hoping to find a route to India, in order to trade for spices; instead, he reached Americas. Columbus is commonly credited as the first European to see the Americas and his voyage marked the beginning of European exploration and colonization of the Americas. However, this new continent was named as "America" after Amerigo Vespucci s first name, another Italian voyager. Pedro Álvares Cabral, Vasco da Gama and Ferdinand Magellan were Portuguese navigators. Cabral is generally regarded as first European discoverer of Brazil. Vasco da Gama discovered an ocean route from Portugal to the East and was the first person to sail directly from Europe to India, through Cape of Good Hope in Africa. Ferdinand Magellan led the first successful attempt to circumnavigate the Earth and became the first person to lead an expedition sailing westward from Europe to Asia and to cross the Pacific Ocean. The Dutch navigators explored the parts of Europe and Oceania. Among these, Willem Barents was a leader of early expeditions to the far north, after whom the Barents Sea is named. Another Dutch navigator, Willem Janszoon or Willem Jansz was the first European known to have seen the coast of Australia. Abel Janszoon Tasman, also Dutch by birth, is best known for his voyages to reach the islands of Tasmania, New Zealand and Fiji. He also mapped substantial portions of Australia. Among other explorers were Juan Ponce de León (Spanish), Yermak Timofeyevich (Russian) and Captain James Cook (English). Juan Ponce de León was the first European to set foot in Florida. He also established the oldest European settlement in Puerto Rico and discovered the Gulf Stream, a current in the Atlantic Ocean. Yermak Timofeyevich was the explorer of Siberia. Captain James Cook made several voyages to the Pacific Ocean, accurately charting many areas and recording several islands and coastlines on European maps for the first time. His most notable accomplishments were the British discovery and claiming of the east coast of Australia; the European discovery of the Hawaiian Islands; and the first circum-navigation and mapping of Newfoundland and New Zealand. Cook's sailing around the Pacific Ocean contributed much to European knowledge of the area. Several islands such as Easter Island and the Sandwich Islands (Hawaii) were encountered for the first time by Europeans. With the discovery of new lands across the Atlantic, Pacific and Indian Oceans, the Age of Exploration is generally said to have ended in early eighteenth century. Almost all the expeditionary leaders relied on Ptolemy's calculations about latitudes and circumference of the earth. As a result of these voyages huge knowledge was accumulated. There was a huge mass of literature that had to be tested, to be made accessible to others. Mainly the invention of printing machine and the works of the scholars like Copernicus, Kapler, Galileo and Newton brought the scientific and technological developments in geography during the Age of Discovery. A German printer, Johannes Gutenberg is credited for the invention of 'Printing Machine' in Europe; it aided the scientific developments by publishing the accounts of explorations. The Polish Scholar, Nicholus Copernicus, carried out numerous experiments on "celestial mechanics", to explain planetary motion, and provided the first modern formulation of a heliocentric theory of the solar system. He also proved that planetary motions are elliptical. Galileo Galilei, the Italian astronomer, presented the proof that Copernicus was right about the heliocentric Universe. Galileo's achievements include improvements to the telescope, a variety of astronomical observations, the laws of motion, and effective support for Copernicanism. Isaac Newton, the English physicist and astronomer, presented the 'Laws of Gravitation'. He was the first to show that the motion of bodies on Earth and of celestial bodies is governed by the same set of natural laws. The unifying and deterministic power of his laws was integral to the scientific revolution and the advancement of heliocentrism.

The field of geography that was most benefited one during Age of Discovery was Cartography. Ptolemy's maps were unearthed and restudied. Ptolemy's calculations about the latitudes, longitudes and circumference of the earth were corrected. The shape of the earth, i.e. elliptical, was finally ascertained. With the filling up of missing details and the removal of discrepancies, the world map became almost complete. On this basis the first accurate 'globe' of the world was made by Martin Behaim, a Portuguese navigator, depicting the Earth in its true three-dimensional form. Besides, Mercator's Projection, based on the concept of the line of constant bearing, was devised, especially, for navigators. With all the above developments, there came a scientific revolution not only in materialistic terms but also in the ideas and thinking of man. The quality of geographic literature, as such, improved. The impact of the developments during Age of Discovery on geography was tremendous. The positive gains of this period may be summarized as: the Discovery of new lands and completion of world-map; the revival of interest in geography from a scientific perspective; 'Academic Freedom' of geography, i.e. identification as an independent field of study; Accumulation of huge amount of data and knowledge of geographical nature and its publication the form of *Geographia Generalis* by Bernard Varenius, the German Scholar; Advancements in Cartographic technique of data representation and map-making; Beginning of scientific or speculative investigations and generalizations in geography; Emergence of various modern schools of thought in geography (as German, French, American and British); Geography acquiring a new definition and methodology; and Split of geography in various specialized fields of study, e.g. physical, human etc. Thus, the foundations of Modern period of geographical thought were already laid before the Medieval period came to a close after the Age of Discovery. This was in the form of the writings of Bernard Varenius, a German scholar.

III
MODERN

10
Dominance of German School of Thought (Paradigm Phase III)

The knowledge of geography was not organized in late medieval and early modern period. In the absence of an appropriate methodology, it was no more than encyclopaedic compilations of places and their descriptions. The principles behind spatial arrangement of phenomena, their causal connections and inter-relationships that could give geography a true scientific status were yet to be established. The credit for this goes to the German scholars. It is generally said that Greeks laid the foundation and Germans built the structure of geography. For about next 300 years after the Age of Discovery geography was purely a German science. Lot of research was carried on towards defining geography and deciding upon its methodology (approach, scope, content, method of study). Ultimately the place of geography was seen established among scientific disciplines. The Paradigm Phase III represents developments especially during 1650s–1850s when the German School of Geographic Thought was at its dominance. The school got established and popularized mainly through the writings of Varenius, Kant, Humboldt and Ritter. Right up to the opening of twentieth century all significant developments were related to the works of these scholars.

The initial impetus to German School came from the works of Bernard Varenius (1622–1650). The time of Varenius was the one close to Medieval period and beginning of Modern period.[1] It was the time when a large mass of accumulated empirical knowledge was at disposal in the discipline. Such unprecedented expansion in knowledge paved the way for a new advance in geography. Varenius had studied mathematics and medicine, and intended to practice medicine; but, the recent discoveries by various navigators, attracted him to geography.[2] He is best known for his work entitled '*Geographia Generalis*', considered as the first modern literature in geography. In this, he endeavoured to lay down the general principles of the subject on a wide scientific basis according to the knowledge of his day. His work long held its position as the best treatise in existence on scientific and comparative geography. Varenius brought together contemporary knowledge of astronomy and cartography and subjected the different theories of his day to sound critical analysis. Through this he tried to provide a scientific methodology to geography by advocating two approaches for its study, viz. 'General' (systematic) and 'Special' (regional). The former approach emphasizes on one aspect of study throughout all the regions. The regional studies, on the other hand, deal with a particular area in detail covering all its relevant and inter-related elements. The systematic studies lead to the formulation of generalizations, laws and theories. General geography, on the other hand, provides the fundamentals that need to be applied to regional studies. These generalizations help to understand the regional geographies. All these ideas are summarized in '*Geographia Generalis*'. The book itself is divided into two parts, viz. General Geography and Regional Geography. General Geography consists of description of three forms of earth features, viz. Absolute, Relative and Comparative. The First investigates mathematical facts relating to the earth as a whole, its figure, dimensions, motions, their measurement etc.; the Second part considers the earth as affected by the heavenly bodies, climates, seasons, the difference of apparent time at different places, variations in the length of the day etc.; and the Third treats with divisions of the earth surface, their relative positions, globe and map-construction, longitude, navigation etc. Similarly, Regional Geography also consists of descriptions of three types of properties of the constituent areal units. These include Celestial, Terrestrial and Human. The Celestial Properties are related to the study of climate, atmosphere etc.; Terrestrial Properties deal with the location, relief, soil and land characteristics; and Human Properties are concerned with population, political divisions, communication, socio-cultural traits, religion, diet etc. Thus, the geography of this period included both 'general' studies of particular kinds of phenomena of the earth's surface, and descriptions of many kinds of phenomena found in particular areas. The terms that Varenius used: 'General' and 'Special' geography became standard in the discipline. Among later geographers, Humboldt and Ritter, popularly known as the fathers of modern geography, drew attention to Varenius' genius and services to science. They did it respectively in the form of a natural scientist and a human geographer. In the meanwhile, writings of Emmanuel Kant (1724-1794) guided the research in Germany.

Emmanuel Kant, an armchair geographer, was an outstanding example of a professional philosopher in Western Thought.[3] During the last years of the eighteenth and early years of the nineteenth centuries, the work of Kant served to anchor geography within the methods and philosophy of science. Kant called for the organization of all knowledge with three categories: the grouping of facts systematically, the examination of information and events temporally, and the evaluation of phenomena as they relate to one another in space. In doing this, he provided the organization foundation for most modern intellectual pursuits as well as the basic model under which most institutions of higher learning in Europe and North America now function. His call for the grouping of facts systematically gave rise to the notion of academic disciplines organized around transcendent themes. Furthermore, in recognizing that all observable phenomena happen in time and space, he effectively identified the unifying themes and importance of the modern disciplines of History and Geography. Kant was also instrumental in freeing geography from its previous close Relationship with Theology. However, for Kant, Geography was a chorographic and mainly a descriptive science, distinct from the systematic sciences. He, thus, considerably modified the scientific structure of geography as proposed by Bernard Varenius. He laid an exceptionalist claim not only for geography, but also for history. Therefore, he has been labelled as 'the father of exceptionalism' in geography by F.K. Schaefer, a twentieth century American scholar. Schaefer considered Kant a poor and immature geographer. However, in spite of all the criticisms put forward against Kantian philosophy, the basic tradition of the field of geography has persisted in his writings, i.e. "geography is a description of the whole earth, where the earth is understood to mean the surface of the earth and the dwelling place of the man". The exposition of geography as 'chorology' by Alfred Hettner and Richard Hartshorne in the late nineteenth and early twentieth century had closely followed the Kantian philosophy. Hartshorne's

"*Nature of Geography*" (1939) also followed the line of thought first formalized by Kant. Through his book, it became the dominant concept of geography as a field of learning in the mid twentieth century. Kant agreed that there are two ways to acquire knowledge, viz. through 'sense perceptions' and through 'pure reasoning'. The former approach is 'inductive' and the latter is 'deductive'. The sense perceptions, in turn, are of two types, viz. 'inner' and 'outer'. However, Kant's approach to geographical studies was purely deductive. Kant also pointed out that there are basically two ways of classifying knowledge for the various phenomena on the surface of earth. They are: 'Logical Classification', i.e. grouping according to the nature of the inherent characteristics of the phenomena; and 'Physical Classification', i.e. grouping according to the position or occurrence of phenomena in time or space. The Logical Classification of Kant deals with organizing knowledge according to the type of objects studied, and accordingly covers a series of systematic (pure) sciences that are devoted to the study of particular kinds (sets) of phenomena or activities, e.g. Botany, Zoology, Physics, Chemistry etc. The Physical Classification, on the other hand, deals with organization of knowledge in its temporal or spatial dimensions and, therefore, includes historical and spatial sciences. The historical sciences are concerned with the arrangement of phenomena in terms of time dimension, the chronological order; and the spatial sciences are concerned with the arrangement of phenomena in the space dimension, the chorological order. The disciplines falling within this category, according to Kant, are different from those in the previous classification. In simple words, History is the description in terms of time; and Geography is the description in terms of space. Kant considered both history and geography as descriptive in their nature. He also regarded both of these disciplines as complementary to each other. According to him, history and geography are both essential sciences, standing alongside the systematic sciences. Without them, man cannot achieve a full understanding of the world. Besides, Kant laid the greatest emphasis on the study of Physical Geography. Along with it, he also identified some other branches of geography namely (1) Mathematical Geography—studying the form, size and movement of the earth, and its place in the Solar system; (2) Political Geography—the study of relationship between the political units and their physical background; (3) Moral Geography—the study of customs and characteristics of man in relation to environment; (4) Mercantile Geography—the study of trade and commerce; and (5) Theological Geography—the study of distribution of religious traits in different types of environment, e.g. Christians in Europe, Muslims in Central Asia, Confucians in Hwang Ho Valley in China, and Vaishnavs in Sutlej-Ganga Plains etc. After Kant, Alexander von Humboldt and Carl Ritter revived the tradition of Bernard Varenius.

The writings of Humboldt and Ritter characterize the Classical Phase of German School. Both were contemporary; they lived and worked at Berlin for more than three decades. They together tried to mould the subject matter of geography into scientific form, following the tradition of Bernard Varenius. The prevalent philosophies at the time of Humboldt were 'natural' and 'moral'. The natural philosophies were related to the physical world, envisaging the earth as an organic whole, inseparable and mutually interdependent; whereas, the moral philosophies were related to man's inner world and human thinking. Humboldt's basic inclination was towards natural, whereas Ritter had anthropocentric view.

Alexander von Humboldt was a great traveller, explorer, naturalist and cosmographer. His travels covered the regions around Mediterranean in Europe, Mexico and Caribbean Islands in the West and Siberia in the East. He envisaged the earth as a whole, an organic whole, of all parts of which being mutually interdependent. Like Kant, Humboldt also attempted on a division of knowledge, but the classification he contemplated fell into three groups, viz.: (1) Systematic Sciences—studying the phenomena in terms of their similar characteristics, e.g. Botany, Geology, Physics, Chemistry—etc.; (2) Historical Sciences—studying the history of development of existing groups of phenomena, e.g. History, Anthropology etc.; and (3) Earth Sciences—concerned with the distribution and arrangement of phenomena on the earth's surface. The objective of such sciences, e.g. Geography, is to study the phenomena in their spatial distribution, spatial relationships and inter-dependence. Although, the basis of this classification was close to that of Kant's, but, Humboldt treated all kinds of subject-matter (including history and geography) as scientific and accorded them the status of sciences. While dealing with the subject matter of geography, Humboldt coined a term 'cosmography', the study of the Universe; and divided it into Uranography and Geography. In his opinion Uranography is the descriptive astronomy that is concerned with mapping the stars, galaxies or other celestial bodies. Geography, on the other hand, deals with the terrestrial part. Geography, according to Humboldt, is the description of the earth, which deals with the inter-relationships of phenomena that exist together in an area. Humboldt's approach was holistic. Unlike Kant, he followed empirical and inductive method of study. It was based on self-observation and experimentation, because he himself collected enormous data related to plants, animals, rocks and other aspects. Humboldt's methods of recording and describing in the field are important in themselves. He noted, measured and mapped changes in relief, plant life, crops, tree and snowline, in a way that had never been done before. Since no related maps were available, he himself drew cross-sections of the areas he travelled to generalize on these features. He used to travel with 40 different types of instruments for observation and recording. They included, e.g. telescope, sextants, quadrants, barometer etc. As far as the analysis of the recorded information was concerned, he used systematic and comparative methods. Because the purpose of his travels was not only to measure and record various kinds of phenomena, but also to find out the ways in which the great variety of observable phenomena of the landscape are associated and inter-connected with each other at different places. In order to bring about precision and accuracy in the descriptions he also used various cartographic techniques. Humboldt conceived man and his activities as part of the nature. *'Cosmos'*, the ordered Universe, is Humboldt's best-known work. It is a comprehensive survey of the world. In *'Cosmos'*, Humboldt regarded man as an integral and constituent part of the nature. Through this work, particularly, Humboldt was able to make geography "an original science" rather than a collection of facts from the physical and biological sciences. Although Humboldt included man and his works in the concept of nature and natural areas, but he did not consider man as a primary determinant.

Carl Ritter's work was not only contemporary to that of Humboldt, but it was also complementary to it. He exercised profound influence on geography by adopting a different methodology. Where Humboldt's inclination was towards nature, Ritter emphasized more on human. Humboldt followed systematic approach and Ritter was in favour of regional or synthetic approach. Although Ritter did make some travels in Europe, he was basically an armchair geographer. Just like *'Cosmos'* of Humboldt, Ritter also authored a great work entitled *Die-Erdkunde*. Ritter was deeply a religious man. He thought that divine has created the earth as a home of man. Unlike Humboldt, Ritter was an anthropo-geographer who had an inclination towards human and regional geography. He also had a vision of an ordered and harmonious universe. But, his approach was teleological. Since, Ritter was deeply a religious man, he gave teleological interpretations to man-nature relationships. Being a teleologist, Ritter asserted that all phenomena are spatially distributed according to the plan of God for mankind. The main problem of the teleological philosophy is that such a philosophy cannot be tested empirically and therefore, does not qualify as a scientific explanation. Nevertheless, Ritter's teleological outlook did form an active school of geography, and had the characteristics of a paradigm.[4] Ritter's teleological approach is generally taken to mean that a phenomenon is explained in relation to the purpose it is believed to serve. The 'holistic-synthesis' of organic relationships is strongly related to teleological explanatory models. This approach is reflected in most of the religions and their philosophies. However, in spite of being an armchair scholar himself Ritter did identify geography an empirical and descriptive science. According to him, geography deals with local conditions and embraces the attributes of a place in the form of topographical, formal and material characteristics. Topographical attributes deal with the natural divisions of the earth; Formal are concerned with systematic or general features as atmosphere, movement of water etc.; and Material attributes describe the geographical aspects of natural history and cover distribution of human beings, minerals, plants, animals etc. These three attributes constitute the basic content of his most important work, *Die-Erdkunde*, as well. '*Die Erdkunde*' is Ritter's best-known work and is based on the regional descriptions of the earth covering largely Asia, Africa and Europe. Like *Cosmos* of Humboldt, Ritter's *Die-Erdkunde* is also a professional work—a product of his lifetime reading, selection, interpretation and organization of a mass of available data. This work was written with a distinct theme, i.e. comparative and inter-relationships. On the basis of this he proved that geography also has a right to be distinctively considered as a science with the same dignity as the other sciences. Ritter claimed that the central principle of geography is 'the relation of all phenomena and forms of nature to the human race, examined and organized within the framework of the unique geographical associations of land and man on the earth's surface'. Through his *Erdkunde*, Ritter made a rhetorical claim for geography as the science of the earth. He believed "the earth to be an organism made, even in its smallest details, with divine intent, to fit the needs of the man to perfection; as the body is made for the soul, so is the physical globe made for mankind".

Ritter is also popular among geographers for his Regional Concept. His division of the Earth consists of four levels in hierarchical order: Earth or '*Erdkunde*' (1st Order); Continental Units or '*Erdteile*' (2nd Order); Division of each Continent into various reliefs or Orographic regions (3rd Order); and Smaller Areal or terrestrial units arrived at inductively (4th Order). After the Earth, the second order units, according to Ritter, are arrived at deductively. The conceptual basis of this division is traditional, i.e. each continent to be sub-divided by its orography (relief characteristics). He held that the continent is not a generic complex, but a complex of individual units. The fourth order areal units are arrived at inductively from the detailed configuration and terrestrial mosaic. Ritter appears to have envisaged each continent as a complex of individual units whose whole structure and content could be arrived at by inductive procedures from the smallest areas of associated phenomena (synthetic approach). In this way one can classify the regions of various orders (micro to macro and *vice versa*). Thus, Ritter sought to develop the concept of terrestrial or spatial unity. He divided the earth into major continental units (*Erdteile*). According to him the major continental units were to be arrived at deductively; and the smaller units within each continent to be derived and studied with an inductive approach, i.e. to divide the individual continental units into smaller areal units on the basis of empirical observations. But, he could not attempt this himself due to the lack of maps and related data. Therefore, he was forced to involve himself in teleological speculations (imaginations). To conclude, Ritter's main contribution to geography was his concept of regional associations of terrestrial phenomena at various levels on the earth's surface. One has to overlook teleological approach of Ritter in order to discover a treasure of geographic insights in his works. His plan of study, based on the concept of regional associations of terrestrial phenomena, became the model for geographical presentations and had far-reaching influences. This classical period of German geography terminated with the death of both Humboldt and Ritter in the year 1859.

Both Humboldt and Ritter worked towards providing geography as the status of a scientific discipline and appeared, as such, inspired by Bernard Varenius. Although, they did not lay any exceptionalist claim for geography, the Kantian influence can also be seen in the classification of sciences by Humboldt and in the chorological perspective of Ritter. Both had a holistic approach to geography, as they believed in the Unity of Nature. It was axiomatic that the unity of nature included organic as well as inorganic, human as well as non-human, immaterial as well as material. The exclusion of any part would be not only arbitrary but would destroy the coherence and unity of the whole. The writings of Humboldt and Ritter had a thrust on inter-relationships, spatial associations or spatial integration of phenomena. The Physical Geography was considered as the base of all geographical studies by Humboldt and Ritter. Both emphasized on the importance of empirical method of research in geography. However, the sphere of their research differed. Humboldt worked on general or systematic studies, with a major emphasis on physical. Ritter, on the other hand, mainly explored the human world with regional approach. Humboldt was a student of science and we know him as a natural scientist; whereas, Ritter was basically a religious man, a human and regional geographer, an anthropogeographer. Although both believed in the Unity of Nature, they differed in their philosophical approach to geographical studies. For Humboldt, this unity of nature was a balanced unity of the whole of nature, of which man was a part; whereas for Ritter, it was teleological, as he spoke of discovering the 'divine secrets'. In spite of all the differences, the works of Humboldt and Ritter together, particularly '*Cosmos*' and '*Erdkunde*' did provide an almost complete explanation for geography, in terms of its definition, content, scope, approach and method of study.

Thus, Humboldt and Ritter were not only contemporary; their works were also complementary to each other. None of the personalities were complete in themselves as far as their contribution to geography is concerned. The great difference in temperament and in general outlook on life gave different colour to the work of the two founders of modern geography, and in the works of their followers this difference became fundamental. Because, they could not leave a clear and unmistakable framework for our subject. Therefore, after their death in 1859 geography entered a long phase of dualism. Coincidently, both of them died in the same year. The year 1859 marked not only the death of Humboldt and Ritter but also the publication of Darwin's "On the Origin of Species". Charles Darwin (1809-1882) was an English naturalist who established the theory of organic evolution.

NOTES

1. Rana, L. (2008).

2. Chisholm, Hugh, ed. (1911), p. 905.
3. Holt-Jenson, A. (2009) p. 112.
4. Hussain, M. (1995), pp. 307.

REFERENCES

Chisholm, Hugh, ed. (1911). "Varenius, Bernhardus". *Encyclopædia Britannica.* 27 (11th ed.). Cambridge University Press.

Holt-Jenson, A. (2009). "Paradigms and Revolutions", *Geography: History and Concepts: A Systematic Guide,* Fourth Edition, Sage, London, pp. 100-124.

Hussain, M. (1995). "Paradigm in Geography", *Evolution of Geographic Thought,* Rawat Publications, New Delhi, pp. 300-340.

Rana, L. (2008). *Geographical Thought: A Systematic Record of Evolution,* Concept Publishing Company, New Delhi, ISBN-13:978-81-8069-536-0.

11
Darwinism and Dualistic Tendencies (Crisis Phase III)

From the early nineteenth to mid twentieth century, geography was developing in several directions under the influence of distinct schools of thought. The influence of Humboldt and Ritter was very strong on geography in the early part of nineteenth century, and there were supporters of each. Ritter's followers emphasized on historical and human aspect in regional studies, and those of Humboldt on natural or physical aspects in systematic manner. The geographical studies were, thus, bifurcated in two directions: study of earth as a natural body and study of earth as a dwelling place of man. Since the followers of Humboldt and Ritter exaggerated certain aspects of views of each of the founders, a number of dualisms overshadowed the discipline for about 100 years after the emergence of German School of Thought. However, the more expanding frontiers were those of physical and biological sciences than human in the late nineteenth century. Now, the main impact was that of Darwin.[1] Darwinism, dualistic tendencies and a physical bias in geographical studies brought a turning point in the history of the discipline during this phase of development that may be marked as the Crisis Phase III in Geography.

Darwinism

The year '1859' has a special significance in the history of geographical thought. The year marks not only the death of both Humboldt and Ritter but also the publication of *"On the Origin of Species"*, the most popular work of Charles Darwin (1809-1882), the English natural scientist who revolutionized the entire philosophy of science. The theory of evolution by natural selection, as devised by Charles Darwin, caused considerable intellectual ferment in mid nineteenth century. The philosophy of Darwin also had a tremendous sway on geographic thought. This influence came through his work and his theory of 'survival of the fittest'. Post-Ritterian period was dominated by the thinking of Charles Darwin.[2] It was during this period that geographers started thinking seriously about nature of the discipline and concentrated on the issue, whether geography can be regarded as a science. Darwin's ideas on evolution stimulated social scientists to reappraise old concepts. New observations in the biological sphere seemed to provide the key to scientific understandings of differences in cultural levels and economic activities. After Darwin, the scientists looked for the laws which controlled nature and for materially conditioned social laws and so on.[3] It was at this point in time that the cosmographic way of thinking was replaced by the causal explanations characteristic of developing natural sciences.

Darwinism was primarily concerned with the mechanism of the change. The Darwinian evolution provided fresh incentive to concepts of biological origin which date back to Ritter and before. Through his evolutionary biology, Darwin made four-fold contribution. The first is the non-constancy of species, or the modern conception of evolution itself; the second is the notion of branching evolution, implying the common descent of all species of living things on earth from a single unique origin; thirdly, Darwin noted that evolution is gradual, with no major breaks or discontinuities; and finally, he reasoned that the mechanism of evolution was natural selection. Four themes in Darwin's writings, in turn, are significant in influencing the development of geographical thought:[4] the idea of development through time strongly influencing the progress of geomorphology, pedology, ecology, and to some extent the social sciences; Darwin's stress on the intimate relationships between organic life and habitat gaving impetus to organismic interpretations of regions; the themes of selection and struggle being deterministically applied in both human and political geography; and lastly, the theme of the random nature of original variations. Finally, Darwin's work so changed the nineteenth century world view that the development of geography as a science itself became possible. The philosophy of science and the modern spirit of geography were highly revolutionalised.[5] Darwinism rejected all supernatural phenomena and causations. The theory of evolution by natural selection explains the adaptedness and diversity of the world solely materialistically. Eliminating God from science made room for strictly scientific explanations of all natural phenomena. It not only gave rise to positivism, but also produced a powerful intellectual revolution, the effects of which have lasted to this day.[6] Until 1859, all evolutionary theories were teleological. Darwin questioned this orthodox view. He saw no final purpose of evolution. His doctrine of "survival of the fittest" (a term later coined by Darwin's contemporary, Herbert Spencer) treats the world as an incessant struggle for survival. Darwin's theory of evolution thereby undermined the value of traditional religion and morality, which had been accepted for centuries as the guiding principle of mankind. Darwin made it necessary to re-evaluate the most essential concepts which humanity had created for the last 2000 years: man, nature, consciousness, God, soul and so on. Mankind had been proud of these concepts because they put man in a superior position in relation to the world of nature, but Darwin shattered them by one theory. As a result, traditional natural philosophy professed by amateurs became transformed into modern science developed by professional scientists who base their competence on rigid theoretical and practical research. A new term 'Social Darwinism' emerged in the second half of nineteenth century to refer to various ways of thinking and theories that tried to apply the evolutionary concept of natural selection to human society.[7] Social Darwinism is the view that believes human groups and races are subject to the same laws of natural selection as Charles Darwin had perceived in plants and animals in nature. Darwin believed that this struggle for natural resources allowed individuals with certain physical and mental traits to succeed more frequently than others, and that these traits accumulated in the population over time, which under certain conditions could lead to the descendants being so different that they would be defined as a new species. Social Darwinism is often cited as an ideological justification for much of eighteenth and nineteenth century European enslavement and colonization of Third World countries. Darwin was highly impressed by the publication of Thomas Malthus who argued that human beings are involved in an intense 'struggle for existence', competing for the limited resources.[8] This idea helped Darwin uncover the mechanism he needed. Combining the idea of competition with his other observations, Darwin explained how evolution could occur. He stated that variation exists among individuals of a species and the scarcity of resources in a burgeoning population would lead to competition between individuals of the same species because all use the same limited resources. Such competition would lead to the death of some individuals, while others would survive. From this reasoning Darwin concluded that individuals having advantageous variations are more likely to survive and reproduce than those without the advantageous variations. Darwin coined the term 'Natural Selection' to describe the process by which organisms with favourable variations survive and reproduce at a higher rate. An inherited variation that increases an organism's chance of survival in a particular environment is called an 'adaptation'. Over many generations, an adaptation could spread throughout the entire species. In this way, according to Darwin, evolution by natural selection would occur.

Dualism

The philosophy of Darwin had a tremendous influence on geographic thinking in the latter half of the nineteenth century. Geography during this phase of development became highly dualistic in nature and a number of dualisms overshadowed the discipline for about a century ahead. It is because of the cumulative effect of the works of Darwin and the followers of Humboldt and Ritter who overemphasized views of the founders. The dualistic concept to which the geographers particularly clanged was concerned with the study of the earth as a natural body on the one hand, and as the dwelling place of man on the other. Thus, geography bifurcated in two main directions, and its position as a branch of knowledge was thereby brought into serious question. And, the geographers all around the world remained busy, at least up to Second World War, in seeking to establish the status of geography as an independent discipline, distinct in its concepts and procedures from other disciplines. Because, the foundations of geography as secured by Humboldt and Ritter did not provide a clearly unified field, the differences in the viewpoints persisted for a long time in the discipline. The emphasis of research shifted fast from one viewpoint to the other. Ritter's followers tended to emphasize the 'historical and human' aspects even more than he himself did. However, soon interest was deflected from Erdkunde and Ritterian philosophy was criticized. It was focused on systematic studies in a manner reminiscent of Humboldt. Thus, the dualistic tendencies that have always been part of geographical studies since beginning became more conspicuous after the death of Humboldt and Ritter. The major controversies that emerged in geography during modern period were related to its subject matter and approach to study, viz. Physical *versus* Human and Systematic *versus* Regional respectively. And, very soon the discipline entered a phase of double-dualism, i.e. 'systematic-physical' *versus* 'regional-human'.

It is particularly with the writings of Kant in the initial stage and of that of Humboldt later that physical aspect of geographical studies got greater emphasis than human side. Kant clearly stated that 'physical geography' is the basic of all geographical studies and categorically separated it from the other branches of geographical study. He considered 'physical geography' as the only systematic part of the discipline. Humboldt conceived man as the part of nature, dependent upon nature-minus-man, which existed before man and which could exist independent of man, whereas the reverse cannot hold true. He believed that the natural laws govern human beings as well. Since man was conceived as the product of nature, it was advocated that the study of man-nature relationships should begin with nature. Moreover, the knowledge in the natural sciences was more complete than the social sciences, and the former were on a firm footing comparatively. And in geography, it was physical geography that was considered as the basic to all geographical studies. The German scholars, particularly, have given this kind of understanding. Humboldt, for instance, considered 'physical geography' as 'the science of laws like physics'. His 'physical geography' was directly comparable to the 'general geography' or the 'systematic geography' of Varenius. Carl Ritter, on the other hand, had an anthropocentric approach to geographical studies. Ritter conceived Earth as the home of man. According to him, the physical surroundings get meaning only with reference to human beings. They are required to be known only in relation to human beings. Because it is the man who interacts with the physical world. The relation between the world of man and the non-human world is of prime concern in geography. Since the earlier times, the human and the non-human aspects have contributed to the content of geography. The dichotomy is what aspect is more important to study in geography, physical or human. Originally the controversy between man and physical (nature) was external to geography. It appears to have entered into the discipline first with the writings of German scholars, and was later strenghthened by the rise of natural sciences and Darwinism. The important sources of problem may, therefore, be identified as: the works of German scholars; Environmentalism (Determinism and Possibilism); and Systematic *versus* Regional Dualism.

The studies in geography have always been conducted in either of the ways, viz. to study all the aspects in one region (chorological); or to study one aspect in detail throughout all the regions (systematic). The chorological or regional approach deals with complete study of all the elements within an area, specifically demarcated for this purpose. The parallel approach advocated was systematic or topical. It leads to a series of studies focused on the character, distribution, process and behaviour of one element or phenomenon throughout the world. The simplest form of study in systematic geography is the consideration of the differential character of the earth surface in terms of any single geographic factor. Such studies have generally been related to the natural factors—the climatic factors, landforms, soils etc. However, any study in systematic geography, whether of an element or an element-complex, concentrates on one particular kind of phenomena or phenomenon-associations. It naturally leads, therefore, to the establishment of generic concepts. The 'Systematic *versus* Regional dualism' had been there in the ancient and medieval periods as well, but became prominent particularly after the classic work of Bernard Varenius (1622-1650), a German scholar. He divided Geography into two: 'General' and 'Specific'. According to him, 'General Geography' studies the earth in general, describing its various phenomena. Such a study provides the discipline with general laws and theories. 'Specific Geography', on the other hand, is the study of individual regions, countries or areas. The terms like 'General' and 'Specific' were caught by various geographers later and the dichotomy between the two widened to the extent that there had been periods when there existed double dualism in geography, i.e. 'Systematic and Physical' *versus* 'Regional and Human". One such time was that of Emmanuel Kant (1724-1804). Kant substituted the word 'physical' for 'general', and classified all the general or systematic studies under 'physical'. It was under his influence that the geographers in the late eighteenth century concentrated on systematic or general studies at the expense of regional or specific studies.

But, during late nineteenth century the interest of some of geographers, like Ritter and Ratzel, shifted towards human aspects. Then these human geographers and their followers emphasized the concept of region. They considered any law or generalization to be empty and unpredictable without its application to regional or specific geography. Therefore, they gave more emphasis on the regional studies, and conceived general studies as only the honourable occupation of the people without some practical or empirical base. The development of geography during the first half of the twentieth century has been marked by an increasing interest in regional geography. The climax was witnessed in the views of Richard Hartshorne (1899-1992) and Alfred Hettner (1859-1942), when the concept of geography as a chorological science was advocated by full force. This led to the development of concept of geography as a science of 'areal differentiation'. The idea of region was further boosted by its usage in different forms, e.g., 'Paysage' in France (Blache), 'Landschaft' in Germany (Otto Schluter) and 'Landscape' in America (Carl Sauer). This was a period when the 'Possibilistic' thinking was at its peak. At this climax phase, this argument took the form of the old 'hen and egg story'. One group believed that the core of geography was regional; whereas the others believed that systematic geography was really indispensable.

Physical Bias

The close of nineteenth century found an increased interest on physical or natural geography; and the systematic studies in their new guise were highly specialized. They were prosecuted by scientists who tended to be geomorphologists and climatologists. This was a time when 'systematic' and 'physical' became synonymous and the study of landforms emerged as the dominant and most attractive part of geography. This attracted the researchers not only from Germany but also from outside, especially U.S.A. The development that started after 1859 culminated in the works of W.M. Davis (1850-1934). In comparison, Human Geography lagged behind, because, some geographers wanted to make geography a preserve of pure science. Since human behaviour did not appear to be subject to general laws, they proposed to exclude man entirely from the field of geography.

The early twentieth century saw a preponderance of physically-biased studies in geography, particularly brought in by the German and American Schools. The shift in geographical thinking during this period came primarily due to the works of Oscar Peschel, George Gerland, Ferdinand Richthofen, Friedrich Ratzel, Albrecht Penck, Alfred Hettner, Otto Schluter, Alfred Wagener and Walther Penck in Germany. The physical bias in geographical studies is clearly evident in the works of the American scholars prior to Davis as well. The prominent ones include, for instance, M.F. Faury, Louis Agassiz, Arnold Guyot, Samuel Forry, Elias Loomis, John Wesley Powell and Grove Karl Gilbert. Most of these scholars, except Ratzel, Hettner and Schluter, promoted physically biased studies in geography. Ratzel was an anthropogeographer, whereas, Hettner and Schluter advocated for chorological (regional) approach in geography.[9]

Peschel (1879) attempted to establish physical geography as a science. He believed that physical and human geography constituted two entirely separate domains of knowledge, and as such the two could not form part of a single science. He was strongly of the view that geography should be pursued primarily as a study of the physical phenomena of the earth[10]. He was critical of Ritter for having neglected physical geography, and he criticized Humboldt for not having attempted a scientific classification of landforms. He was also critical of Humboldt for creating the impression that general geography could be equated with the entirely of natural science. He further decried Ritter for holding a teleological view about the nature and origins of earth phenomena, and for having subordinated geography to history. Peschel's exhortation that in order to project its status as a science, geography should identify itself more and more as physical geography was further advanced by two other contemporary scholars, namely George Gerland (1833-1919) and Ferdinand Richthofen (1833-1905). Gerland had gone so far as to suggest that since man could not be put to scientific analysis, study of man should be taken outside the purview of geography[11]. Richthofen was in greater sympathy with Humboldt than Ritter, and like Humboldt, he endeavoured to restore the close connection of geography with the natural sciences. He also emphasized that in order to reach useful and reliable conclusions, geographical study of any part of the earth's surface must start with a careful description of its physical features, and from there the student should move on to examine the interrelationships of other features of the earth's surface, and before anything else the physical geographic framework should be described at the outset. But, Ratzel, a younger contemporary of Richthofen, was not in much agreement to this view. He again brought human component in the fold of geographical studies. Ratzel and his work continued to be influential in the studies of modern geography.

Friedrich Ratzel was a remarkable Human geographer. The thinkers like Charles Darwin and the zoologist like E.H. Haekel, in turn, influenced Ratzel's writings. He became popular among geographers because of his work on: Man-Nature relationships and Political Geography, making the two most important literary contributions in this direction, viz. 'Anthropogeographie' and 'Political Geography'. Ratzel explored human relationships with the physical surroundings in a systematic manner. The findings of his work are based on empirical observations as he approached all human-nature relationships through travel and direct contact with the realities. In his Anthropogeographie, Ratzel explores the influence of physical environment on mankind systematically and empirically. The work runs in two volumes. The basic theme of the work in First volume, published in 1882, is that the man and his actions are not free from the influences of the surroundings. They are controlled by nature. He believed that man is the creature of his environment in much the same way as Darwin had demonstrated the adaptation and survival of the fittest in the animal world. The second volume of the book, published in 1891, is written with more liberal views on man-nature relationships. Because, he added that the survival of man and his adaptation to the environment also depends on fact that how fit and skilful he is in moulding the natural circumstances. Man's reactions to the nature differ sharply accordingly to his culture and historical background. If belonging to primitive culture then nature has a stronghold; and if advanced technologically, then the nature's hold becomes weaker. But, even then the nature is dominant. However, according to Ratzel, whether man adapts or modifies the surroundings, in any case, is the creature of his environment. Because, even the act of moulding is a kind of adaptation prompted by the nature in an indirect manner. Ratzel's work, in a way, was reworking of the theme developed by Ritter in his '*Erdkunde*', but with a newly established scientific method. But, his work was different in two respects, viz. First, Ratzel treated Human Geography systematically and not regionally as done by Ritter and Second, Ratzel's views were highly influenced by Charles Darwin. Ratzel saw man as the end product of evolution, in which the mainspring was natural selection of types according to the human capacity to adjust to the physical environment. Thus, where Ritter considered man and nature as part of harmonious whole serving the creative purpose of God, Ratzel tended to see man as a product of his environment, moulded by physical forces that surround man, and succeeding only in so far as he made the correct adjustment to their demands; the ultimate purpose of these adjustments, if any, lay outside the core of his enquiry. A deterministic tint, thus, colours most of his writing.

In the name of Penck, there were two scholars, namely Albrecht Penck (1858-1945) and Walther Penck (1888-1923), former being the father of latter. Albrecht Penck was basically a geologist. He is noted for his study of glaciation (especially in the Alps), for his pioneer classification of landforms, and for his work in the development of modern regional geography. Penck particularly dedicated itself and geography to international meaning. Walther Penck's life was cut extremely short when in 1923 he died of cancer. After his death, his father, Albrecht Penck, gathered together his research and published the book 'Morphological Analysis of Landforms'. Walther Penck has contributed a considerable amount to the field of geomorphology and more specifically to the idea of slope development. The main opponent to Penck's view of landform development was William Morris Davis. Penck's ideas were very controversial and, thus, were never extremely popular. But, the physical geographer who emerged after the death of Penck became highly popular among geogrphers. He was Alfred Wagener, who gained popularity among geographers for postulating the theory of Continental Drift.

In American School, on the other hand, Mathew Fontaine Maury (1806-1873) recorded information on the ocean currents and the winds. This information was the basis for many of his publications that earned him the title "Pathfinder of the Seas." It was Maury who proposed a world conference for developing a universal system of weather observations. The primary work of Louis Agassiz was on classification of living and fossil fish—ichthyology. He also promoted the concept of a great "Ice Age" in the recent geological past. Agassiz was the first to scientifically propose that the Earth had been subject to a past Ice Age. In his publication entitled '*Etudes sur les glaciers*' ("Study on Glaciers", 1840), he discussed the movements of the glaciers, their moraines, their influence in grooving and rounding the rocks over which they travelled, and in producing the striations seen in Alpine-style landscapes. The publication of this work gave a fresh impetus to the study of glacial phenomena in all parts of the world. Agassiz was the world's foremost authority on fossil and living fish. Arnold Guyot (1807-1884), a friend and contemporary of Agassiz, studied the flow of ice and the distribution of glacial erratic and served to underpin the theory of glaciation that had been advanced and championed by his close associate, Agassiz. Guyot was intimately involved in the formative years of weather forecasting in the United States and was responsible for selecting and equipping some fifty meteorological observation stations for the network. Arnold Guyot made fundamental contributions to the study of geology, glaciology and meteorology on two continents, namely Europe and America. The contributions of Samuel Forry (1811-1844) and Elias Loomis (1811-1889) to atmospheric science and geomagnetism are important components in the development of the earth sciences. Loomis encouraged the use of the growing national telegraph system in the 1850s, especially as used by the military, to record weather conditions around the United States. He was the first to use these synoptic data to map the air pressure differences across parts of the country. Loomis, in a sense, was the originator of experimental as well as observational meteorology. John Wesley Powell (1834–1902) is the most important name of the American and world geomorphology of nineteenth century. He is well known for exploration of the Great Colorado Canyon (1869), emphasizing river erosion, for the theory of absolute and temporal erosional base, classification and terminology of landforms, drainage basins, genetic types of rivers and structural types of valleys. Grove Karl Gilbert (1843–1918) was another important geologist of the late nineteenth century. He is known for his geological study on mechanics of river processes. But he also has merits for many new ideas, e.g. the theory on dynamic equilibrium of river profile and concept of grade, which were crucial in constructing of Davisian geographical cycle. His inventions were also the explanation of many other aspects of geomorphology, as for example, the upwarping of the shore-lines of ancient lakes, the study of climatic influence on land forming, the law of uniformity of the slopes, the laws of declivities and structure, the law of divides, the idea of lateral planation and origin of pediments etc. All these scholars who wrote prior to W.M. Davis covered numerous aspects of physical geography in their studies, ranging from meteorology, oceanography, glaciology and ichthyology. However, the influence of Davis became stronger than any one else in the late nineteenth and early twentieth century.

Thus, after the death of Humboldt and Ritter in 1859 till the First World War, the emphasis of geographical studies kept on shifting. Although this shift was evident in the form of Physical and Systematic (Peschel, Gerland), to Physical and Regional (Richthofen, Penck), to Systematic and Human (Ratzel), to primarily Regional (Hettner, Schluter) and predominantly Physical (Wegener, Penck) during the late nineteenth and early twentieth century, the physical bias was very strong in most of these studies. It was largely the Darwinian tradition that led geography into next phase of professionalism.

NOTES

1. Charles Robert Darwin (1809-1882) is best known for his work as a naturalist, developing a theory of evolution to explain biological change. In 1831, he embarked on a five-year survey voyage around the world on the HMS *Beagle*. His studies of specimens around the globe led him to formulate his theory of evolution and his views on the process of natural selection. In 1859, he published *On the Origin of Species*. His scientific theory of evolution by natural selection became the foundation of modern evolutionary studies.
2. Hussain, M. (1995), pp. 307.
3. Holt-Jenson, A. (2009), p. 113.
4. Stoddart, D.R. (1966).
5. Ernst Mayr, (2000).
6. Andrzej Diniejko, Warsaw University.
7. Crook, Paul (1999).
8. In 1838 Darwin read a book called 'Essay on the Principle of Population' by a British economist, Thomas Malthus (1776-1834). Malthus stated that the human population growing unchecked would double every 25 years. Resources such as food, air and water cannot increase at the same rate, Malthus argued.
9. Rana, L. (2008), pp. 144-179.
10. Dikshit, (1997), 67.
11. *Ibid.*

REFERENCES

Andrzej Diniejko, "Charles Darwin's Theory of Evolution and the Intellectual Ferment of the Mid-and Late Victorian Periods" (Dr. Andrzej Diniejko, Senior Lecturer in English Literature and Culture, Warsaw University; Contributing Editor, Poland), *Internet*.

Carroll, Joseph, (2004). *Literary Darwinism: Evolution, Human Nature, and Literature,* Routledge, New York.

Crook, Paul (1999). "Social Darwinism in European and American Thought 1860–1945". *The Australian Journal of Politics and History*. 45.
Darwin, Charles (1859). "*On the Origin of Species by Means of Natural Selection*, John Murray, London.
Ernst Mayr, (*2000*). "Darwin's Influence on Modern Thought', *Scientific American*, July Issue.
Himmelfarb, Gertrude, (1959). *Darwin and the Darwinian Revolution*, Garden City, NY: Doubleday.
Holt-Jenson, A. (2009). "Paradigms and Revolutions", *Geography: History and Concepts: A Systematic Guide*, Fourth Edition, SAGE, London, pp. 100-124.
Hussain, M. (1995). "Paradigm in Geography", *Evolution of Geographic Thought*, Rawat Publications, New Delhi, pp. 300-340.
Rana, L. (2008). *Geographical Thought: A systematic Record of Evolution*, Concept Publishing Company, New Delhi.
Stoddart, D.R. (1966). 'Darwin's Impact on Geography', *Annals of the Association of American Geographers*, Volume 56, Issue 4, pp. 683-698.

12
Rise of Determinism (Paradigm Phase IV)

Rise of Determinism was the major philosophical outcome of the developments that took place in late nineteenth century and the trend remained so up to 1920s[1]. It constitutes Paradigm Phase IV in Geography. The theory of evolution by natural selection, devised by Charles Darwin, caused considerable intellectual stir in late nineteenth century onwards. Darwin suggested that the physical world had been and still was subject to continuous change through the action of natural forces, and man is the product of these forces. No book has so profoundly affected the modern view of man than Darwin's *'The Origin of the Species'* (1859). This publication prompted a sharp reorientation of philosophical and moral attitudes. As discussed earlier, during this period physical geography dominated the scene particularly through "German and American Schools of thought". The maximum contribution came from geologists and geomorphologists and the emphasis was on morphological research and landform studies. Physical geography had a phase in the late nineteenth century, when it concentrated on systematic work at the expense of regional studies.[2]

Darwin's paradigm in conjugation with that of Davis' replaced all the Old traditions till the emergence of Chorological perspective in Geography in the twentieth century. The writings of W.M. Davis played a great role in further promoting professionalization in the discipline during this period. Davis sought to integrate the two elements, viz. Nature and Man. This he did after a period in which a marked dualism prevailed in the discipline, i.e. between exclusively physical geography (including plants and animals) and the geography of man. Davis found the key at this time in the Darwinian evolutionary approach that regarded the facts of nature and man as existing in a 'harmonious relationship' in a continuous chain of cause and effect. W.M. Davis remained a source of inspiration for a long chain of geographers. His 'Theory of Cycle of Erosion' is the most important contribution to Physical Geography. The core of geography, Davis stated, is the study of the relationship between the physical environment (the control) and the human behaviour (the response). This kind of explanation represents the influence of the prevailing deterministic thinking in the first decades of the twentieth century. Davis sought to classify the facts of man in reference to the facts of the physical earth, calling them respectively 'Ontography' and 'Physical Geography'. The physical earth required to be studied by physical geographers with respect to origins and processes. The resultant forms of land, water and air were to be examined by the geographers in relation to the 'human responses' and 'controls'. The human facts on the earth were called 'Ontography' as opposed to the facts of the physical earth, 'Physiography'. This was an extension of the evolutionary interpretation of man on the earth in so far as human characteristics show direct results of geographical (physical) influences and controls. Davis believed that such a system would form the framework of an integrated science. This viewpoint very clearly set the pattern for the pursuit of man-land relationships on a deterministic basis. Davis identified 'Ontography' as the other half of geography, and defined it as 'geographical description of life forms on the earth'. Davis' ideas on geomorphology remained largely unchallenged for a long time in America and outside. His 'Ontography' paradigm, developed into a major paradigm, i.e. ecological, in the late 1960s, when the emphasis of geographical studies was attached to the ecosystem as a geographical principle and method.

Besides Davis, the contemporary thinking was also dominated by the ideas of his pupils like Ellen Churchill Semple (1863-1932), Ellsworth Huntington (1876-1947) and Harlan Barrows (1877-1960). The environmentalist approach was taken up in their diligent and far ranging researches. Semple was an environmental determinist. Huntington was known for his studies on climatic determinism. He, particularly, sought to measure the 'influences' of the 'physical environment' on human activities and historical development through his work 'Principles of Human Geography' (1934). He examined the influence of climate and climatic change on men and the course of history. He linked the natural and cultural or historical phenomena in causal relations. Harlan Barrows presented the concept of geography as human ecology for the first time. He stated that the physical environment does not cause man's adaptation to the conditions of his habitat. This adaptation, in fact, is the function of human choice. He proposed human ecology as the unifying theme for geographical work.[3]

At this point of time Friedrich Ratzel (1844–1904), the German scholar, corrected the prevailing tendency of overstressing the physical aspects of geography and established a more balanced viewpoint. Ratzel was a remarkable human geographer. The thinkers like Charles Darwin and the zoologist like E.H. Haekel, in turn, had influenced Ratzel's writings. 'Anthropogeographie' (1882 and 1891) and 'Political Geography' (1897) were his two most important literary contributions. In these works, Ratzel treated human geography systematically and from Darwinian viewpoint. He is also notable for coining the term *'lebensraum'* ("living space"). His work served to fix interest still more firmly on systematic studies, and regional geography continued to receive scant attention. Among geographers influenced by Darwin, it was Ratzel, above all, who explored the influences of physical environment on mankind.[4] Ratzel's 'Anthropogeographie' was a seminal work in this direction and it created intellectual debate on the both side of Atlantic to make it a paradigm. Scholars like Huntington, Semple and Davis extended and elaborated his work into what become known as Environmental Determinism; a view that dominated geography for decades. In Anthropogeographie, Ratzel explored the influence of physical environment on mankind systematically and empirically. The basic theme of the work in its first volume is that the man and his actions are not free from the influences of the surroundings. They are controlled by nature. He believed that man is the creature of his environment in much the same way as Darwin had demonstrated the adaptation and survival of the fittest in the animal world. Accordingly, he attempted to show in this volume of 'Anthropogeographie' as how the distribution of man on earth had been more or less controlled by natural forces depending on the level of his cultural advancement. However, in his second volume of Anthropogeographie, written with more liberal views on man-nature relationships, he showed a different spirit, where the emphasis was on distribution and density of population, forms of settlements, migrations, and diffusion of cultural traits. Here, he added that the survival of man and his adaptation to the environment also depends on fact that how fit and skilful he is in moulding the natural circumstances. Man's reactions to the nature differ sharply accordingly to his culture and historical background. If belonging to primitive culture then nature has a stronghold; and if advanced technologically, then the nature's hold becomes weaker. But, even then the nature is dominant. However, according to Ratzel, whether man adapts or modifies the surroundings, in any case, is the creature of his environment. Because, even the act of moulding is a kind of adaptation prompted by the nature in an indirect manner. For the explanation of these features Ratzel did not merely resort to environmental influences but equally, or more so, on historical-cultural factors. Ratzel saw man as the end product of evolution, in which the mainspring was the natural selection of types according to their capacity to adjust themselves to the physical environment. A deterministic tint thus coloured most of his writings. Ratzel advocated the concept of cultural geography, a connection between the natural sciences and the study of man, as a central theme and use of scientific method in the study of human geography. In this environment-dominated-man-response approach, he used deductive perspective to present the first systematic study of geography of man.

In 1897, Ratzel published 'Politische Geographie', the first work written with systematic approach to political geography.[5] In this book, Ratzel develops the concept that views the state as "a particular spatial grouping on the earth's surface." The state, as defined by Ratzel, consists of "a human group with definite organization and distribution"[6]. From these ideas, Ratzel developed the concept of *'lebensraum'* or living space, Ratzel hypothesized that the state naturally seeks to increase its size. If the state's neighbours are weak, the state will grow larger and spread into other states. As evidenced, Ratzel believed that space was a great political force. Through his 'Political Geography', Ratzel reshaped the old topic along the lines conformable with the principles he had enunciated in his Anthropogeographie. Through his work, he tried to make a comparative investigation of the relation between 'the political state' and 'the earth's surface'. The influence of revolutionary biology of Darwin led Ratzel to adopt the organic theory of state and society. The political entities (nations or states) expand or shrink like organisms. Based on this viewpoint, he postulated the 'Organic Theory of State', the concept of state as an organism, a piece of humanity on the surface of the earth. He believed that a nation or state is not only a spatial unit, but also a human entity, and therefore, it is a living unit. It has spatial dimensions in terms of location, boundaries etc. At the same time, as a human unit, it has a particular history of mankind, culture, power and economy. Besides, the state-capital has been called its 'heart' and the lines of communication as its 'arteries'. The state evolves, grows and dies like an organism. The healthy or powerful states are fit to survive; they grow and develop fast, whereas the poor or weak states have a retarded development. They may die, vanish or may be over-powered by the stronger states. However, during the World War Second, the 'Organic Theory of State' attained its extreme form of explanation, and out of Political Geography of Ratzel emerged a new concept, i.e. the concept of 'Geopolitik'. Rudolf Kjellen, a Swedish student of Ratzel, further elaborated on the 'Organic theory of State' and coined the term 'geopolitics'[7]. It became the governing principle in Germany during the Second World War, the period of Hitler, when Germans considered themselves as the supreme race. Ratzel's publications influenced the development of geography for at least half a century. Ratzel's impact, particularly on American geographers, was quite strong owing to the interpretation of his ideas by Ellen Churchill Semple, his American pupil.

Thus, Darwin laid the foundation of the deterministic approach to geography.[8] In his opinion, the natural conditions determine the socio-cultural developments of a society.

Ratzel systematically postulated the viewpoint of Environmental Determinism in the later half of the nineteenth century. According to this viewpoint, man is the product of nature. It is the nature, which governs man and his activities, and determines his way of life, living styles and also physical appearance. Because of this postulation, Ratzel became very popular particularly in Germany and also in U.S.A. The relationship of man with nature and the resultant pattern or description over space in a particular time is the core of all geographical studies. Determinism has emerged as one of the ways to study this relationship. Much of the geographical work since the mid nineteenth century, however, has either explicitly or implicitly taken its inspiration from biology and in particular from Charles Darwin.[9] Darwin established a sphere of scientific enquiry free from *a priori* theological ideas, and freed natural science from the argument of natural theology (science of God). And finally, Darwin established man's place in nature, and in doing so he made man a fit object for the scientific study. Modern geography is inconceivable without these general advances.

The philosophy of environmental determinism rose to prominence in the late nineteenth and early twentieth century when it was taken up as a central theme by the discipline of geography. The view persisted up to the Second World War in one form or the other. Supporters of the determinism favour the view that environment controls the patterns of man's behaviour, so that all his actions, his settlements, his modes of life are controlled, directly or indirectly, by his surroundings. They believed that nature is superior to man. According to the determinists, man is a passive agent on whom the physical factors are acting and determining his attitudes and the process of his decision-making. The essence of the deterministic school of thought is that the history, culture, living style and the stage of development of a social group are exclusively or largely governed by the physical factors of the environment. Those who believe in this view say that humans are strictly defined by stimulus-response (environment-behaviour) and cannot deviate. In crude environmental determinism, only the physical environment has been given importance. The physical environment, in turn, has two aspects, viz. 'climatic' and 'physiographic'. The basic argument of the environmental determinists was that aspects of physical geography, particularly climate, influenced the psychological temperament of individuals, which in turn shaped the culture and society that those individuals formed.

However, Determinism is one of the oldest schools of thought.[9] The origin of deterministic thinking goes back to antiquity, when the Greeks and Roman scholars like Hippocrates, Aristotle, Strabo and few others made first attempts to explain the physical features and character traits of various people and their culture with reference to the influence of natural conditions. Hippocrates (born in 460 B.C.), for instance, discussing his work "on Airs, Waters and Places", commented on the easy going life of Asians endowed by nature with very favourable physical environment, thus making them inactive, relaxed and slackened. In comparing life of Asians with those of Europeans, who live in relatively harsh physical environment, he found the latter as hard working. Aristotle repeated the same generalization about the national character and added the political element. According to him, the people inhibiting the cold places (of Europe or North America) are full of spirit but somewhat deficient in intelligence. The people of Asia, on the other hand, are intelligent and skilful in temperament, but lack spirit. Therefore, the latter have been in continuous slavery and subjection. Similarly, Strabo, in his geographical accounts established a close relationship between the shape, relief, climate and space and the rise and strength of Roman Empire. The geographical determinism continued to reflect the writings of Arab Scholars in the medieval period as well. The Arab Scholars divided the habitable world into several 'terrestrial zones' (*Kishwars*) and highlighted the physical and cultural characteristics of races and nations living in these zones. Al-Battani, Al-Masudi, Ibn-Hauqal, Al-Idrisi, Ibn-Khaldun etc. attempted to correlate the environment with human activity and their mode of life. Al-Masudi, for example, asserted that the land where water is abundant, the people are gay and humorous, while people of dry and arid lands are short tempered. The nomads who live in open air are marked by strength and resolution, wisdom and physical fitness. Determinism was also strongly observed in the work of German Scholars. For example, Emmanuel Kant (1724-1794), stated that the people of New Holland (East-Indies) have a half closed eyes and cannot see to any distance without bending their heads back until they touch their backs. This is due to innumerable flies that are always flying in the eyes. Kant further stressed that all the inhabitants of hot lands are exceptionally lazy and timid. Timidity engenders superstition and in lands ruled by kings leads to slavery. In support of his hypothesis of influence of climate, he stated that animals and men that migrate to other countries are gradually affected by their environment. Humboldt and Ritter, on the other hand, looked on this 'man-environment relationship' more scientifically, i.e. 'cause and effect relationship'. Therefore, the development of astronomy in Arab, they attributed to the physical environment—open skies and desert lands where astronomical observations are easier to be made.

Later the philosophy of determinism was greatly influenced by the high profile of evolutionary biology of Charles Darwin. After the publication of the theory of Darwin, a change was seen in the development of deterministic viewpoint as well. Because, the pre-Darwinian views on Determinism were found relatively more modest in comparison to that in post-Darwinian period, when they attained more extreme form. The maximum influence of deterministic thinking was infact, exercised after the popularity of Darwin's Theory. The scientific developments in geography that followed Darwin were deeply influenced by the role of nature on organisms, including man. This role of nature was deeply studied in human affairs. In other words, it was tried to prove that man's houses, communications, agriculture, his physiographic characteristics, nature, attitudes, capacities, general way of life and beliefs all are determined by the nature or environment (particularly including climate, relief, soil and vegetation). Gradually the term 'environmental influence' was replaced by the 'environmental control', i.e. the environment does not simply influence but determines. The observations of various scholars caused them to conclude that 'man's immediate surroundings have a decisive (deterministic) effect not only on his occupations, dwellings and the materials he uses in everyday life, but also on his personal traits, his general character, and even his religious beliefs. They found environment to be superior to man and mastering all of his actions. Such generalizations were called 'extreme' and not as the result of 'cause and effect' relationship between man and environment. Thus, there emerged extreme form of determinism in the late nineteenth century.

The scientific milieu in the later half of the nineteenth and the early twentieth centuries was dominated in part by Darwin's idea, deductive approaches and an acceptance of the Newtonian cause and effect relationships. Fitting well into this intellectual environment, the theme of environmental determinism was prevailing view not only in Germany but also in American geography. Ratzel supplemented 'classical' geographical determinism with the elements of 'Social Darwinism' and developed a 'theory of the state as an organism' that owed its life to the earth, which was ever striving to seize more and more territory. In the opinion of Ratzel "similar locations lead to similar mode of life". He cited the example of British Isles and Japan and asserted that both these countries have insular locations, which provide natural defence against the invaders. Consequently, the people of these countries are making rapid progress. Ratzel, a follower of Darwin, believed in 'the survival of the fittest' and saw 'man' as the end product of the evolution, an evolution in which the mainspring was the natural selection of types according to their capacity to adjust themselves to the physical environment. He was convinced that the physical features and location of a place in relation to mountains and plains closely influence the course of history, the mode of life of the people and the stage of development. In his deterministic approach he gave more weightage to location in relation to topographic features.

At the beginning of the twentieth century 'environmentalism' became particularly widespread in the United States, where its leading proponents were Ellen Churchill Semple (1863-1932) and Elsworth Huntington (1876-1947). Semple was the direct descendent of Ratzel. She preached the philosophy of her teacher and thus was a staunch supporter of determinism. Based on the philosophy of determinism she wrote two books, namely, *"American History and its Geographic Conditions"* in 1903 and *"Influences of Geographic Environment"* in 1911. Semple in her book distinguishes the attitudinal characteristics of the people living in different physical settings and stresses that the dwellers of mountains are essentially conservative. There is little in his environment to stimulate him to change and little reaches him from outside world. Hence, innovation is repugnant to him. As a matter of fact, the process of diffusion of new ideas and innovations in the hilly tracts of isolation is slow as compared to the well-linked plain parts of the world. This relative isolation of the hill dwellers leads to orthodoxy, conservativism and suspicious attitude towards strangers. They are extremely sensitive in their tradition and do not like criticism. They have strong religious feelings and an intense love for family. The bitter struggle for existence makes the hill man industrious, frugal, provident and honest. Contrary to this, the people of plain parts in Europe are energetic, serious, thoughtful, rather than emotional, and cautious rather than impulsive. The people of Mediterranean region, where the climate is temperate and mild, are gay, imaginative and life is easy going. According to Semple, there are four ways in which the environment can influence the man and they are: Physiological Influences: on the features, stature and physical appearance of the human beings; Psychological Influences: on the level of thinking, beliefs, knowledge of the human beings; Influences on the Economy and Society: on the occupation, culture, traditions of the people; and Influences on the Human Mobility: on the mode of living, as sedentary or practicing transhumance. Elsworth Huntington, an American geographer, was another protagonist of 'Environmental Determinism'. He explained more in terms of 'Climatic Determinism'. It was between 1915 and 1924 that Huntington was most actively promulgating the doctrine of the effects of climatic variations, based on his early field studies in Central Asia. Huntington assessed the role of climate on man at three levels, viz. Global, Regional and Personal. He made the most decisive step since the time of Hippocrates towards something new and conclusive in environmental causation thinking. So over many years he was engaged in developing the idea of climate's leading role in the advancement of civilization. The basic philosophy of Huntington was that the supreme achievements of civilization in any region were always bound up with a particular type of climate and variation in climate led to 'pulsations' in the history of culture. He associated with the climatic cycles the 'Golden Age' in ancient Greece, the Renaissance in Western Europe, and cyclical fluctuations in iron production or the price of shares. Huntington divided the world in the 'mild' and 'harsh' climatic zones and established that the ancient civilizations (Egyptian,

Mesopotamian, Chinese and Indus) flourished in the fertile river valleys of mild climates. He also established the hypothesis of invasion and tribal warfare. The great outpouring of nomadic people from Central Asia which led to Mongol Conquest of Turan, Turkistan, China and India and the raids in Eastern Europe in the thirteenth century could be explained by the dying of pastures on which the nomads were dependent. According to Huntington, the religion and racial character are the products of climate. A temperature of about 20°C with variable atmospheric conditions is good for the high mental and physical efficiencies. Such a climatic condition is found in the countries of North-West Europe. Huntington has attributed the advancement of Europeans in the field of science and technology to climatic conditions. The underdevelopment of Tropics, he explained, is owing to the humid, hot, oppressive weather that makes the people lethargic, lazy, inefficient and timid. The subsequent geographers like H.J. Mackinder, Chislomn, Davies, Bowman, Robert Mill, Geddes, Herbertson, Taylor etc. interpreted the progress of societies with a deterministic approach. Many of the scholars made it vividly clear that the climate influences the physical properties of soil that ultimately determines the cropping patterns, dietary habits, physique and attitudes of human societies.

In the 1920s and 1930s, environmental determinism came under attack as its claims were found to be hopelessly inadequate at best, and often destructively wrong. The examples above represent the more extreme opinions of a school of determinism which has few followers today, and which went beyond the ideas of the early exponents of environmental control. As a result, Determinism started fading away since the beginning of the twentieth century. Particularly after Second World War, the philosophy of determinism was attacked vigorously. Many geographers in the USA, France, Britain, Canada and other countries were drawing attention to the one-sided approach adopted by the environmentalists in their interpretation of historical reality to their exaggeration of nature's active role and the fact that they only acknowledge man as capable of passive attempts at adaptation. Works of man reveal many facts for which environmental forces alone give no satisfactory explanation. The natural environments may be modified by the technological improvements in order to make the inhospitable and harsh physical situations worthy of human occupancy.

As a consequence of the above criticism, the doctrine of crude Determinism began weakening away since the second quarter of the twentieth century, but after few decades it resurged in the form of 'Neo-Determinism'. In the meanwhile there emerged the doctrine of "Possibilism" in France, as an antithesis to "Determinism". In spite of its vigourous criticism the school of Determinism did not die away completely, its central theme continued to survive in the shape of Neo-Determinism. Where some of the early environments were apt to make unsound generalizations, the modern determinists were more careful, to seek and record, wherever possible, a true sequence from cause to effect. They were aware that man may be subjected to many surrounding influences that are not purely geographical ones. Sometimes geographical influences are not strong, nor their effects obvious, but to the environmentalists these geographical influences are always present as part of the general environment which determines man's course of action. The modern determinists differed most radically from the 'possibilists' in their view on the nature of human 'choice'. They did not allow that man has complete freedom to choose, but rather, that he narrows down alternative courses of action in his mind, and, in selecting one alternative rather than another is, he is consciously or subconsciously influenced by some part of his general environment. The supporters of this new school of determinism basically believed in determinism but not to the extent as earlier determinists did. They did not use unsound generalizations; they believed in cause and effect relationships between man and nature; they were more careful, and applied theoretical and empirical examination of the facts with the help of modern techniques. Griffith Taylor used the term 'Stop-Go-Determinism' for Neo-Determinism (Taylor, 1951). Taylor maintained that it may be that the well-endowed parts of the world offer a number of different possibilities for making a living; but in most of the earth's land area, nature speaks out clearly—"this land is too dry, or too cold, or too wet, or too rugged". He wrote that the settlers who fail to heed this nature-given limitation must face disaster. According to Taylor, the role of nature was far from crudely deterministic, but the environment was nevertheless a powerful force in human action which man could ameliorate but not escape. It is like the case of the flow of traffic on a busy road. The traffic policeman cannot wish away the traffic that must necessarily flow. All that he does is to regulate the traffic through the temporary stop-go-method. Similarly, the human agency through the use of technology can modify the force of nature but it cannot escape it. The role of human agency is similar to that of the traffic regulator.[10] The widespread criticism to Determinism also brought a paradigmatic shift in geographical thinking.

NOTES

1. Himmelfarb, Gertrude (1959).
2. Schaefer, Fred K. (1953), p. 228.
3. Dikshit, R.D. (1997), p. 107.
4. Rana, L. (2008), pp. 400-413.
5. Pearcy, (1957), p. 22.
6. (Dickinson, 1969, 68).
7. *Geopolitiks* is the method of political analysis, popular in Central Europe during the first half of the twentieth century, that emphasized the role played by geography in international relations. The term was first used (1916) by Rudolf Kjellen, a Swedish scientist, and was later borrowed by Karl Haushofer, a German geographer and follower of Freidrich Ratzel. Haushofer's writings found favour with the Nazi leadership, and his ideas were used to justify German expansion during the Nazi era. Many expasionist justifications, including the German 'Lebensraum', are based on geopolitical considerations. 'Geopolitics' was one of the chief features of the Nazi propaganda. 'Geopolitics' has been defined as "a science of the earth's relationships of political processes. It is based on political geography and views space from the standpoint of the 'state'. In its extreme German form it considers the state as a necessarily expanding organism, in which the individual is entirely sub-ordinated to this organic state, which is to be aided in its growth by all powers of the military and civilian population." (Taylor, 1951., 587) It is clearly an extreme form of determinism. Geopolitiks is different from political geography, a branch of geography concerned with the relationship between politics and the environment. The difference between the concept of 'Political Geography' and 'Geopolitiks' is that the 'Political Geography' studies the evolution and distribution of political units over the surface of the earth; whereas, 'Geopolitiks' is the geography of political power or political supremacy. According to the latter view, as interpreted by the Germans, the stronger state has every right to grow and expand, even at the cost of the others. This was an extreme form of explanation for the study of relationships in between the political power and the geographical space.
8. Hussain, M. (1995), pp. 307-308.
9. Rana, L. (2008), pp. 387-398.
10. Davies, (1972), pp. 52-68.
11. Dikshit, (1997), 246.

REFERENCES

Davies, W.K.D. (1972). *Conceptual Revolution in Geography*, Rowman & Littlefield, New Jersey.

Dickinson, R.E. (1969). *The Makers of Modern Geography*, Praeger Publishers, New York.

Dikshit, R.D. (1997). *Geographical Thought: A Contextual History of Ideas*, Prentice Hall of India.

Himmelfarb, Gertrude (1959). *Darwin and the Darwinian Revolution*. Garden City, NY: Doubleday.

Hussain, M. (1995). "Paradigm in Geography", *Evolution of Geographic Thought*, Rawat Publications, New Delhi, pp. 300-340.

Pearcy, E. (1957). *World Political Geography*, Crowell Company, New York.

Rana, L. (2008). *Geographical Thought: A systematic Record of Evolution*, Concept Publishing Company, New Delhi.

Schaefer, Fred K. (1953). "Exceptionalism in Geography: A Methodological Examination", *Annals of the Association of American Geographers*, Vol. 43 (3) : 226-249.

13
Shift to Human Studies (Crisis Phase IV)

Geography encountered another crisis phase (IV) in early twentieth century when there was an extensive dissapproval with the doctrine of Determinism. Resultantly, there took place a paradigmatic shift from Physical to Human. However, this turned out to be relatively a very short period in transition as the discipline soon witnessed the next Paradigm phase. As the dominating paradigm in human geography, determinism had a short life, being challenged by the possibilists and the French School of human geographers[1].

Criticism to Determinism

As a result of the development during Paradigm Phase IV, geography acquired recognition in the community of sciences. But human geography showed a stunted growth. By 1900, the philosophy of determinism had become particularly widespread in Germany. The thinking also percolated in U.S.A. through the works of Allen Churchill Semple, a pupil of Ratzel and a staunch supporter of determinism. But, from the beginning of twentieth century this doctrine came under attack.

The chief faults in the assumptions of the extreme environmentalists lay in over-simplication in the approach dealing with the issues related to the study of man-environment relationships. Man is undoubtedly profoundly affected by physical conditions. Environment does strongly affect the nomad, and has almost certainly contributed to the nature of religious beliefs of such peoples. But assertions of the kind quoted above are too sweeping, and were not, in the main, backed by careful investigations that modern environmentalists employ in their search for a sequence from cause and effect. From the perspective of the late twentieth century, it appeared that the determinists over-emphasized the role of environment in human affairs. This does not imply that environmental influence is inconsequential or that the human geographer should not study such an influence. Rather, it suggests that the physical environment is only one of the many forces affecting human culture and rarely the sole determinant of human behaviour and beliefs. The crude determinism was a one-sided approach, as there was a lack of cause-and-effect relationships.

The determinists over-emphasized the role of physical environment in human affairs. Man was seen only as a passive element of the environment. Such works of man were overlooked for which environmental forces alone give no satisfactory explanation. O.H.K. Spate (1952), for instance, criticized the fanatic approach of environmental determinists.[2],

He stressed that "environment taken by itself and without man is a meaningless phrase. Equally important is his indication of the need to consider the psycho-physiological influence of the geographical environment *via* social structure. Geographical environment is only one of the factors of territorial differentiation and it acts through society and cultural tradition." Hartshorne[3] put forward similar argument. He rejected determinism purely on the grounds that the latter separated nature and man, and this he found disruptive to the fundamental unity of the field, contradicting the concept of geography as an integrating science.

Although, the school of determinism, as such, had a short span of life, its central theme did not die out completely. It survived and continued to persist in the form of 'Neo-Determinism' or 'Stop-Go-Determinism'. Where the early determinants were apt to make unsound generalizations, the neo-determinants were more careful, to seek and record a true sequence from cause to effect. They were aware that man might be subjected to many surrounding influences that are not purely physical ones. Where the neo-determinists differed most radically from the 'possibilists', they did not allow that man has complete freedom to choose, but rather that he narrows down alternative courses of action in his mind, and, in selecting one alternative rather than another, is, influenced, consciously or subconsciously, by his general environment. They were more careful and believed in theoretical and empirical examination of facts. In spite of the emergence of Neo-determinism, the move towards human studies in geography grew stronger. Largely, as a reaction to the exclusive concentration on systematic studies on the part of physical geographers, the interest began to shift towards social or human geography.[4] The French scholars owe the credit for this change.

Rise of French School

The growth of geography in France, unlike that in Germany with its several distinct schools of thought, has been shaped by the work of one man.[5] For about 4 to 5 decades in the late nineteenth and early twentieth century, geography in France was largely a one-man-show. Vidal de La Blache (1845-1918) was the leader of French School of Geographical Thought. For generations his school was dominant and nearly all the professors in geography and all the occupants of chair in geography were the pupils or the pupils of pupils of Blache. In no other country in the world so far the development of geography has been centred on one man as was in France in the modern period. Many generations of scholars after Blache were inspired by him, directly or indirectly, till Second World War. They altogether have made outstanding contributions to geography. During his academic career, Blache was able to inspire many students in geography, who later formed a long genealogy of geographers in France. Blache dominated the scene in the University of Paris for nealy 40 years. The first half of his career covers the last quarter of the nineteenth century. In this period, although a contemporary of Richthofen, Ratzel and Wagner, he was junior and learned much from them. He was particularly influenced by the second volume of 'Anthropogeographie' (1891) of Ratzel. In the latter half of his career, in the first quarter of twentieth century, Blache was virtually contemporary with Penck and Hettner. This period also saw the florescence of his school with the publication of massive doctorate studies by his pupils of particular areas in France. This tradition also continued even after his death. In the First quarter of the twentieth century there emerged the first generation of scholars with the inspiration of Blache. The distinguished scholars of this phase included the geologists like Elisee Reclus

(1830-1905) and De Margerie (1862-1953), the cartographer Franz Schrader (1824-1924), Lucien Gallois (1857-1941) and the historian Auguste Himly (1823-1906). The outstanding students of Blache who dominated the professional scene in the second quarter of the twentieth century, make up the second generation in France. They include Jean Bruhnes (1869-1932), Albert Demangeon (1872-1940), Em de Martonne (1873-1955), Raoul Blanchard (1877-1965), Jules Sion (1879-1940), Henri Baulig (1877-1962), Camille Vallaux (1870-1945), Andre Seigfried (1875-1950), Rene Musset (1881-1977) etc. The younger scholars emerging after Second World War belong to third generation, including Andre Cholley (1886-1968), Pierre George (1909-2006), Roger Dion (1886-1981) and Pierre Deffontaines (1894-1978). The fourth generation represents the still younger geographers. The chief of these are Jean Gottmann (1915-1994), Philippe Pinchemel (1923-2008), Max. Derruau (1920-2004), Pierre Birot (1909-1984), Pierre Vilar (1906-2003), Jean Gottamann (1915-1994) etc. Thus, the influence of Blache penetrated right into the late twentieth century.

Humanistic Tilt

Blache insisted on the study of the Man and the physical earth that he referred to as the geographical environment. He considered, however, that the peculiar task of the geographer lies in the correlation of physical and human conditions on their spatial basis. The physical (natural) environment provides a range of possibilities which Man turns to his use according to his needs, wishes and capacities, in creating his habitat. Hence, as opposed to the "environmental determinism" that was prevalent at the time as a Darwinian heritage. In its place, he established a conceptual framework of 'Possibilism' that was fully supported by a critical historian, Lucien Febvre (1828-1956), who was also the friend of Blache. The major thrust of the work of Blache was on the study of man-environment relationships from the point of view of man. We know him as a great human geographer. Among the other scholars, working on the lines of Blache, the contributions of Himly, Reclus, Bruhnes, Demangeon, Seigfried, Vilar and Gottamann are noteworthy. Their attempts proved instrumental in bringing a humanistic tilt in geographical studies.

Auguste Louis Himly A historian and geographer published, in two volumes, his remarkable *'Histoire de la formation territoriale des etats de l'Europe centrale'*, in which he showed the reciprocal influence exerted by geography and history. In his work, 'l'Homme et la terre', Élisée Reclus set the crown on his previous greater works by considering man in his development relative to geographical environment. Jean Bruhnes was a leading exponent of French systematic (as opposed to regional) geography. He studied human artifacts in the context of environment. He authored many texts, including *'Human Geography'* (1910) and *'Human Geography of France'* (2 vol., 1920-26). Bruhnes' outlook was inspired by the teaching of Blache, but it inevitably shifted the emphasis from the focal study of "geographical conditions and social groupings" to "the study of surface features of the earth". Bruhnes elaborated a system of "Human Geography" with its raw material in "the visible and tangible facts" of the human activity on the earth's surface. He has called these the 'essential facts' of the human geography. Similarly, if one adds the physical facts of the landscape–surface configuration, vegetation, soils in their surface expressions as forms of terrain, then we have 'the essential facts of geography'. The essential facts of geography he divided into three: (1) Facts of unproductive occupation of the soil (houses, roads, settlements etc.); (2) Facts of plant and animal conquest (e.g. cultivation of crops, animal husbandry, plantations etc.); and (3) Facts of destructive economy or exploitation (i.e. negative use of land, e.g. clearing of forests, hunting and mining etc.). He examined these in particular type of areas or 'regional studies' to show the interdependence of these phenomena in contrasted physical and social environments. According to Bruhnes, study of these three sets of observable phenomena does not set the limit of geography. Beyond the essential facts, but in association with them, is an area of wider study. This is described as "the geography of history", and it embraces five aspects, viz. The distribution of population (static conditions and migrations); Geography of work or economic activities (related to production, transport, exchange etc.); Geography of political control or power of man (territory, routes and frontiers, groups of states); Geography of civilization or social geography (nationalities, races, languages, religions, intellectual, artistic and technical phenomena) and social attitudes (mentality, organization, judicial, social) where these are associated with essential facts; and Regional Geography (i.e. a synthesis of all the above). Albert Demangeon was an outstanding promoter of regional and human geography. In spite of his dedication to regional geography initially, later he established a systematic approach in the fields of economic geography, political geography and the studies on the habitat and the rural house. He wrote great number of articles in the Annales de Géographie. Andre Seigfried was the French political scientist and educator. Siegfried's work was dedicated to the study of the geopolitics of its time, to the history and the daily life of the countries like United States, England, New Zealand and France, and to economic and political geography. After accepting a position at the Collège de France in Paris in 1933, he continuously wrote numerous economic, political, and cultural analyses, particularly of Latin America, Canada, Switzerland, South Africa, Great Britain, and the United States. Through these Siegfried outlined his ideas about "electoral geography," a system of political research that attempted to explain regional political and voter participation through an analysis of economic, social, and geographical factors. Pierre Vilar was one of the greatest historians of twentieth century. He was committed to an idea of history as a human science. Geography was his first calling, but his approach to the discipline was always decidedly historical. He made early contact with Lucien Febvre, to whose Annales he contributed, almost from the beginning. His books, written in a style that combined intellectual passion and persuasiveness with a regard for rhetorical elegance, in the best French tradition, are extremely readable, but also intellectually demanding.

In spite of the fact that German School is famous for its physically-biased studies in geography in the late nineth and early twentieth centuries, by the mid of the twentieth century, this School also showed a gradual inclination towards the use of humanistic and regional (chorological) viewpoint in the discipline. Interestingly, this development was parallel to the one going in other modern schools of geography as well, and more particularly the French and American. Two German scholars are known to have brought this change in Germany and they were Alfred Hettner and Otto Schluter; and in America it was largely due to Isaiah Bowman and Carl Sauer.

Alfred Hettner (1859-1941) vigorously encouraged the development of regional studies through publication in the *Geographische Zeitschrift*, the journal he founded in 1895. Hettner's methodology and philosophy was grounded in the work of Immanuel Kant and had a great influence. Under the leadership of Hettner, chorology became the guiding principle of Geography in Germany, and soon all over the English-speaking world[6]. Hettner subscribed to a view of classification of sciences and the placing of geography as a science that was similar to Kant's view on the subject. Hettner noted that on the one hand we have a 'logical classification' of the fields of learning, according to which we have a series of subject–sciences each concerned with the study and analysis of a logically defined circle of facts. On the other, we have a 'physical classification' of fields of knowledge according to which fields of study are defined not on the basis of logical unity of the subject matter concerned, but on the basis of the physical association of phenomena of diverse origin existing together. The physical association (or arrangement) of phenomena may be viewed in two different dimensions: Time and Space. Accordingly, physical classification of knowledge gives us two groups of fields of study, namely, 'historical sciences', which study integrations of diverse phenomena occurring together in the context of time (i.e. historical period); and 'chorological sciences' which study diverse phenomena existing together in segments of the earth's surface. Hettner vigorously encouraged the development of regional studies through publication in the Geographische Zeitschrift. Otto Schluter (1872-1952) had suggested that geographers should concentrate on the study of phenomena on the surface of the earth that could be perceived through the senses; and the focus should be on the totality of perception in each area. This totality of "visual perception of area" was termed as landscape. According to him, acceptance of landscape as the subject matter of geography had raised the discipline to the level of the other logically defined fields of science. Under the influence of Richthofen, Schluter developed a strong interest in questions of scope and method of geography, an interest that led him to his famous "Objectives of the Geography of Man" (1906), which projected the alternative concept of geography as the study of landscapes. Schluter incorporated the concept of process, i.e. development through time, in his concept of geography as the science of landscape by means of the derivative concepts of cultural and natural landscapes. He called them as *'Kulturlandschaft'* and *'Naturlandschaft'* respectively. For him the purpose of the study of landscapes was "not only of classifying categories of phenomena and determining their distribution and associations, but of examining their characteristics through the process of change through time". The concept of landscape and the related methodological principles of Schluter have also been widely used by geographers other than German before the Second World War.

In America, the physical bias in geographical studies is clearly evident in the works of the scholars prior to Davis. This bias continued even in the thinking of Davis and his followers, with an inclination towards environmentalism. However, since the times of Isaiah Bowman (1878-1950) and Carl Sauer (1889-1975), a chorological tilt has replaced everything. Bowman and Sauer became the active advocates of Possibilism in U.S.A.[7] Bowman was one of the outstanding geographers of the American scholarship of the early twentieth century. Based on his scientific expeditions and observations, he wrote his doctoral thesis, "Geography in the Central Andes"—a landmark in the contemporary conceptual structure of geography in U.S.A. His expeditions, in fact, led him to discard the Davisian paradigm which was the ruling paradigm for the geographical community.[8] In his later publications, he sought for an effective way to generalize the many detailed observations regarding the terrain and people for which he had colled information during his expeditions. He applied regional approach in his work, which in itself set forth the distinct paradigm for the research community. It strengthened the conceptual structure of applied geography in America and elsewhere. He treated the territorial problems within the framework of the regional paradigm. According to him, the geographical elements are fixed only in the narrow sense of the term. Earth facts do not determine the form and nature of human society and development. They condition it. New earth facts are continually being discovered and the old Earth facts give new significance as human knowledge, promoted thought and social action develop. The relations are reciprocal. In this manner, Bowman promoted professional geography in U.S.A. which was markedly different from that of Davis and others. Just as William Morris Davis, along with his associates, was the main promoter and exponent of geography in the first quarter of the twentieth century, so Carl Sauer (1889-1975) was the dominant academic figure in its second quarter. Carl Sauer led the field of cultural geography at the University of California Berkeley, where he taught for over thirty years. Sauer's influence in the field became known as the 'Berkeley School' of geography. Carl Sauer is famous as a human geographer and a chorographer. The best-known work of Carl Sauer is '*Morphology of Landscape*' published in 1925. Sauer's main interest was in the study of the impact of physical environment on man. Sauer insisted that instead of going into the field with a set of a priori principles concerning the effect of the physical environment on man, on should seek to observe facts himself and draw conclusions later. In other words, he favoured the use of inductive approach and emphasized on the use and importance of field studies and surveys in acquiring knowledge of geographical nature. Thus, he advocated for the formulation of a *posteriori* theories, i.e. in which the conclusions are drawn on the basis of the facts empirically observed and collected on the field. According to Sauer, geography relies on a purely evidential system, without prepossessions regarding the meaning of its evidence. "Geography is concerned only to establish the connections of phenomena in a visible landscape; and these connections are ones of spatial associations and not of some hidden causality". Although Sauer was a supporter of Possibilism and also remained an activist of the 'possibilistic' movement in U.S.A. during the inter-war period, he was not rigid about the 'possibilistic' viewpoint. Instead his ideas appeared more close to 'neo-determinism', i.e. acknowledging the superiority of nature on man rationally. Moreover, according to Sauer, the field of geography is based on areal characterization, rather than on environmental relationships, which he considered too general and wider field to be enfolded within the content of geography. This was the main philosophy of the school of chorographers, which Sauer actually belonged to. The Concept of Landscape of Sauer served as an important Paradigm in the Second Quarter of twentieth century. The roots of the concept lie in German term 'Landschaft', which emerged in the writings of Otto Schluter. The French equivalent is 'Paysage'. 'Landschaft' means "a piece of land (region) with certain physical and human attributes". The term 'Landscape' was accepted by many of the followers of Sauer as the basic term in geography. The human occupation and use of the natural area results in various forms, patterns and associations of phenomena; and Cultural forms and patterns are explainable in terms of historical antecedents, racial characteristics or any other force or agency.

However, notable among the American writers after Sauer who seems to be mainly interested in defining the nature of geography and its development over the years in general was Richard Hartshorne (1899-1992). The real change in American geography to chorology as the basic paradigm for geographical work came only after Hartshorne. In 1939, there emerged the monumental work of Hartshorne, '*The Nature of Geography*' that investigated the literature of several countries to synthesize what had been thought and written concerning the nature of geography. This book presented an authoritative account of developments in modern geography, and had soon become essential reading for graduate students in every geography department. From then on, the concept of geography as a chorological science became enshrined as the mainstream concept of the discipline, and it continued to be so until the late 1950s[9]. In 1959 he published *Perspective on the Nature of Geography*, which provided the benefit of 20 years' further thought and reflection by the author.

Thus, the various American scholars, from M.F. Faury to Hartshorne enriched geography in several ways. However, in general, the writings of the scholars in the early nineteenth century (Faury, Agassiz, Guyot, Forry, Loomis etc.) showed high inclination towards physical geography. In the late nineteenth century this physical bias continued, but there also emerged a deterministic tilt, particularly in the works of Davis, Semple and Huntington. Thereafter, since the beginning of the twentieth century, the American geography has been shaped by chorological perspective, a trend in consonance with the German and French schools of thought. The American scholars continued to dominate the geographical scene even after the Second World War.

The above scholars in the modern period have contributed to human studies with regional approach, particularly emphasizing on the historical, economic, political, colonial and electoral aspects. Their attempts have been successful in bringing a shift in geographical studies from physical to human, to the extent that a new paradigm was established, getting popularity as 'Videlian Tradition'.

NOTES

1. Hussain, M. (1995), pp. 308.
2. Oskar Hermann Khristian Spate (1911–2000) was a geographer best known for his role in strengthening geography as a discipline in Australia and the Pacific.
3. Richard Hartshorne (December 12, 1899–November 5, 1992) was a prominent American geographer, and professor at the University of Wisconsin-Madison, who specialized in economic and political geography and the philosophy of geography. He is known in particular for his methodological work *The Nature of Geography*, published in 1939.
4. Schaefer, Fred K. (1953), p. 228.
5. Rana, L. (2008), pp. 180-199.
6. Dikshit, (1997), p. 74.
7. Holt-Jenson, Arild (2009), p. 66.
8. Adhikari, Sudeepta (2015), pp. 161-63.
9. Dikshit, 1997. *op. cit.*

REFERENCES

Adhikari, Sudeepta (2015). *Fundamentals of Geographic Thought*, Orient Blackswan, pp. 161-163.

Dikshit, R.D. (1997). *Geographical Thought: A Contextual History of Ideas*, Prentice Hall of India.

Holt-Jenson, Arild (2009). *Geography: History and Concepts: A Systematic Guide*, 4th Edition, Sage, p. 66.

Hussain, M. (1995). "Paradigm in Geography", *Evolution of Geographic Thought*, Rawat Publications, New Delhi, pp. 300-340.

Rana, L. (2008), *Geographical Thought: A systematic Record of Evolution*, Concept Publishing Company, New Delhi.

Schaefer, Fred K. (1953). "Exceptionalism in Geography: A Methodological Examination", *Annals of the Association of American Geographers*, Vol. 43 (3) : pp. 226-249.

14
The Videlian Tradition (Paradigm Phase V)

Paradigm Phase V was strictly devoted to research in human and regional (chorological) studies in geography in the Second quarter of twentieth Century (1930s–W.W. II). During this period, the deterministic thinking was replaced by French school of the thought. This stream of geography was installed and emphasized particularly after the First World War, when a sociological trend was put forward through the writings of Vidal de la Blache, Lucien Febvre and Jeans Bruhnes in France, and Isaiah Bowman and Carl Sauer in United States. In spite of the fact that German School was famous for its physically-biased studies in geography in the late nineth and early twentieth centuries, it also showed a gradual inclination towards the use of chorological viewpoint in the discipline by mid twentieth century. Interestingly, this development was parallel to the one going in other modern schools of geography as well, and more particularly in France and America. The scholars are known to have brought this change in Germany were Alfred Hettner and Otto Schluter.

As to Greeks, Arabs and Germans, the geographers are also indebted to French, who made valuable contributions particularly in the fields of regional, human, historical, political and colonial geography. In the twentieth century geography emerged as an independent field in modern sciences in France; and after the First World War geography was well represented in all its Universities and nearly all the professors were pupils of Paul Vidal de la Blache (1845-1918). France entered the modern era of geography shortly behind Germany. While the German tradition quickly filled with new scholars, the French owe much of their tradition to one man, Blache. As the founder of modern geography in France, Blache is credited for establishing an entire generation of geographers here. He dominated the scene in the University of Paris for nearly 40 years. The first half of his career covers the last quarter of nineteenth century, when he was a contemporary of Richthofen, Ratzel and Wagner. He was particularly influenced by the second volume of 'Anthropogeographie' (1891) of Ratzel. The latter half of Blache's career was in the first quarter of twentieth century and he was then virtually contemporary with Penck and Hettner. This period also saw the florescence of his school with the publication of massive doctorate studies by his pupils of particular areas in France. This tradition also continued even after his death. Geography in France had a stultifying place in historical studies. Blache proceeded to direct geography away from this subsidiary position. His approach to geography was empirical encouraging detailed field studies. Under his leadership, the various university geography departments undertook field studies. In fact, after Ritter, it was Blache who re-emphasized the regional studies and synthetic approach in geography. Blache insisted on the study of the Man and Nature that he referred to as the geographical environment. The physical environment provides a range of possibilities which Man turns to his use according to his needs, wishes and capacities, in creating his habitat. Hence, he opposed the "environmental determinism" that was prevalent at the time as a Darwinian heritage. The major thrust of the work of Blache was on the study of man-environment relationships from the point of view of man. We know him as a great human geographer. With the fusion of regional and human approaches to geographical studies, Blache was able to put forward three interrelated doctrines, viz. 'Possibilism', '*Genre de Vie*' and 'Paysage'. These viewpoints constitute the 'Videlian Tradition'.

As a proponent of 'Possibilism', Blache considered that the main task of geographer is to correlate the physical and human conditions on the basis of their spatial inter-relations. He viewed this man-nature relationship from the point of view of man and, thus, rejected the doctrine of 'Environmental Determinism'. In its place, he established a conceptual framework of 'Possibilism' that was fully supported by Lucien Febvre (1828-1956), a critical French historian. He stressed the importance in viewing man in relation to his natural environment. Laying stress on Possibilism, Blache and his followers declared that man is not a passive agent ruled by the forces of nature which play their role and determine man's destiny and shape the human society. The main theme of 'Possibilism' is that the physical surroundings provide a range of possibilities to man and man makes a choice. This choice depends on his needs, aspirations and capacities. Secondly, man establishes relationships with nature not as an individual, but as a group or society. In other words, man views this relationship through the prism of his culture. 'Culture' here includes his beliefs, institutions, traditions, attitudes, religion, language, diet, habits, customs, skills and technology etc. In this way, man creates his own habitat, the local modified environment within which he lives, works or acts. This habitat is a distinct landscape created or developed because of man's occupancy, imprint and usage. This landscape has been termed as 'pays' at micro (smaller) level and 'paysage' at macro (larger) level by Blache. Thirdly, with the cultural advancement in time, the level of interaction of man and environment increases. This brings changes in human life-styles ('*Genre de Vie*') as well.

The possibilists developed the approach that in order to understand a society and its habitat, field study is most important. These micro-level studies were suggested to promote the regional approach, and thus making geography an idiographic or regional science. These geographers emphasized that humans have free will to evaluate different options and participate in the development of each landscape through unique historical processes. Methodologically, geographers were trained to concentrate their study on the unique single region. This inevitably limited the development of theory (as normally understood in science). The appropriate methodology under the approach would be to try to understand a society and its habitat through field study of the ways of life and attitudes of mind of the inhabitants. Fieldwork was regarded as of the utmost importance in the French School of regional geography. This fieldwork was, however, mainly concerned with the collection of factual material ways of life in the regions and had to be supplemented with the collection of factual material from statistical, historical and archaeological sources.

As a regional geographer, Blache preferred to view man within a physical milieu, or 'pays', and within his cultural milieu, or '*genre de vie*' together. A 'pays' is a natural region with some homogeneous physical characteristics, such as the mountainous Massif Central region in France. A '*genre de vie*', on the other hand, is a way of living or local culture. It includes the traditions, institutions, language, habits, foods etc. of a people. In other words, when one studies geography regionally, one may identify a region with regard to both, its physical characteristics and its human characteristics. According to the 'Determinists', human response to the environment is the same in similar physical circumstances. Contradicting this viewpoint the 'Possibilists' believe that even the similar physical circumstances can generate the different life-styles (*Genre de Vie*) depending upon the culture, economy and system of occupancy of land by the concerned society. The nature sets limits and offers possibilities for development, but the way man adjusts to the natural conditions of the area of his inhabitance is largely a function of his own tradition and mental structuring. Therefore, the same environment carries different meanings to people with different '*Genres de Vie*' (ways of life or culture). According to Blache, culture (i.e. inherited traits) is the basic factor in determining which of the many possibilities in natural environment should be selected or exploited by a given community. He drew attention to the fact that the same environment frequently leads to different responses from people with different '*Genres de Vie*', since the complex of institutions, traditions, attitudes and technical skills which make up a '*Genre de Vie*', largely determine which of the many possibilities offered by nature in any given area shall be selected for development. Blache was opposed to the dichotomy between 'natural' and 'cultural' aspects of the earth's surface; instead, he considered them as the two ends of the same continuum. To him, natural and cultural aspects of the earth's surface cannot be separated from each other. It is because, he reasoned, in every inhabited part of the earth's surface, the original landscape is significantly transformed as a result of human habitation. Such changes are greater in the case of culturally advanced societies where, owing to the more developed technology, the degree of man's intervention in nature is more far-reaching[1]. It is, therefore, impossible to study landscapes meaningfully without due reference to the interlocking roles of the nature and culture. Thus, the possibilistic approach to the study of man-environment relationships demands a three-fold interpretation, i.e. in terms of spatial, cultural and historical (temporal) perspectives. In other words, the present ways of life of a society, its cultural, socio-economic and overall development, is the result of the history of its occupation of the physical surroundings.

Possiblism continued to grow and spread both in France and outside it. Jeans Brunhes, who was one of the major proponents of the possibilists viewpoint, started formulation of human geography as a systematic approach to the study of geography. Outside France also the idea given by Vidal was accepted by a large group of geographers and sister disciplines. Harlan Barrow seriously took this idea in ecological conceptualization of geography. He stressed upon non-recursive feedback relationship between man and the environment.

Blache developed the concept that environment contains a number of possibilities, and their utilization is dependent almost entirely on human selection. This concept gave rise to the school of Possibilism. According to possibilists it is man who is the primary architect of his environment. It depends on the choices he makes among the numerous possibilities offered by the environment. In order to use these possibilities he makes two types of adjustments with the environment, i.e. in the form of 'adaptation' and 'modification'. For the possibilists, the works of man, not the earth, are the starting point and the most important is the freedom of man to choose. There are no necessities, but everywhere possibilities and man as a master of these possibilities is the judge of their use. In spite of the fact that man has numerous possibilities in a given physical setting, he cannot go against the directions laid by the physical environment. On this ground, the possibilistic approach was also criticized, and the possibilism, too, did not hold for many years.

There were two alternative views build around the French Concept of 'Paysage', viz. 'Landscape' in American) and 'Landschaft' in Germany, respectively by Carl Ortwin Sauer (1889-1975) and Otto Schluter (1872-59). The relationship of a community to the physical landscape of the area of its inhabitance is so intimate that it is difficult to think of the one without the other. As such, each segment of the cultural landscape, with the passage of time, acquires a unique personality of its own. Such areas were named as 'paysage'. According to Blache, the study of such regions constitutes the primary task of geography as a professional field. Thus, in his scheme of geography, Blache asserted that human occupancy creates homogeneity even where natural features are not homogeneous. The 'Cosmopolitan Urban Culture' may be cited here as an example. Blache sought to establish geography as a distinct discipline. Its field of study is the ensemble of phenomena that occurs in the zone of contact of solid, liquid, and gaseous masses (geosphere) that make up the planet. These phenomena are studied in relation to 'place, localization and distribution'. Thus, the field of study has a double aspect—Nature and Man. To explain their expression on the earth's surface one needs to borrow from both the natural sciences and humanities. He outlines the 'distinctive characteristics of geography' and claims that study has its goal in the characteristics and groupings of the phenomena of the landscape as the expression of man's presence and works in it.

The Concept of Landscape of Carl Sauer served as an important Paradigm in the Second Quarter of twentieth century, its roots lying in German term 'Landschaft', which emerged in the writings of Otto Schluter. Sauer has provided the detailed exposition of 'Landscape Concept' in his book entitled "The Morphology of Landscape". The term 'Landscape' was accepted by many of the followers of Sauer as the basic term in geography. The views of Carl Sauer, related to the Landscape concept, may be summarized in brief as follows: Geography is a comparative study of the landscapes that comprise the earth's surface; Description in an essential part of geography, and is equal in importance to explanation; Man's activities, as such, are not of prime importance to the geographers but rather those material features, resulting from man's activities; Features of the natural environment give character to an area, independent of human and cultural influence; the human occupation and use of the natural area results in various forms, patterns and associations of phenomena; and cultural forms and patterns are explainable in terms of historical antecedents, racial characteristics or any other force or agency. Sauer considered that geography must find its entire expression in the 'landscape' on the basis of the significant reality of chorological relations. The facts of geography, he maintains, are place facts. Their associations in area create this 'landscape'. Rejecting the terms 'region' and 'area' Sauer adopts 'landscape' to designate such terrestrial associations. This landscape is an area made up of a distinct association of forms (physical and cultural). The design of a landscape includes both: the features of natural area; and the forms superimposed over it by the activities of man, converting it into cultural landscape (or 'Kulturlandschaft' as identified by the Germans). Originally the 'landscape' is a natural landscape, and later it turns into cultural landscape or human habitat. Thus, a landscape at a later stage is the creation and workplace of human beings. The natural landscape is the assembly of natural features of an area that are relevant to human occupancy. Geography, Sauer claims, therefore, is 'distinctly anthropocentric'. The study of natural landscape, however defined, is directed to understanding its significance as habitat. The landscape is studied in terms of the process of human activity in time and in area. Various elements of this cultural landscape are related to: Habitation (type and grouping); Production (land use and workshops); and Communication (mobility). The relationships of human beings with their habitat are conditioned by their group habits and cultural habits. There may be a succession of these landscapes with a succession of cultures (e.g. Primitive-to-Ancient-to-Medieval-to-Modern). Hence, the idea of landscape may be defined in terms of both: its time relations and its space relations, i.e. in its spatio-temporal framework. It is a continuous process of development or dissolution and replacement. The landscape, therefore, changes continuously. The landscape elements are studied in terms of their form, function and pattern. These are interpreted in terms of the way in which they are associated in area as the spatial impact of cultural processes. In addition, Sauer considers landscape as an individual unit of area, i.e. geographical unit of study. The various forms, which characterize this landscape, therefore, may be identified, its structure can be recognized, and its origin and growth functions can also be understood. Thus, Sauer represented geography as a science that finds its entrifield in the concept of landscape. Accordingly any area is geographically distinctive and individual because of the nature and arrangement of two sets of features that cover its surface: those which are provided by nature (climate, surface features, soils etc.); and those which man has added (houses, cultivations, lines of communication etc.). Geographer systematically observes and records both the natural and cultural features, describes them in their associations and patterns, and seeks to explain the growth and development of the cultural panorama in its natural setting. He does not limit his study to search for relationships between man and earth. Rather, he shifts the emphasis from a study of relationships to a study of the material features, natural and cultural, within an area. The 'characteristics', 'associations' and 'patterns' of these surface forms, together with the bonds of relationships that connect them, are his chief lines of investigation. The region thus studied is viewed as a functioning organism composed of natural and cultural groups. Since, it is the observable material features that give chorographic individuality to any area on our planet, the chorographer does not directly include man's activities among his primary objects of study. When such activities express themselves in material forms, which can be observed and recorded, they (the forms) are then a part of the landscape, and as such open to geographic analysis. Further, the material features of an area are the 'primary facts' with which a geographer deals, and all those other facts used to explain and interpret the observable, areal scene are classed as 'secondary'. Thus, the geographic analysis is both 'descriptive' and 'explanatory' of the man-made landscape. The concept of landscape was based on visual perception of the surface features in given areas.

A parallel view emerging with 'the concept of landscape' was that of 'Landschaft' by Otto Schluter (1872-1952), the German scholar, the French equivalent being 'Paysage'. All three viewpoints meant 'a piece of land (region) with certain physical and human attributes'. The German word 'landschaft' literally and most generally means 'landscape'. In his 1906 methodological paper, Schluter had suggested that geographers should concentrate on the study of phenomena on the surface of the earth that could be perceived through the senses; and the focus should be on the totality of perception in each area. This totality of 'visual perception of area' was termed as landscape. According to him, acceptance of landscape as the subject matter of geography had raised the discipline to the level of the other logically defined fields of science. Schluter incorporated the concept of process, i.e. development through time, in his concept of geography as the science of landscape by means of the derivative concepts of cultural and natural landscapes. He respectively called them as 'Kulturlandschaft' and 'Naturlandschaft'. For him the purpose of the study of landscapes was "not only of classifying categories of phenomena and determining their distribution and associations, but of examining their characteristics through the process of change through time." The concept of landscape and the related methodological principles of Schluter have also been widely used by geographers other than German before the Second World War.

The word 'landschaft' has two distinct meanings: (i) 'An extent of area/territory distinguished by a more or less uniform aspect', and (ii) 'The aspect of the face of the Earth as seen in perspective but without connotation of areal extent.' Otto Schluter noted that accepting the landscape as the subject matter of geography would give the field a logical definition, like that of any other academic field. The non-material content of areas, such as political organization, religious-cultural beliefs, economic institutions and the like, were believed to lay outside the central focus of geography as a science. Schluter sought to distinguish between the landscape that predated the arrival of 'cultural' man, and the landscape created by the human culture, i.e. the cultural landscape. Schluter's 'kulturlandschaft' (cultural landscape) appears to have a greater resemblance with the French 'Pays'—small homogeneous well-defined areas/regions manifesting intimate and reciprocal relationships between man and his immediate surroundings (milieu) through centuries. A 'pay' is, therefore, an areal expression of the *Genre de Vie* of a human community whose occupation of the area/region over centuries seems to have created the landscape.

Chorological Tilt

As part of Videlian tradition, the major concern of geographers became to study region. Blache and French School of regional geography had produced scientific works that served as exemplars for a large group of students, and thus could be said to have functioned as a paradigm. Regional descriptive geography appeared to them to be the only honourable occupation. Resultantly, there emerged a bulk of descriptive literature. In whatever methodological writing they did, the geographers aligned themselves with the opponents of scientific method.[2] The concept of 'region' or 'regional' concept has for long been a major concern to geographers, particularly to those who believe in the 'geographical synthesis' of the earth-bound phenomena. But, in the second half of twentieth century, it strongly reflected in the works of certain prominent American and German scholars as well, as, e.g. that of Richard Hartshorne (1899-1992), Alferd Hettner (1859-1941) and their supporters. They made use of regional approach to study geography. These geographers expressed their concern about the variations in the character of the face of the Earth, which became known later as 'areal differentiation'. They recognized that there were distinctly different kinds of areas on the Earth which were distinguished from their surroundings and which showed a certain degree of homogeneity within boundaries that could be defined. Extending the Kantian philosophy, Hartshsorne and Hettner, both regarded the regional synthesis as central to geography. In the writings of these scholars, geography emerged as the 'integrating science', completely different from other disciplines, whose unique importance finds its expression in the special methods which it must use to reach its profound results. Following the Greek geographers, this viewpoint has been called the 'chorological' or 'chorographic' one, depending on the level of abstraction. Hartshorne's

'Nature of Geography' (1939) that remained a central textbook and codified this chorological viewpoint into one of areal differentiation.

Besides the emergence of chorological perspective in geographical studies, the man-environment debate that had continued in the discipline for about a century after the death of Humboldt and Ritter, also got resolved. This was due to a compromise being reached in the Ecological School of Thought followed by Quantitative Revolution. The ecological approach no longer accepted the sharp distinction between man and the external biophysical surroundings. It considered man not as the same creature everywhere and at all times, but diverse in outlook and capability, depending on the society to which he belongs. And each society perceives and interprets its biophysical setting through the selective prism of its culture. Thus, nature and culture are interwoven and complementary parts of whole rather than opposing forces.

One distinct development that was observed commonly among various modern schools of thought (particularly, French, German and American) in geography showed a tilt towards the adoptition and dominance of chological (regional) approach in general.[3] For a long time, geographers continued to stress the central position of Regional geography. Regional geography flourished in such countries as France, where in school and University teaching, geography was closely associated with history and the educational system fostered a national self-image of study, peasantry and cultured townsfolk. Regional studies were also important to the academic leaders of the emergent nations in central and Eastern Europe, who were seeking to establish and preserve the uniqueness of their national heritage. The German geographers also recognized as the major concern of geography being the manner in which the natural phenomena, including man, were distributed in space. This implied that geographers must describe and explain the manner in which 'things combine to fill an area'. These combinations change, of course, from area to area. Different areas contain different factors or the same factors in different combinations. These differences, either in the combination of factors or in their arrangement from place to place, underlie the common sense notion that areas differ. Geography, thus, must pay attention to the spatial arrangement of phenomena in an area and not so much to the phenomena themselves. These spatial relations are the ones that matter in geography and no others[4].

The goal of the chorological point of view is to know the character of regions and places through comprehension of existence together and interrelation among the different realms of reality and their varied manifestations and to comprehend the earth surface as whole in its actual arrangement of continents, larger and smaller regions and places (Hartshorne, 1961). According to Taaffe (1974), Hartshorne used the term "areal differentiation' to characterize the way in which geographers dealt with the wide variety of phenomena physical, economic and social, which exist together in the area and distinguish them from other areas."

The chorological viewpoint, in turn, presented geography with a two-dimensional investigation. The geographer's investigations, be it a physical, economic or political, are of two different types: either 'systematic' or 'regional'.[5] A region contains, to be sure, a special, unique, yet in some ways uniform combination of kinds or categories of phenomena. The detail with which the regional geographer describes, lists, or catalogues these features at the outset of his investigation depends, of course, on the size of the region considered. Next he will want to gather information as to the spatial distribution of the individuals in each class. First, he must find those relations obtaining among individuals and the classes by virtue of which the area considered has that unitary character that makes it a region.

Second, he must identify the relations which obtain in this particular area as instances of the causal interrelations that hold, by virtue of general laws among such features, individuals, classes, or what have you, in all known circumstances. This second step amounts, therefore, to an application of systematic geography to the area in question. Only after both these steps have taken can one say that a scientific understanding of the region has been achieved. This brings us to systematic geography. Its procedure is in principle not different from that any other social or natural science which searches for law or, what amounts to the same thing, has reached the systematic stage. Spatial relations among two or more selected classes of phenomena must be studied all over the earth's surface in order to obtain a generalization or law. Regional geography flourished in France and got diffused in its neighbouring countries. But, later on this approach also became inadequate to explain the regional personality and therefore, a period of crisis in the discipline emerged.[6]

NOTES

1. Dikshit R.D. (1997), pp. 85-86, 92-93, 245.
2. F.K. Schaefer (1953), p. 230.
3. Rana, L. (2008), p. 234.
4. F.K. Schaefer (1953), p. 226.
5. *Ibid.*, pp. 227-228.
6. Holt-Jenson, A. (2009), pp. 113-115.

REFERENCES

Dikshit, R.D. (1997), *Geographical Thought: A Contextual History of Ideas*, Prentice Hall of India.

Rana, L. (2008), *Geographical Thought: A Systematic Record of Evolution*, Concept Publishing Company, New Delhi.

Hartshorne, R. (1939), *The Nature of Geography: A Critical Survey of Current Thought in the Light of the Past*, Association of American Geographers, Lancaster.

Holt-Jenson, A. (2009), "Paradigms and Revolutions", *Geography: History and Concepts: A Systematic Guide*, Fourth Edition, Sage, London, pp. 100-124.

Hussain, M. (1995), "Paradigm in Geography", *Evolution of Geographic Thought*, Rawat Publications, New Delhi, pp. 300-340.

Schaefer, Fred K. (1953), "Exceptionalism in Geography: A Methodological Examination", *Annals of the Association of American Geographers*, Vol. 43, No. 3, pp. 226-249.

15
Modern Tradition

The Modern period witnessed three major and rapid paradigmatic shifts (III, IV and V) in geography. It was as a result of the rise and dominance of German School inducting the studies with Systematic and Regional approach; origin of Determinism focusing the role of environment in human affairs; and emergence of Videlian tradition as an antithesis to Determinism, with a thrust on man in man-environment study. Major paradigms of this phase included Determinism, Possibilism and Chorological (Human and Regional) view. The intervening Crisis phases developed various dualistic tendencies addressed to these dotrines in geography. Particularly, three modern schools of thought have been instrumental behind this development, viz. German, French and American.

The Paradigm Phase III represents developments during the next 300 years after the Age of Discovery when the German School of geographic thought was at its peak and geography was purely a German science. Lot of research was carried on towards defining geography and deciding upon its methodology (approach, scope, content, method of study). In late medieval and early modern period, Geography was no more than encyclopedic compilations of places and their descriptions. The principles behind spatial arrangement of phenomena, their causal connections and inter-relationships that could give geography a true scientific status were yet to be established. The credit for this goes to the German scholars particularly to the writings of Varenius, Kant, Humboldt and Ritter. The initial impetus to German School came from the works of Bernard Varenius, best known for his work Geographia Generalis, considered to be the first modern literature on geography. Varenius brought together contemporary knowledge of astronomy and cartography and subjected the different theories of his day to sound critical analysis. Through this he tried to provide a scientific methodology to geography by advocating two approaches for its study, viz. 'General' (systematic) and 'Special' (regional). General Geography included description of three forms of earth features, viz. Absolute, Relative and Comparative. Similarly, Regional Geography consisted of descriptions of three types of properties of the constituent areal units, viz. Celestial, Terrestrial and Human. Thus, the geography of this period included both 'general' studies of particular kinds of phenomena of the earth's surface, and descriptions of many kinds of phenomena found in particular areas. The terms, 'General' and 'Special', that Varenius used became standard in the discipline. Among later geographers, Humboldt and Ritter, popularly known as the fathers of modern geography, drew attention to Varenius' genius and services to science. They did it respectively in the form of a natural scientist and a human geographer. In the meanwhile, writings of Emmanuel Kant guided the research in Germany. For Kant, Geography was a chorographic and mainly a descriptive science, distinct from the systematic sciences. He considerably modified the scientific structure of geography as proposed by Bernard Varenius. Kant agreed on the two ways of acquiring knowledge, viz. through 'sense perceptions' and through 'pure reasoning'; the former approach is 'inductive' and the latter is 'deductive'. The sense perceptions, in turn, are of two types, viz. 'inner' and 'outer'. However, Kant's approach to geographical studies was purely deductive. Kant also pointed out that there are basically two ways of classifying knowledge for the various phenomena on the surface of earth. They are: 'Logical Classification', i.e. grouping according to the nature of the inherent characteristics of the phenomena; and 'Physical Classification', i.e. grouping according to the position or occurrence of phenomena in time or space. The Logical Classification of Kant deals with organizing knowledge according to the type of objects studied, and accordingly covers a series of systematic or pure sciences that are devoted to the study of particular kinds. The Physical Classification, on the other hand, deals with organization of knowledge in its temporal or spatial dimensions and, therefore, includes historical and spatial sciences, viz. History and Geography. Considering both history and geography as descriptive in nature, Kant also regarded both of these disciplines as complementary to each other. According to him, history and geography are both essential sciences, standing alongside the systematic sciences. Without them, man cannot achieve a full understanding of the world. Besides, Kant laid the greatest emphasis on the study of Physical Geography. Along with it, he also identified some other branches of geography, namely: Mathematical Geography, Political Geography, Moral Geography, Mercantile Geography and Theological Geography. Kant has been labelled as 'the father of exceptionalism' in geography by F.K. Schaefer, a twentieth century American scholar, for laying an exceptionalist claim not only for geography, but also for history. However, in spite of all the criticisms put forward against Kantian philosophy, the basic tradition of the field of geography has persisted in his writings, i.e. "geography is a description of the whole earth, where the earth is understood to mean the surface of the earth and the dwelling place of the man".

The exposition of geography as 'chorological' in nature by Alfred Hettner and Richard Hartshorne in the late nineteenth and early twentieth century has closely followed the Kantian philosophy. After Kant, Alexander von Humboldt and Carl Ritter revived the tradition of Bernard Varenius. Both were contemporary, and they together tried to mould the subject matter of geography into scientific form, following the tradition of Varenius. The prevalent philosophies at their time were 'natural' and 'moral'. The natural philosophies were related to the physical world, envisaging the earth as an organic whole, inseparable and mutually interdependent; whereas, the moral philosophies were related to man's inner world and human thinking. Humboldt's basic inclination was towards natural, whereas Ritter had anthropocentric view. He envisaged the earth as a whole, an organic whole, all parts of which being mutually interdependent. Like Kant, he also attempted on a division of knowledge, but the classification he contemplated fell into three groups, viz. Systematic, Historical and Earth Sciences. Although, the basis of this classification was close to that of Kant's, but, Humboldt accorded the status of sciences to all kind of subject-matter (including history and geography). While dealing with the subject matter of geography, Humboldt coined a term 'cosmography', the study of the Universe; and divided it into Uranography, concerned with mapping the stars, galaxies, or other celestial bodies; and Geography, studying the terrestrial part with the interrelationships of phenomena that exist together in an area. Humboldt followed empirical, inductive systematic and comparative methods of study. Humboldt conceived man and his activities as an integral part of the nature. This theme runs through the 'Cosmos', his best-known work. Humboldt included man and his works in the concept of nature and natural areas, but he did not consider man as a primary determinant. Carl Ritter's work was not only contemporary but also complementary to that of Humboldt. He exercised profound influence on geography by adopting a different methodology. Contrary to and complementing Humdoldt, Ritter emphasized more on human, favoured regional or synthetic approach, and worked basically as an armchair geographer. Just like 'Cosmos' of Humboldt, Ritter also authored a great work entitled '*Die-Erdkunde*'. Ritter was deeply a religious man. He thought that divine has created the earth as a home of man. He also had a vision of an ordered and harmonious universe, but, his approach was teleological. He asserted that all phenomena are spatially distributed according to the plan of God for mankind. In spite of being an armchair scholar himself, Ritter did identify geography an empirical and descriptive science. According to him, geography deals with local conditions and embraces the attributes of a place in the form of topographical, formal and material characteristics. These three attributes constitute the basic content of his most important work, *Die-Erdkunde*, as well. '*Die Erdkunde*' is based on the regional descriptions of the earth covering largely Asia, Africa and Europe. This work was written with a distinct theme, i.e. comparative and inter-relationships. Ritter claimed that the central principle of geography is "the relation of all phenomena and forms of nature to the human race, examined and organized within the framework of the unique geographical associations of land and man on the earth's surface." Through his Erdkunde, Ritter made a rhetorical claim for geography as the science of the earth. He believed "the earth to be an organism made, even in its smallest details, with divine intent, to fit the needs of the man to perfection; as the body is made for the soul, so is the physical globe made for mankind." Besides, Ritter is also popular among geographers for his Regional Concept. His division of the Earth consists of four levels in hierarchical order: Earth or Erdkunde (1st Order); Continental Units or Erdteile (2nd Order); Division of each Continent into various relief or Orographic regions (3rd Order); and Smaller Areal or terrestrial units arrived at inductively (4th Order). His plan of study, based on the concept of regional associations of terrestrial phenomena, became the model for geographical presentations and had far-reaching influences. This classical period of German geography terminated with the death of both Humboldt and Ritter in the year 1859. Both Humboldt and Ritter worked towards providing geography as the status of a scientific discipline and appeared, as such, inspired by Bernard Varenius. The writings of Humboldt and Ritter had a thrust on inter-relationships, spatial associations or spatial integration of phenomena. The Physical Geography was considered as the base of all geographical studies by Humboldt and Ritter. Both emphasized on the importance of empirical method of research in geography. Although both believed in the Unity of Nature, they differed in their philosophical approach to geographical studies. For Humboldt, this unity of nature was a balanced unity of the whole of nature, of which man was a part; whereas for Ritter, it was teleological, as he spoke of discovering the 'divine secrets'. In spite of all the differences, the works of Humboldt and Ritter together, particularly '*Cosmos*' and '*Erdkunde*' did provide an almost complete explanation for geography, in terms of its definition, content, scope, approach and method of study. Coincidently, both of them died in the same year. The year 1859 marked not only the death of Humboldt and Ritter but also the publication of Darwin's "*On the Origin of Species*". Charles Darwin was an English naturalist who established the theory of organic evolution.

From the early nineteenth century to mid twentieth century, geography was developing in several directions due to distinct schools of thought developed under the influence of Humboldt, Ritter and Darwin, and there were followers of each. The geographical studies were basically bifurcated in two directions: Study of earth as a natural body and Study of earth as a dwelling place of man. At the same time a number of dualisms also overshadowed the discipline for about 100 years after the emergence of German School of Thought. It is because, in the latter half of the nineteenth century, the followers of Humboldt and Ritter exaggerated certain aspects of views of each of the founders. However, in the late nineteenth century, the more expanding frontiers were those of physical and biological sciences than human under the impact of Darwin. Darwinism, dualistic tendencies and a physical bias in geographical studies brought a turning point (Crisis Phase III) in the history of the discipline. The philosophy of Darwin caused considerable intellectual ferment and had a tremendous influence on geographic thought. This influence came through his theory of 'evolution by natural selection' and the concept of 'survival of the fittest'. The cosmographic way of thinking was replaced by the causal explanations characteristic of the developing natural sciences. Darwin was primarily concerned with the mechanism of the change and he rejected all supernatural phenomena and causations. Eliminating God from science made room for strictly scientific explanations of all natural phenomena; it gave rise to positivism. It produced a powerful intellectual and spiritual revolution, the effects of which have lasted to this day. This intellectual ferment led to a great reappraisal of thinking in both the natural and social sciences. As a result, traditional natural philosophy professed by amateurs became transformed into modern science developed by professional scientists. A new term 'Social Darwinism' emerged in the second half of the nineteenth century to refer to various ways of thinking and theories that tried to apply the evolutionary concept of natural selection to human society. Since the foundations of geography as secured by Humboldt and Ritter did not provide a clearly unified field and, therefore, differences in the opinions and viewpoints persisted for a long time in the discipline.

The dualistic tendencies that have always been part of geographical studies since beginning became more conspicuous, after the death of Humboldt and Ritter. The major controversies that emerged in geography during modern period were related to its subject matter and approach to study, viz. 'Physical *versus* Human' and 'Systematic *versus* Regional' respectively. It is particularly with the writings of German scholars like Kant in the initial stage and of that of Humboldt later that physical aspect of geographical studies got greater emphasis than human side. Humboldt conceived man as the part of nature, dependent upon nature-*minus*-man, which existed before man and which could exist independent of man, whereas the reverse cannot hold true. His 'physical geography' was directly comparable to the 'general geography' or the 'systematic geography' of Varenius. Carl Ritter, a contemporary of Humboldt, on the other hand, had an anthropocentric approach to geographical studies. Ritter conceived Earth as the home of man. According to him, the physical surroundings get meaning only with reference to human beings. They are required to be known only in relation to human beings. Because it is the man who interacts with the physical world. The relation between the world of man and the non-human world is of prime concern in geography. Since the earlier times, the human and the non-human aspects have contributed to the content of geography. The dichotomy was what aspect is more important to study in geography, physical or human. Originally the controversy between man and physical (nature) was external to geography. It appears to have entered into the discipline first with the writings of German scholars, and was later supported by the rise of natural sciences and Darwinism. The important sources of problem may, therefore, be identified as: the works of German scholars; Environmentalism (Determinism and Possibilism); and Systematic *versus* Regional Dualism.

The terms 'General' and 'Specific', as suggested by Varenius, were caught by various geographers later and the dichotomy between the two widened to the extent that there had been periods when there existed double dualism in geography, i.e. 'Systematic and Physical' *versus* 'Regional and Human". One such time was that of Emmanuel Kant himself. Kant substituted the word 'physical' for 'general', and classified all the general or systematic studies under 'physical'. It was under his influence that the geographers in the late eighteenth century concentrated on systematic or general studies at the expense of regional or specific studies. The close of nineteenth century found an increased interest on physical or natural geography; and the systematic studies in their new guise were highly specialized. They were prosecuted by scientists who tended to be geomorphologists and climatologists. This was a time when 'systematic' and 'physical' became synonymous and the study of landforms emerged as the dominant and most attractive part of geography. This attracted the researchers not only from Germany but also from outside, especially U.S.A. The development that started after 1859 culminated in the works of W.M. Davis. In comparison, Human Geography lagged behind, because, some geographers wanted to make geography a preserve of pure science. Since human behaviour did not appear to be subject to general laws, they proposed to exclude man entirely from the field of geography. The early twentieth century saw a preponderance of physically-biased studies in geography, particularly brought in by the German and American Schools. Thus, after the death of Humboldt and Ritter in 1859 till the First World War, the emphasis of geographical studies kept on shifting. Although this shift was evident in the form of Physical and Systematic (Peschel, Gerland), to Physical and Regional (Richthofen, Penck), to Systematic and Human (Ratzel), to primarily Regional (Hettner, Schluter) and predominantly Physical (Wegener, Penck) during the late nineteenth and early twentieth century, the physical bias was very strong in most of these studies. The development of geography during the first half of the twentieth century, on the other hand, has been marked by an increasing interest in regional geography. The climax was witnessed in the views of Richard Hartshorne and Alfred Hettner, when the concept of geography as a chorological science was advocated by full force. This led to the development of concept of geography as a science of 'areal differentiation'. The idea of region was further strengthened by its usage in different forms, e.g. 'Paysage' in France (Blache), 'Landschaft' in Germany (Otto Schluter) and 'Landscape' in America (Carl Sauer). This was a period when the 'Possibilistic' thinking was at its peak. At this climax phase, the argument took the form of the old 'hen and egg story'. One group believed that the core of geography was regional, whereas the others believed that systematic geography was really indispensable.

It was the Darwinian tradition that led geography into next phase of professionalism, i.e. Paradigm Phase IV. Rise of Determinism was the major philosophical outcome of this development which happened in late nineteenth and early twentieth Century. Darwin's paradigm in conjugation with that of Davis' replaced all the Old traditions. Davis sought to integrate the two elements, viz. Nature and Man. This he did after a period in which a marked dualism prevailed in the discipline, i.e. between exclusively physical geography (including plants and animals) and the geography of man. Davis found the key at this time in the Darwinian evolutionary approach that regarded the facts of nature and man as existing in a 'harmonious relationship' in a continuous chain of cause and effect. The writings of W.M. Davis played a great role in bringing out professionalization in the discipline during this period. The 'Theory of Cycle of Erosion' is the most important contribution to Physical Geography by Davis. This revolutionary concept was later applied to the study of evolution of Glacial Landforms, Limestone Region, and Islands bordered by Coral Reefs as well. Representing the prevailing deterministic thinking, Davis restated the core of geography as the study of the relationship between the physical environment (the control) and the human behaviour (the response). Davis sought to classify the facts of man in reference to the facts of the physical earth, calling them respectively 'Ontography' and 'Physical Geography'. Davis identified 'Ontography' as the other half of geography, and defined it as 'geographical description of life forms on the earth'. This was an extension of the evolutionary interpretation of man on the earth in so far as human characteristics show direct results of geographical (physical) influences and controls. Davis believed that such a system would form the framework of an integrated science. This viewpoint very clearly set the pattern for the pursuit of man-land relationships on a deterministic basis. Besides Davis, the contemporary thinking was also dominated by the ideas of his pupils like Ellen Churchill Semple, Ellsworth Huntington and Harlan Barrows. The environmentalist approach was taken up in their diligent and far ranging researches. Semple was an environmental determinist. Huntington was known for his studies on climatic determinism. He examined the influence of climate and climatic change on men and the course of history. Harlan Barrows presented the concept of geography as human ecology for the first time. He expressed that the physical environment does not cause man's adaptation to the conditions of his habitat. This adaptation, in fact, is the function of human choice. He proposed human ecology as the unifying theme for geographical work.

Among geographers influenced by Darwinism, it was Friedrich Ratzel, the German scholar, who explored the influences of physical environment on mankind. At this point of time, Ratzel corrected the prevailing tendency of overstressing the physical aspects of geography and established a more balanced viewpoint. The thinkers like Charles Darwin and the zoologist like E.H. Haekel, in turn, influenced Ratzel's writings. In his two most important literary contributions, *'Anthropogeographie'* and *'Politische Geographie'*, Ratzel treated human geography systematically and from Darwinian's point of view. Indeed his work served to fix interest still more firmly on systematic studies, and regional geography continued to receive scant attention. In *Anthropogeographie*, Ratzel explored the influence of physical environment on mankind systematically and empirically. The basic theme of the work in its first volume is that the man and his actions are not free from the influences of the surroundings. They are controlled by nature. He believed that man is the creature of his environment in much the same way as Darwin had demonstrated the adaptation and survival of the fittest in the animal world. However, in his second volume of *Anthropogeographie*, written with more liberal views on man-nature relationships, he showed a different spirit, where the emphasis was on distribution and density of population, forms of settlements, migrations, and diffusion of cultural traits. Here, he added that the survival of man and his adaptation to the environment also depends on fact that how fit and skilful he is in moulding the natural circumstances. Man's reactions to the nature differ sharply accordingly to his culture and historical background. If belonging to primitive culture then nature has a stronghold; and if advanced technologically, then the nature's hold becomes weaker. But, even then the nature is dominant. However, according to Ratzel, whether man adapts or modifies the surroundings, in any case, is the creature of his environment. Because, even the act of moulding is a kind of adaptation prompted by the nature in an indirect manner. For the explanation of these features Ratzel did not merely resort to environmental influences but equally, or more so, on historical-cultural factors. Ratzel saw man as the end product of evolution, in which the mainspring was the natural selection of types according to their capacity to adjust themselves to the physical environment. Through Politische Geographie, his other book, Ratzel reshaped the old topic along the lines conformable with the principles he had enunciated in his Anthropogeographie. In this work, he tried to make a comparative investigation of the relation between 'the political state' and 'the earth's surface'. The revolutionary biology led Ratzel to adopt the organic theory of state and society, i.e. the 'Concept of state as an Organism', a piece of humanity and a piece of earth. The political entities (nations or states) expand or shrink like organisms. Based on this viewpoint, he postulated the 'Organic Theory of State', the concept of state as an organism, a piece of humanity on the surface of the earth. He believed that a nation or state is not only a spatial unit, but also a human entity, and therefore, it is a living unit. From these ideas, Ratzel developed the concept of 'lebensraum' or living space and hypothesized that the state naturally seeks to increase its size. If the state's neighbours are weak, the state will grow larger and spread into other states. As evidenced, Ratzel believed that space was a great political force.

Ratzel is known as the propounder of Environmental Determinism in geography. According to this viewpoint, man is the product of nature. It is the nature, which governs man and his activities, and determines his way of life, living styles and also physical appearance. Because of this postulation, Ratzel became very popular particularly in Germany and also in U.S.A. The relationship of man with nature and the resultant pattern or description over space in a particular time is the core of all geographical studies. Determinism has emerged as one of the ways to study this relationship. Although, Determinism is one of the oldest schools of thought, the philosophy of environmental determinism rose to prominence in the late nineteenth and early twentieth century when it was taken up as a central theme by the discipline of geography. The view persisted up to the Second World War in one form or the other. Supporters of the determinism favour the view that environment controls the patterns of man's behaviour, so that all his actions, his settlements, his modes of life are controlled, directly or indirectly, by his surroundings. They believed that nature is superior to man. According to the determinists, man is a passive agent on whom the physical factors are acting and determining his attitudes and the process of his decision-making. The essence of the deterministic school of thought is that the history, culture, living style and the stage of development of a social group are exclusively or largely governed by the physical factors of the environment. The physical environment, in turn, has two aspects, viz. 'climatic' and 'physiographic'. The basic argument of the environmental determinists was that aspects of physical geography, particularly climate, influenced the psychological temperament of individuals, which in turn shaped the culture and society that those individuals formed. The origin of deterministic thinking goes back to antiquity, when various Greek and Roman scholars made first attempts to explain the physical features and character traits of various people and their culture with reference to the influence of natural conditions. The geographical determinism continued to reflect the writings of Arab Scholars in the medieval period as well. Determinism was also strongly observed in the work of German Scholars. The philosophy of determinism, in turn, was greatly influenced by the evolutionary biology of Charles Darwin later. The pre-Darwinian views on Determinism were found relatively modest in comparison to that in post-Darwinian period, when they attained more extreme form. The maximum influence of deterministic thinking was exercised after the popularity of Darwin's Theory. The scientific developments in geography that followed Darwin were deeply influenced by the role of nature on organisms, including man. This role of nature was deeply studied in human affairs. In other words, it was tried to prove that man's houses, communications, agriculture, his physiographic characteristics, nature, attitudes, capacities, general way of life and beliefs all are determined by the nature or environment (particularly including climate, relief, soil and vegetation). Gradually the term 'environmental influence' was replaced by the 'environmental control', i.e. the environment does not simply influence but determines. The observations of various scholars caused them to conclude that 'man's immediate surroundings have a decisive (deterministic) effect not only on his occupations, dwellings and the materials he uses in everyday life, but also on his personal traits, his general character, and even his religious beliefs. They found environment to be superior to man and mastering all of his actions. Such generalizations were called 'extreme' and not as the result of 'cause and effect' relationship between man and environment. Thus, there emerged extreme form of determinism in the late nineteenth century. The scientific milieu in the later half of the nineteenth and the early twentieth centuries was dominated by Darwin's idea and the deductive approaches. Fitting well into this intellectual environment, the theme of environmental determinism was prevailing view not only in Germany but also in American geography. Ratzel supplemented 'classical' geographical determinism with the elements of 'Social Darwinism' and developed a 'theory of the state as an organism' that owed its life to the earth, which was ever striving to seize more and more territory.

In spite of its vigourous criticism the school of Determinism did not die away completely, its central theme continued to survive in the form of Neo-Determinism. Where some of the early environments were apt to make unsound generalizations, the modern determinists were more careful, to seek and record a true sequence from cause to effect. The modern determinists differed most radically from the 'possibilists' in their view on the nature of human 'choice'. They did not allow that man has complete freedom to choose, but rather, that he narrows down alternative courses of action in his mind, and, in selecting one alternative rather than another is, he is consciously or subconsciously influenced by some part of his general environment. The supporters of this new school of determinism basically believed in determinism but not to the extent as earlier determinists did. They did not use unsound generalizations; they believed in cause and effect relationships between man and nature; they were more careful, and applied theoretical and empirical examination of the facts with the help of modern techniques. Griffith Taylor used the term 'Stop-Go-Determinism' for Neo-Determinism.

Geography encountered another crisis phase in early twentieth century when there was widespread criticism to Determinism. Resultantly, there took place a paradigmatic shift to Human from Physical. However, this turned out to be relatively a very short period in transition as the discipline soon witnessed the next Paradigm phase. By 1900, the philosophy of determinism had become particularly widespread in Germany and the thinking also percolated in U.S.A. But, from the beginning of twentieth century this doctrine came under attack. The chief fault in the assumption of the extreme determinists was found in their over-simplication of the man-environment relationships. Their generalizations were accused of being too-sweeping, without being backed by careful investigations and cause-and-effect relationships. The determinists were blamed for over-emphasizing the role of physical environment in human affairs. Man was seen only as a passive element of the environment. Such works of man were overlooked for which environmental forces alone give no satisfactory explanation. Although, the school of determinism, as such, had a short span of life, its central theme did not die out completely. It survived and continued to persist in the form of 'Neo-Determinism' or 'Stop-Go-Determinism'. In spite of the emergence of Neo-determinism, the move towards human studies in geography continued to grow stronger. Largely, as a reaction to the exclusive concentration on systematic studies on the part of physical geographers, the interest began to shift towards social or human geography. The French scholars owe the credit for this change. Vidal de La Blache was the leader of French School of Geographical Thought. He was particularly influenced by the second volume of '*Anthropogeographie*' of Ratzel. Blache insisted on the study of the geographical environment and Man. He considered, however, that the peculiar task of the geographer lies in the correlation of physical and human conditions on their spatial basis. The physical (natural) environment provides a range of possibilities which Man turns to his use according to his needs, wishes and capacities, in creating his habitat. Hence, as opposed to the "environmental determinism" that was prevalent at the time as a Darwinian heritage. In its place, he established a conceptual framework of 'Possibilism'. The major thrust of the work of Blache was on the study of man-environment relationships from the point of view of man. We know him as a great human geographer. Among the other scholars, working on the lines of Blache, the contributions of Himly, Reclus, Bruhnes, Demangeon, Seigfried, Vilar and Gottaman are noteworthy. Their attempts also proved instrumental in bringing a humanistic tilt in geographical studies. Along with the French, the German and American schools also showed a gradual inclination towards the use of humanistic viewpoint in geography. Two German scholars are known to have brought this change in Germany and they are Alfred Hettner and Otto Schluter; and in America it was largely due to the efforts of Isaiah Bowman and Carl Sauer. The Berkeley School of Carl Sauer, for instance, is based on areal characterization, rather than on environmental relationships, which he considered too general and wider field to be enfolded within the content of geography. All the above scholars in the modern period have contributed to human studies with regional approach, particularly emphasizing on the historical, economic, political, colonial and electoral aspects. Their attempts have been successful in bringing a shift in geographical studies from physical to human, to the extent that a new paradigm was established, getting popularity as 'Videlian Tradition'.

Paradigm Phase V was devoted to research in human and regional (chorological) studies in geography in the Second quarter of twentieth century. During this period, the deterministic thinking was replaced by French school of the thought. As the dominating paradigm in human geography, determinism had a short life, being challenged by the French School of human geographers. Blache and his followers stressed the idea that humans have free will to evaluate different options and participate in the development of each landscape through unique historical processes. With the fusion of regional and human approaches to geographical studies, Blache was able to put forward three interrelated doctrines, viz. 'Possibilism', '*Genre de Vie*' and 'Paysage'. These viewpoints constitute the 'Videlian Tradition'. The main theme of 'Possibilism' is that the nature or physical surroundings provide a range of possibilities to man and man makes a choice. This choice depends on his needs, aspirations and capacities. Secondly, man establishes relationships with nature not as an individual, but as a group or society. In other words, man views this relationship through the prism of his culture. 'Culture' here includes his beliefs, institutions, traditions, attitudes, religion, language, diet, habits, customs, skills and technology etc. In this way, man creates his own habitat, the local modified environment within which the man lives, works or acts. This habitat is a distinct landscape created or developed because of man's occupancy, imprint and usage. This landscape has been termed as 'pays' at micro level and 'paysage' at macro level in French School. Then, with the cultural advancement in time the level of interaction of man and land or environment increases. This brings changes in human life-styles as well. Thus, the possibilistic approach to the study of man-environment relationships demands a three-fold interpretation, i.e. in terms of 'spatial', 'cultural' and 'historical (temporal)' perspectives. In other words, the present ways of life of a society, its cultural, socio-economic and overall development, is the result of the history of its occupation of the physical surroundings. As a regional geographer, Blache preferred to view man within a physical milieu, or 'pays', and within his cultural milieu, or '*genre de vie*'. A 'pays' is considered as a natural region with some homogeneous physical characteristics. A '*genre de vie*', on the other hand is a way of living or local culture. It includes the traditions, institutions, language, habits, foods etc. of a people. In other words, when ones studies geography regionally, one may identify a region with regard to both, its physical characteristics and its human characteristics. The same environment carries different meanings to people with different ways of life (*Genres de Vie*). He drew attention to the fact that the same environment frequently leads to different responses from people with different *Genres de Vie*, since the complex of institutions, traditions, attitudes and technical skills which make up a *Genre de Vie*, largely determine which of the many possibilities offered by nature in any given area shall be selected for development. Blache was opposed to the concept of dichotomy between 'natural' and 'cultural' aspects of the earth's surface; instead, he considered them as the two ends of the same continuum. To him, natural and cultural aspects of the earth's surface cannot be separated from each other. It is because, he reasoned, in every inhabited part of the earth's surface, the original landscape is significantly transformed as a result of human habitation. Such changes are greater in the case of culturally advanced societies where, owing to the more developed technology, the degree of man's intervention in nature is more far-reaching. It is, therefore, impossible to study landscapes meaningfully without due reference to the interlocking roles of the nature and culture. Possiblism continued to grow and spread on the both in France and outside France. According to possibilists it is man who is the primary architect of his environment. It depends on the choices he makes among the numerous possibilities offered by the environment. In order to use these possibilities he makes two types of adjustments with the environment, i.e. in the form of 'adaptation' and 'modification'. For the possibilists, the works of man, not the earth, are the starting point and the most important is the freedom of man to choose. There are no necessities, but everywhere possibilities and man as a master of these possibilities is the judge of their use. In spite of the fact that man has numerous possibilities in a given physical setting, he cannot go against the directions laid by the physical environment. On this ground, the possibilistic approach was also criticized, and the possibilism, too, did not hold for many years.

The concepts of 'Landscape' and 'Landschaft' emerged, respectively in America and Germany, as parallel viewpoints of the time. The Concept of Landscape of Sauer served as an important Paradigm in the Second Quarter of twentieth century. The term Landscape was accepted by many of his followers as the basic term in geography. This 'landscape' is an area made up of a distinct association of forms (physical and cultural). The design of a landscape includes both: the features of natural area; and the forms superimposed over it by the activities of man, converting it into cultural landscape or *'Kulturlandschaft'* as identified by the Germans. Originally the 'landscape' is a natural landscape, and later it turns into cultural landscape or human habitat. Thus, a landscape at a later stage is the creation and workplace of human beings. The relationships of human beings with their habitat are conditioned by their group habits and cultural habits. There may be a succession of these landscapes with a succession of cultures. An alternative view built around 'the concept of landscape' was that by Otto Schluter, the German scholar. As the French equivalent of 'Paysage', 'Landschaft' also means a piece of land (region) with certain physical and human attributes. Schluter's 'kulturlandschaft (cultural landscape) appears to have a greater resemblance with the French 'Pays'—small homogeneous well-defined areas/regions manifesting intimate and reciprocal relationships between man and his immediate surroundings (milieu) through centuries. A 'pay' is, therefore, an areal expression of the *genre de vie* of a human community whose occupation of the area/region over centuries seems to have created the landscape.

As part of Videlian tradition, the major concern of geographers became to study region. For a long time, geographers continued to stress the central position of Regional geography. Regional geography flourished in France and other European countries. Hettner and Hartshorne, both, regarded the regional geographical synthesis as central to geography, arguing for geography to be regarded as a chorological science. Vidal de la Blache and French School of regional geography, on the other hand, produced scientific works that served as exemplars for a large group of students, and thus could be said to have functioned as a paradigm. In the writings of the modern geographers, geography emerged as the 'integrating science', completely different from other disciplines; and labelled as 'chorological' or 'chorographic one'. The goal of the chorological point of view was to know the character of regions and places through comprehension of existence together and interrelation among the different realms of reality and their varied manifestations, and to comprehend the earth surface as whole in its actual arrangement of continents, larger and smaller regions and places. Hartshorne used the term 'areal differentiation' to characterize the way in which geographers dealt with the wide variety of physical, economic and social phenomena, which exist together in the area and distinguish them from other areas. Regional geography flourished in France and got diffused in its neighbouring countries. But, later on this approach also became inadequate to explain the regional personality and therefore, a period of crisis in the discipline emerged again.

IV
POST-MODERN

16
Schaefer-Hartshorne Debate (Crisis Phase V)

The dissatisfaction against the chorological view forced geography to enter the next Crisis Phase V during which this perspective was highly criticized. In the first half of twentieth century the ideas of Hartshorne[1] dominated geographic thinking and were practiced by various scholars. In 1950s they came under attack. New generation of geographers felt that geography was not truly scientific in its approach. Laws and generalizations were largely restricted to natural sciences (physical geography) and human geography was generally devoid of such practice. The most severe statement of the view that geographers should seek for laws and theories came from F.K. Schaefer[2], who opposed the views of Hartshorne in a paper entitled "Exceptionalism in Geography: a methodological examination" (1953). A vigorous criticism of Hartshorne was presented by Schaefer through his paper. The disappointment with the state of affairs in contemporary geography of 1950s had found voice in the forceful paper by Schaefer. He attacked the 'exceptionalist' view of the Kant-Hettner-Hartshorne tradition—the view that geography is quite different from all other sciences, methodologically unique, and therefore, is an idiographic rather than nomothetic discipline. Schaefer maintained that objects in geography are not more unique than objects in other disciplines and that a science searches for laws.[3] Discontentment developed with chorological or regional approaches. The new generation of geographers was increasingly drawn to the systematic studies. Main reason was that the chorological viewpoint was promoting the concept of uniqueness of areas, neglecting theory and generalizations. Chorological paradigm had become sterile in geography and was not yielding laws. Basically, Schaefer favoured the view that without the formulation of laws a science cannot be recognized. Schaefer criticized the exceptionalist claim made for geography as a chorological science, and put forward a strong case for the discipline to adopt the philosophy and methodology of 'scientific positivism'[4]. Geography is concerned with the spatial arrangement of phenomena and not the phenomena themselves. Geographical laws are the 'pattern laws', as contrasted to the 'process laws' of other social science disciplines. Schaefer's paper posed a challenge to the concept of geography as projected and popularized by Hartshorne's 'Nature of Geography' (1939).

Schaefer-Hartshorne Debate

As stated earlier, the philosophy and methodology of geography were apparently debated most earnestly in the writings of two American scholars, namely Richard Hartshorne (1899-

1992) and Fred K. Schaefer (1904-1953) during mid twentieth century.[5] *'Nature of Geography'* constitutes Hartshorne's major work published in 1939. In this work, he strongly advocates for chorological (regional) approach in geography when he states that[6]:

> "Geography is an integrative discipline to which society has assigned responsibility for the study of areas. It is expected to satisfy human curiosity about how much of what is where, and why it is there, in an organized manner that will facilitate comprehension and retention. The discipline deals with an enormous range of phenomena and must provide a congenial home for many different kinds of practitioners. Geography has survived repeated misguided attempts to "make it into a Science" by amputating vital parts of the discipline. Systematic geography generates theories to facilitate an understanding of regions, and regional geography is the proving ground where theories are tested empirically. The idea of the region provides the essential unifying theme that integrates the diverse sub disciplines of geography. The highest form of the geographer's art is the production of evocative descriptions that facilitate an understanding and an appreciation of regions. Regional geography must be informed by a sense of time, and it cannot ignore the physical environment. It begins with the visible features of the earth's surface, but quickly transcends them and attempts to understand the values that motivate the human behaviour that is related to them. Exploration or fieldwork, is a basic research technique in geography, and the most geographic hypotheses are generated by field observation and by cartographic analysis. Effective communication is a difficult and demanding art that regional geographers must master."

Schaefer's article "Exceptionalism in geography" was not only reciprocation to Hartshorne's chorological viewpoint, it was also a call for scientific approach to geography based upon the search for geographical laws, the ultimate form of a scientific generalization. Schaefer died before his article even appeared in print, and so he was never able to elaborate his argument, nor defend himself from Hartshorne's subsequent attack in 1959. But, the article became a rallying point for the younger generation who were keen on reinventing the discipline as a science, or spatial science as it was called later. Schaefer's paper was the first to challenge Hartshorne's presentation and also interpretation of the works of Hettner and others who had tremendous support for chorological viewpoint in geography. This paper was published 14 years after Hartshorne's "Nature of Geography". His purpose was to criticize the 'exceptionalist' claims made for regional geography, and to present the case for geography adopting the philosophy and methods of the positivist school of science. The exchanges between Richard Hartshorne and F.K. Schaefer, provided fuel for continue in questioning the nature of geography[7]. Hartshorne, according to Johnston[8], described geography as:

> "...a science that interprets the realities of areal differentiation of the world as they are found, not only in terms of the differences in certain things from place to place, but also in terms of the total combination of phenomena in each place, different from those at every other place."

Hartshorne tried to prove that the nature of geography is regional and its evolution can be historically traced. This was the paradigm accepted by geographers until contested by Schaefer, during the mid twentieth century. Schaefer's view of geography, on the other hand, as indicated by Johnston, was that:

> "...geography has to be conceived as the science concerned with the formulation of the laws governing the spatial distribution of certain feature on the surface of the earth and these spatial arrangements of phenomena, not the phenomena themselves, should be the subject of geographers' search for law-like statements."

Due to controversies related to systematic and regional approach in geography, the discipline witnessed a small measure of progress during 1950s. The relative importance of systematic and regional work was continuously debated. Hettner, for instance, thought that the core of geography was regional. Hartshorne, on the other hand, believed that systematic geography was really indispensable to regional work. The heart of geography was, nevertheless, considered regional. Earlier too, Humboldt and Ritter clearly saw systematic geography attempting to formulate the rules and laws which could be applied in regional geography. Humboldt felt that the formulation and testing of laws is the highest target a scientist could aim at. The systematic geographer, studying the spatial relations among a limited number of classes of phenomena, arrives by a process of abstraction at laws representing ideal or model situations. But, how much the known laws will explain and how complex a situation the scientists can tackle is a matter of degree, depending on the stage of development of the field. Conversely, it was also felt that, there was no need for regional geography to feel inferior to the systematic geography. For, systematic geography will always have to obtain its data from regional geography, just as the theoretical physicist has to rely on the laboratory for his. Furthermore, systematic geography receives a good deal of guidance as to what kind of laws it should look for from regional geography. For, again, regional geography is like a laboratory in which the theoretical physicists' generalizations must stand the test of use and truth. It appeared, then, in conclusion, that regional and systematic geography are condign, inseparable, and equally indispensable aspects of the field.[7]

One of the causes of the unnecessary argument between systematic and regional views was discovered in 'Kantian philosophy' by Schaefer. Kant claimed geography as a methodologically unique discipline. And Schaefer labelled this situation as 'exceptional' and Kant as 'father of exceptionalism'.[9] Kant was considered, by Schaefer, a poor geographer when compared to his contemporaries or even Bernard Varenius who died more than 150 years before him. Kant made the exceptionalist claim not only for geography but also for history. This was undoubtedly one of the roots of historicist variant of the claim to uniqueness which Schaefer dealt elaborately in his paper.

The criticism against Kantian philosophy proceeds along two main lines in Schaefer's paper.[10] First, the distinction (between the logical and the physical groups of sciences) as intended is untenable in itself. It is simply not true that such systematic disciplines, as, say physics, abstract from or otherwise neglect the socio-temporal coordinates of the objects they study. According to Schaefer, the reason for this is historical. When Kant wrote the passage in his youth he had probably not yet undergone the full impact of Newtonian science. Accordingly, Kant thought of systematic lawfulness as extremely classificatory. Second, it was noticed that the resulting notion of geography is descriptive in the narrowest sense of the term. The social sciences were virtually non-existent at that time. Their place was taken by narrative history. The biological disciplines were at that time still largely classificatory or, as one says in this case, taxonomic. So, it was not unnatural after all, that Kant, in eighteenth century, conceived geography exclusively as a catalogue of the spatial arrangement and distribution of taxonomic features. Hettner's great prestige helped to perpetuate the confusion (between the systematic and regional) that has just been unraveled. Involving the formidable authority of Kant, Hettner successfully impressed upon geography the exceptionalist claim in analogy to history. According to Hettner, both history and geography are essentially chorological. History arranges phenomena in time, geography in space. Both, in contrast to the other disciplines, integrate phenomena heterogeneous in themselves. Also, these phenomena are unique. No historical event and historical period is like any other. In geography, no two phenomena and two regions are alike. Thus, both fields face the task of explaining the unique. Such explanation is, therefore, unlike all scientific explanation which 'explains' by subsumption under laws. Thus, the antiscientific spirit of historicism was one of the major intellectual forces of the nineteenth century. Through Hettner it has penetrated geographical thought and powerfully affected its course. The argument for the uniqueness of the geographical material stems both logically and historically from historicism. The main protagonist of this line of thought in America was Hartshorne.[11] Among American geographers, Carl Sauer is considered as the outstanding representative of historicism, building his geography consistently on Hettner's premise stated above.

The main difficulty of the uniqueness argument is that, it proves too much. There are, for instance, no two stones completely alike in all minute details of shape, colour and chemical composition. Yet Galileo's law of falling bodies holds equally for both. Similarly, limited as our present psychological knowledge is, it seems safe to say that no two people would register identical scores on all tests as yet devised. It does not mean that psychologists have so far not discovered a single law. What it comes down to is a matter of degree. With respect to uniqueness, Hartshorne says, "the degree to which phenomena are unique is not only greater in geography than in many other sciences, but the unique is the very first practical importance. "[12] Hence generalizations in the form of laws are useless, if not impossible, and any prediction in geography is of insignificant value. Thus, he comes to the same conclusion as Kant. "Both history and geography might be described as naïve sciences, examining reality from a naïve point of view, looking at things as they are actually arranged and related, in contrast to the more sophisticated but artificial procedure of the systematic sciences which take phenomena of particular kinds out of their real settings." One may say that Hartshorne goes Kant even one step better. For Kant, geography is a description; for Hartshorne it is 'naïve science' or if we accept his meaning of science, naïve description. As one would expect from all this, and as has been mentioned before, regional studies are for Hartshorne the heart of geography. Geography, according to Hartshorne is essentially idiographic. Mainly through Hartshorne's efforts, American geographers have come to look at Hettner as the major authority in support of the idiographic conception of their field during mid twentieth century. With this evolution came continued debate regarding the significance between two main geographic concerns: 'Idiographic-Chorological' and 'Nomothetic-Systematic' perspectives. Hartshorne's second book '*Perspectives on the Nature of Geography*' (1959) was written as a reaction to Schaefer's allegations. This book was a supplement to his previous book. Hartshorne, subsequently, presented a wide-ranging review on the nature of subject, including discussions on and clarifications about many points of criticism regarding the concept of geography as areal differentiation. Although Hartshorne stuck to essentially the chorological viewpoint in his second book as well, he redefined, clarified many of the previously stated concepts.

The essential difference between the positions taken by Hartshorne and Schaefer was that Hartshorne's was a positive view of geography, i.e. geography is what geographers have made it. Schaefer's view, on the other hand, was a normative one, i.e. what geography should be irrespective of what it had been. The fact, however, was that since Hartshorne's view was fast losing ground, geographers in increasing numbers had, by the end of 1950s, turned round to Schaefer's view of geography as a spatial (locational) science. For this they used methods of other systematic sciences and were increasingly concerned with quantification and development of theory, so that by the time Hartshorne's "*Perspectives on the Nature of Geography*" had appeared in 1959, the Quantitative Revolution had reached its zenith.

Outcome

Schaefer's criticism of the regional view brought about a paradigm shift in geography. He initiated what may be called the quantitative and theoretical revolution in geography or spatial organization paradigm. Schaefer advocated the use of spatial laws as a basis for geographic explanation. He wrote: "Description, even if followed by classification, does not explain the manner in which phenomena are distributed over the world. To explain the phenomena one has described means always to recognize them as instances of laws." Schaefer's spatial viewpoint advocated that geography should focus on space and spatial interaction. Schaefer's ideas about spatial laws and geographic explanation were also elaborated in the writings Bunge's *Theoretical Geography* (1962), Haggett's *Locational Analysis in Human Geography* (1966), and David Harvey's *Explanation in Geography* (1969). Both Chorley and Haggett (1967) in their famous book, 'Integrated Models in Geography' also stated that geography should adopt model based paradigm rather than the traditional paradigm. Model building was set up as the aim of geographical investigation and inquiry, a task to be done with the help of quantitative methods.

The net outcome of Schaefer-Hartshorne Debate was that geography, by the end of the 1950s, had come increasingly to be viewed as a science requiring the use of the 'scientific method' so that like other sciences, it could also develop laws and theories relevant to its field of study. This brought about a distinctive shift in emphasis from 'regional' to 'systematic' studies. This meant that geography thereafter began increasingly to be viewed in a Nomothetic perspective, i.e. it required developing the habit of seeking the general in the particular. This also involved a shift from 'areal' to 'locational' studies; from 'absolute' to 'relative' locations; and from 'areal differentiation' to 'spatial interaction'. The point is, very radically, that there is no such thing as a methodological distinction between pure and applied science. There is only one science and science applied. Whatever distinction there is, is practical, a matter of either interest of emphasis. The laws for which the 'pure' scientist looks are in no way different from those which he himself or his 'applied' colleagues use. The regional geographer who explains some features of a region by the use of laws applies the latter in one sense of the term. The statistical techniques prove to be a powerful tool in making the best of the situation. Like all the other social scientists, geographers have come to appreciate this tool. There are many important differences between the natural sciences and the social sciences. Logically, it is submitted, these are differences of degree, not kind.

Schaefer's paper brought a theoretical and methodological revolution in geography.

And geography underwent a radical transformation of spirit and purpose, best described as 'Quantitative Revolution' later. The mechanistic approach of environmentalism, which dominated much of the nineteenth century, was soon replaced by a more nomothetic and scientific approach, setting the stage for the next paradigmatic shift.

NOTES

1. Richard Hartshorne (1899–1992) was a prominent American geographer, and professor at the University of Wisconsin-Madison, who was specialized in economic and political geography and the philosophy of geography. He is known in particular for his methodological work *The Nature of Geography,* published in 1939. Hartshorne's 1939 book *The Nature of Geography: A Critical Survey of Current Thought in the Light of the Past,* reflected his concern that geographers, as scientists and scholars, should familiarize themselves with, and take account of, past work in their field. The book itself became a standard in the field and remained in print for decades; the seventh edition was published in 2000. In the 1950s Hartshorne was part of a key geographical debate over the nature of the subject. Fred K. Schaefer called for the adoption of the 'scientific method' and study of spatial laws and criticised the 'old method' promoted by Hartshorne as the 'Hartshornian orthodoxy'.
2. Fred K. Schaefer was originally an economist, and later he joined the group of geographers at the University of Iowa. Fred K. Schaefer (1904-1953) is considered as one of the pioneers of quantitative revolution. He was born in Berlin. He was involved in politics as a member of Social Democratic party and after the rise of fascism he fled from Nazi Germany. Later in the United States he became an inaugural member of the Department of Geography at Iowa. He died of a heart attack in 1953. He is well-known for his article *Exceptionalism in geography: A Methodological Examination* published in American periodical, Annals of Association of American Geographers.
3. Holt-Jenson, A. (2009), p. 116.
4. Scientific or Logical Positivism is a philosophy that combines positivism—which states that the only authentic knowledge is scientific knowledge—with a version of apriorism—the notion that some propositional knowledge can be had without, or "prior to", experience. Logical positivism denied the soundness of metaphysics and traditional philosophy, and affirmed that statements about metaphysics, religion and ethics are devoid of cognitive meaning and thus nothing but expression of feelings or desires; only statements about mathematics, logic and natural sciences have a definite meaning. Logical positivism originated in 'the Vienna Circle' in the 1920s, where Rudolf Carnap, Otto Neurath and others divided meaningful statements into those which are analytic (true *a priori*), and those which are synthetic (verified by sensory experience, *a posteriori*). Logical positivism refuted synthetic *a priori* knowledge: an evident criticism to Kantian philosophy.
5. Rana, L. (2008), pp. 276-79.
6. Hartshorne, R. (1939), pp. 366-459.
7. Johnston (1979). pp. 50-8.
8. *Ibid.*
9. Schaefer, Fred K. (1953), p. 230.
10. *Ibid.*, pp. 231-232.
11. *Ibid.*, pp. 233-236.
12. *Ibid.*, pp. 238-239.

REFERENCES

Bunge, W., (1962). *Theoretical Geography*. First Edition. Lund Studies in Geography Series C: General and Mathematical Geography. Lund, Sweden: Gleerup.

Chorley, R.J. & P. Hagget (1967). *Integrated Models in Geography,* Part IV, Methuen & Co. Ltd., London.
Haggett, P. (1966). *Locational Analysis in Human Geography,* Edward Arnold.
Hartshorne, R. (1939). *The Nature of Geography: A Critical Survey of Current Thought in the Light of the Past,* Association of American Geographers, Lancaster.
Hartshorne, R. (1959). *Perspective on the Nature of Geography,* Rand Mcnally & Co., Chicago.
Harvey, D. (1969). *Explanation in Geography,* Arnold Publishers, New Delhi.
Holt-Jenson, A. (2009). "Paradigms and Revolutions", *Geography: History and Concepts: A Systematic Guide,* Fourth Edition, Sage, London, pp. 100-124.
Johnston, R.J. (1979). Geography and Geographers: Anglo-American Human Geography since 1945. London (Edward Arnold). ISBN 0-7131-6239-2.
Rana, L. (2008). *Geographical Thought: A Systematic Record of Evolution,* Concept publishing Company, New Delhi.
Schaefer, Fred K. (1953). "Exceptionalism in Geography: A Methodological Examination", *Annals of the Association of American Geographers,* Vol. 43 (3) : 226-249.

17
The Nomothetic Turn (Paradigm Phase VI)

Conspicuously, two traditions prevailed in geographic development during mid-twentieth century, viz. 'Historicism' and 'Positivism'. The former focused on the study of historical and human sciences proceeding through interpretation of the subjective meanings and intentions of individual; and the latter on seeking to discover universal laws, offering causal explanations. Of all the problems of concern in the thinking of geographers during this point in time, the most disturbing was related to the method of obtaining and organizing information in geography. Here the dualism developed between that aspect of science that can be expressed in universals (Nomothetic) and that which is concerned with the individual attribute or object (Idiographic). The major question was whether geography 'like other sciences' can develop 'the knowledge of principles, laws, and general truths'—and thus lay claim to the name of science—or whether its function is merely to describe innumerable unique areas.[1]

As far as geography is concerned, the contrast between that aspect of science that can be expressed in universals and that, which is concerned with the individual object or phenomenon as worthy of study, had received the particular attention, at the turn of the nineteenth century itself. This was because of two German philosophers, Windelband[2] and Rickert[3]. Windelband is mainly remembered for the terms 'nomothetic' and 'idiographic', which he introduced. Closely associated with Windelband was Heinrich Rickert. They appear to have classified the different branches of science on this basis, distinguishing between law-making sciences and those concerned with the unique. In a way this seemed to conform to the conventional division between the natural sciences and the social sciences. These terms were used to describe two distinct approaches to knowledge, each one corresponding to a different intellectual tendency, and each one corresponding to a different branch of academe. The status of geography was under debate and the key issue was the one between the two research methods: nomothetic and idiographic. The debate concerned which method of enquiry is more important and which would allow greater and more valid investigation into the field of geographic investigation. The central axes of these controversies, as stated in the beginning, were 'positivism' and 'historicism', i.e. the search for general causal explanations, and the search for hermeneutic understanding of specific individuals and historical moments. The distinction at this point of time was drawn by Windelband between nomothetic and idiographic sciences.[4] Historicists rejected the idea that the social sciences should aspire to the discovery of universal laws of society. They argued that the human sciences were profoundly different from the natural sciences because they involved interpretive understanding of human action rather than mechanical explanation of a system of causes. Rickert's theories reconsidered the distinction proposed by Windelband, proposing that the distinction between nomothetic and idiographic knowledge does not correspond to the distinction between natural and social knowledge. Our cognitive interest is directed either at the concrete phenomena in their position in space and time and at their concrete interrelationships, or else at the recurrent patterns in which they appear. The former research direction aims at knowledge of the concrete or, more correctly, the individual, the latter at knowledge of the general.

Meaning

Nomothetic means 'stating laws', whereas Idiographic means 'pertaining to the study of individual cases or events'. Often, nomothetic approaches are quantitative, and idiographic approaches are qualitative. Nomothetic is based on a tendency to generalize, and is typical for the natural sciences. It describes the effort to derive laws that explain types or categories of objective phenomena, in general. Nomothetic describes the study of classes or cohorts of individuals. Nomothetic refers to the use of generalization rather than specific properties in the same context. Nomothetic research is about attempting to establish general laws and generalizations. The focus of the nomothetic approach is to obtain objective knowledge through scientific methods. Hence quantitative methods of investigation are used, to try and produce statistically significant results. The nomothetic approach is considered scientific due to its precise measurement, prediction and control of behaviour, investigations of large groups, objective and controlled methods allowing for replication and generalization.

Idiographic, on the other hand, is based on a tendency to specify, and is typical for the humanities. It describes the effort to understand the meaning of contingent, unique, and often cultural or subjective phenomena. The idiographic model focuses on a complete, in-depth understanding of a single case. The idiographic approach, unlike the nomothetic approach, focuses on the individual. It suggests that everyone is unique and therefore, everyone should be studied in an individual way. Due to this, no general laws are possible. The methods of investigation, by this approach tend to collect data, investigating the individual. Case studies are the most common method. Unlike the nomothetic approach, this provides a more complete understanding of the individual. As a result of this focus the individual is more likely to feel valued and unique. Also despite claims, that this approach is unscientific, it does satisfy some of the key aims of a science, i.e. description and understanding. However, as subjective experience cannot be empirically tested, it remains unscientific. In other words, the idiographic method deals with discrete facts or events. Although all branches of science are concerned to some extent with the study of the unique as well as of universals, there are some sciences in which the actual work of many students may be largely, if not completely, absorbed in the search for universals, the study of the unique being left to other divisions of science. For instance, this may be the case in physics and chemistry, particularly in their theoretical aspects. To a lesser extent the same is true of the biological sciences and psychology, and of certain branches of economics and sociology. On the other hand, there are certain branches of most of the sciences in which the students are very largely concerned with the study of the individual object, the unique. For example, this form of study is peculiar in astronomy. Idiographic studies are certainly of great, if not major, importance in geology. This is obviously the case in paleontology. It is true of mineralogy in so far as the student concentrates on certain mineral deposits for securing complete knowledge of those deposits rather than of developing new principles of mineral deposition. Even more does the statement apply to the great and certainly valuable work of the geological survey in mapping the areal geology of areas though this might logically be considered a branch of geography carried on by geologists.

Roots

As far as Geography is concerned, in the writings of scholars particularly after 1900, it has emerged as an 'integrating science', i.e. completely different from other sciences and studying all the physical and human elements in an area. Because of this kind of growth, geography developed a unique methodology, termed as 'idiographic', i.e. studying different areas with their own individual significance. The proponents of this methodology argued that every area or region on the surface of the earth is a unique combination of various elements. This unique combination provides that area a unique outlook or personality. Therefore, there is only one Alps, Himalayas, Amazon, Germany, London, New York, Paris, and so on. They all appear to have their own individual significance. To cite another example, Africa and North America both have a system of lakes, but as a system of connected fresh water lakes the Great Lakes of North America have no counterpart in the world. Thus, geography began with the description of the earth's surface and it has always shown a great concern for individual cases in the situation. Then, for a long time, the bond of geography with history (and human culture) dominated geographical thinking. This also strengthened the belief of geographers in the uniqueness of areas or areal combinations. The most important supporter of this kind of thinking was Emmanuel Kant, popularly known as the 'Father of Exceptionalism in Geography'. Kant's philosophy and writings were largely responsible to support and promote the study of unique cases in geography. Therefore, his philosophy is known by the name of 'Exceptionalism in Geography'. Because, his methodology to the study of geography was not considered scientific and he appears to have profoundly changed the trend of development of geographic thought in his times. According to this view, geography is different from all other disciplines and is methodologically unique. Therefore, it deserves a name of its own. Kant made an exceptionalist claim not only for geography but also for history. Some of the later geographers, namely Ritter, Hettner, Hartshorne etc. made 'idiography' a very important method of geographical studies. Kant wrote that the history and geography are two parts of the whole world of perception of man. Both the disciplines are based on descriptions, one according to the time and the other in terms of space. History is narrative, i.e. reporting the events which fall one after the other in time. Geography, on the other hand, is descriptive, i.e. the study of phenomena which have taken place side by side in space.

The major point taken from the statement made by Kant in the support of the idiographic method is that there is no limit placed in the class or number of objects in geographical studies as in systematic sciences. If there is no limit to the factual content of geography, then the discipline must be defined or recognized in terms of its distinctive method of study, i.e. idiographic. It is the description of unique or specific areas or of elements in unique or specific combinations in an area, rather with developing generalizations about the elements. Kant argued that the locations are unique. If locations are unique, then the descriptions of what existed at those unique locations could not be accomplished by referring to general laws. Rather, it requires the employment of the idiographic method. The idiographic view was very strong in the earlier part of the twentieth century. There were many supporters. The view generated great professional activities and a strong reaction to deterministic thinking. These supporters constructed false similarities between history and geography. For instance, according to Alfred Hettner, the German scholar, as no historical event or period is like each other, so is geography of no two or more phenomena or areas or regions is alike. Thus, both the fields explain the unique.

The Twist

Since 1950s onwards various objections were raised against idiographic method of study in geography. Particularly, the American scholar, F.K. Schaefer, pointed out that Kant's views were immature. According to him, this immaturity was the result of the fact that they were expressed before Newton's scientific laws were put forward. The similarity between geography and history as proposed was not acceptable to him. The concept of uniqueness proved too much to him. For example, though there are not really two stones that are completely alike in all minute details as shape, colour, chemical composition etc. yet Galileo's law of falling bodies holds equally for both. The idiographic method of study suggests considering everything on the surface of the earth. If this is taken as a practice without knowing the importance or significance of the study, then geographers would be lost under the heap of useless and insignificant facts and details and it will become unfruitful. Thus, the idiographic method had adversely affected the course of geographic thought. Kantian influence, however, was a disaster, since it led to Exceptionalism in geography.

The modern trend is to simplify the complexity and predict for the future. Since, there exists a heap of information, it cannot be dealt without following a general approach. The prime concern of geographers in the present times is to form generalizations and establish laws and theories. The idiographic method cannot help in this process. Based on this argument, the idiographic method has been repeatedly criticized in contemporary period. Then, the essential point here is that the uniqueness does not preclude similarity. From these similarities the relationships can be formed to provide the basis for scientific generalizations. To cite an example, although the moon, an apple and a ball are unique in themselves with no formal similarities, yet they share certain common attributes. In a trivial descriptive term, they each have an inside and outside, and are round; whereas Newton recognized a major functional similarity when he proposed his laws of gravitation.

The 'Nomothetic' method emphasized the task to search for similarities and especially the functional similarities. Different cities, for example, may not be the same or similar in terms of size, pattern, structure, shape, density, culture etc., but most of are similar, at least, in terms of their non-agriculture function. Such generalizations or laws may not be true and applicable to the entire world (as such an attempt is rarely possible), but they are true and applicable to a large extent and to a majority. The patterns produced by the individual facts and events in geography are spatial. This implies that geographers must describe and explain the way the phenomena combine to fill an area or are distributed over space. Humboldt, the father of scientific geography, had stressed that geography must pay attention to the spatial arrangement of the phenomena on the earth's surface, and not to the individual characteristics of the phenomena themselves. This is sufficient to call geography a scientific discipline and not the unique. It was agreed that logically everything is unique. But similarities can be found among unique things. General statements can be elicited from these. These general statements after testing may be expressed as formal laws. In spite of the difficulties involved, geography has been able to develop general principles of the relationships among the variable phenomena. Nomothetic method is, therefore, was proposed as the key to the study of geography. There cannot be any kind of compromise between the two absolute opposite methods—Idiographic and Nomothetic. The geographers' task is to study the general character of the individuals and not their uniqueness. As his task is to seek for the most complete presentation possible, his method of study should be such that a majority, if not all, aspects should be brought under investigation. However, in this process of seeking universal truths some individual cases may be left out, which are not explained or described.

Hartshorne admitted to having used the words 'idiographic' and 'nomothetic', but rejected the idea that different sciences can be distinguished as being either idiographic or nomothetic[5]. The two aspects of the scientific approach are present in all branches of knowledge. Both Hettner and Hartshorne have made a distinction between 'Systematic Geography' which seeks to formulate empirical generalizations or laws, and the study of unique in 'Regional Geography' whereby generalizations are tested so that subsequent theories may be improved.[6] Hartshorne suggests that geographical studies show 'a gradational range along a continuum from those which analyze the most elementary complexes in areal variation over the world, to those which analyze the most complex interpretation in areal variation within smaller areas.'[7] With the development of natural sciences in the eighteenth and nineteenth centuries it became apparent that mere description would not do. Description even followed by classification, does not explain the manner in which phenomena are distributed over the world. To explain this one has to describe the means to recognize them as instances of laws. Another way of saying the same is to insist that science is not much so interested in individual facts as in the patterns they exhibit.[8] In geography, the major pattern producing variables are, of course, spatial. It became an accepted proposition that all natural relations and, all spatial relations, are governed by laws. For this type of work tools had to be provided in the form of concepts and laws. Here geography had to be concerned as the science governing the spatial distribution of certain features on the surface of the earth.[9]

The condition during mid-twentieth century indicated a stage of development, which found most geographers still busy with classification rather than looking for laws. Undoubtedly, classification is the first step in any kind of systematic work. But when the other steps, which naturally follow, are not taken and classification becomes the end of investigation, then the field becomes sterile.[10] The strength of historicism in German thought was a matter of record. However, this preponderant historicism in the environment faded away gradually. And the German universities, during the same period, became the centres of the rising natural sciences and the positivist philosophy of Science, stressing the search for laws and the methodological unity of all inquiry.[11] It was also accepted that the laws and concepts of systematic geography can be applied to the regional material. It was found misleading to oppose systematic and regional geography to each other as an instance of pure and applied science.[12] The point is, very radically, that there is no such thing as a methodological distinction between the pure and applied science. There is only one science and science applied. Whatever distinction there is, is practical, a matter of either interest or emphasis. The laws for which the 'pure' scientist looks are in no way different from those which he himself or his 'applied' colleagues use. The regional geographer who explains some features of a region by the use of laws applies the latter in one sense of the term.

Science, to repeat, once more, searches for laws. The laws in geography fall in three categories.[13] Typical to the First are most of the laws of Physical geography. These are applied systematically to various conditions that prevail on the surface of the earth and are analyzed with particular attention to the spatial variables they contain. To be specific, the climatologist uses much physics (meteorology), the agricultural geographer the applied biology (agronomy), for instance. Typical to Second category, are many laws of economic geography. As far as they are morphological, these laws are genuinely geographic. The pioneer work in this area has, in fact, done by economists, or the geographers like Christaller. The Third category laws are Process laws, and the mature social sciences look for them. Knowing such laws one can ideally predict the whole course of history in a region. The future of geography, according to Schaefer, depends on the extent to which the geographers cultivate its systematic aspects.

NOTES

1. Hartshorne, (1959), p. 148.
2. Wilhelm Windelband (1848-1915) was the German philosopher and historian of philosophy. He was a neo-Kantian who argued that philosophy should engage in humanistic dialogue with the natural sciences rather than uncritically appropriating its methodologies. His interests in psychology and cultural sciences represented an opposition to historicism. Closely associated with Windelband was Heinrich Rickert.
3. Heinrich John Rickert (1863–1936) was a German philosopher, one of the leading Neo-Kantians. Rickert, with Windelband, led the so-called Baden School of Neo-Kantians. On the basis of Windelband's distinction between nomothetic (universal) and ideographic (particular) judgments, Rickert developed his logic of the historical disciplines. At both the scientific and the prescientific stages of conceptualization, he claimed, there are two ways of grasping reality: individualizing and generalizing. Individualizing thought is proper to historical thinking. Instead of fabricating a copy of a historical phenomenon in its complex totality, it establishes the essential relationships that bind the phenomenon to its environment and traces the various stages of its development. Philosophy studies the concept of development, while the objects of historical study are unique developments. Generalizing thought, therefore, is proper to the natural sciences but is inapplicable to history.
4. Tim Unwin (2013).
5. Hartshorne, R. (1955), p. 242.
6. Holt-Jenson, A. (2009), p. 116.
7. Hartshorne, R. (1959), p. 121.
8. Schaefer, Fred K. (1953), p. 226.
9. *Ibid.*
10. *Ibid.*, p. 228.
11. *Ibid.*, p. 241.
12. *Ibid.*, p. 242.
13. *Ibid.*, pp. 248-249.

REFERENCES

Hartshorne, R. (1955). "'Exceptionalism in Geography' Re-examined", *Annals of the Association of American Geographers*, Vol. 45, pp. 205-44.

Hartshorne, R. (1959). *Perspective on the Nature of Geography*, Rand Mcnally, Chicago, IL.

Holt-Jenson, A. (2009). "Paradigms and Revolutions", *Geography: History and Concepts: A Systematic Guide*, Fourth Edition, Sage, London, pp. 100-124.

Schaefer, Fred K. (1953). "Exceptionalism in Geography: A Methodological Examination", *Annals of the Association of American Geographers*, Vol. 43 (3) : 226-249.

Tim Unwin (2013). *The Place of Geography*, Routledge.

18
Quantitative Revolution (Crisis Phase VI)

The Quantitative Revolution was one of the major turning points in the history of geography. As a consequence of a long methodological debate upon its scope and content an apparently new perspective had been opened up under the impact of Quantitative Revolution during mid twentieth century. It marked a revolutionary change in the method behind geographical research, from regional geography into a spatial science.[1] It sought to develop a more rigorous and systematic methodology for the discipline that later lead to a paradigm shift. Burton[2] described the Quantitative Revolution as 'a radical transformation of the spirit and purpose of geography'. By this he meant a newfound enthusiasm for the use of numerical techniques of various kinds, directed towards elucidating the details of Earth surface patterns. In the words of Burton (1963) this revolution had set out to discover universals, to build models and to establish methods and theoretical bases for geographical studies. The Quantitative Revolution has been called one of the greatest periods of intellectual achievement in the whole history of the discipline. With all its merits and demerits, the phase of quantitative revolution, however, has been one of the decisive trends in geography's evolution.

Beginnings and End

Traditionally, geography has been a discipline that studied and described the surface of the Earth. But in due course of time, its definition and nature changed; and it began looking for more accurate, systematic, rational descriptions and explanations of the variations in the geographical phenomenon that occurred over the Earth's surface. The Revolution began under the banner of the scientific method. The quantitative revolution was geography's attempt to redefine itself as a science. The revolutionaries declared that the purpose of geography was to test general laws about the spatial arrangement of phenomena. Adopting the philosophy of 'positivism' from natural sciences, they turned to 'mathematics', especially 'statistics', as a way of proving solutions to their problems.

Although its antecedents can be traced far back, the quantitative revolution in geography has begun actually in the late 1940s and early 1950s. It reached its culmination in the period from 1957 to 1960. The revolution had come largely as the result of the impact of work by non-geographers, a process shared by many other disciplines where an established order has been overthrown by a rapid conversion to a mathematical approach. The movement was initiated by natural scientists, specifically physicists and mathematicians. It expanded and led to change physical sciences followed by biological sciences.

In geography, the revolution was set in motion by the preceding modes of thought. Particularly important in this respect was regional geography. The quantitative revolution responded, particularly, to the regional geography paradigm that was dominant at the time. The regional geography was at its apex during the 1940s and 1950s. Following Richard Hartshorne's *The Nature of Geography* (1939), Regional geography was the major philosophical school. It was the central paradigm within geographical studies. In the early 1950s, there was a growing sense that the existing paradigm for geographical research was not adequate in explaining how physical, economic, social and political processes are spatially organized, ecologically related, or how outcomes generated by them are evidence for a given time and place. Regional geography involved the study of unique combinations of characteristics in specific areas. No generalizations were possible (Peet, 1998), except that all areas were unique. It was criticized for its descriptiveness and the lack of theory. Various criticisms were raised against Regional geography[3]. A growing number of geographers started to express their dissatisfaction with the traditional paradigm of discipline and its focus on regional studies. The movement away from the regional geography approach was stimulated by three factors. Firstly, the practical demands of the war led to geographical research; secondly, non-geographers became involved in human geography research and applied natural science methodologies to human geography research; and the above points exacerbated the final factor, the growing frustration that some geographers felt towards the regional approach which was increasingly considered descriptive, fragmented and non-scientific and outdated (Peet, 1998; Unwin, 1992; Cloke *et al.*, 1991). All of these events presented a great threat to geography's position as an academic subject and thus geographers began seeking new methods to counter critique.

The Quantitative Revolution began not only as a response to regional geography paradigm, but also as a result of the 1950s crisis[4]. During the late 1940s and early 1950s a series of events nearly led to the downfall of geography as an academic subject. This is known as the 1950s crisis in geography. The crisis occurred for several reasons as: the closing of many geography departments and courses in universities, e.g. the abolition of the geography programme at Harvard University in 1948; continuing division between Human and Physical geography—general talk of Human geography becoming an autonomous subject; Geography was seen as overly descriptive and unscientific and it was claimed that there was no explanation of 'why' processes or phenomena occurred within the regional framework; Geography was seen as exclusively educational and there were few if any applications of contemporary geography; and continuing question of what geography is—Science, Art, Humanity or Social Science.

Soon thereafter, a series of debates regarding methodological approaches in geography took place. To address these concerns, early critics such as Ackerman[5] suggested the systematization of the discipline. One of the first illustrations of this was the Schaefer *versus* Hartshorne Debate. In 1953, *Exceptionalism in Geography: A Methodological Examination* was published by Schaefer, a professor at the University of Iowa. This led to Schaefer attacking regional geography in 1953. He argued that objects in geography were no more unique than in other disciplines and that a science should search for laws (Peet, 1998). He urged geographers to study systematically, using quantitative methodologies (Holt-Jensen, 1999), providing the stimulus for the quantitative revolution. In his work, Schaefer rejected Hartshorne's exceptionalist interpretations about the discipline of geography and having the region as its central object of study. Instead, Schaefer envisioned the establishment of morphological laws through scientific inquiry as the discipline's main objective, i.e.

incorporating laws and methods from other disciplines that place a greater emphasis on processes. Hartshorne, on the other hand, addressed Schaefer's criticism in a series of publications[6], where he dismissed Schaefer's views as subjective and contradictory. He also stressed the importance of describing and classifying places and phenomena, yet admitted that there was room for employing laws of generic relationships in order to maximize scientific understanding. In his view, however, there should be no hierarchy between these two approaches. Against this backdrop, the geographers launched a changeover in the philosophy and methodology of geography. Quantification was the key to this radical transformation.

The changes introduced during the 1950s and 1960s under the banner of bringing 'scientific thinking' to geography led to an increased use of technique-based practices, including an array of mathematical techniques and computerized statistics that improved precision, and theory-based practices to conceptualize location and space in geographical research.[7] Then, during the Second World War technology became very influential in our society. This culminated in the 1960s and 70s in the Quantitative Revolution, during which science and mathematics enjoyed unrivalled importance. Nomothetic aspects of geography heightened in importance and quantitative methods were applied to virtually all investigation and experimentation.

By the mid-1960s the quantitative revolution had successfully displaced regional geography from its dominant position and the paradigm shift was evident by the myriad of publications in geographical academic journals and geography textbooks. The adoption of the new paradigm allowed the discipline to be more serviceable to the public and private sectors.[8] In geography, the quantitative techniques were first used in physical geography, and they entered later in human studies. In 1950s and 1960s itself, several geomorphologists had started using quantitative methods and the practice spreaded gradually. They were, for instance, Strahler, Chorley (1957), Dury (1954), Mackay (1963), Wolman (1955) etc. As far as the field of climatology is concerned, there has been a little argument about the widespread application of quantitative techniques to it. This branch of geography embraces the most apparently manageable and quantifiable continuum that geographers have been concerned to study. The scholars like Koeppen, Thornthwaite, Mather, Hare, Bryson and others applied quantitative techniques to climatic problems with great effect. By far the greatest struggle for the acceptance of quantitative methods has been in human geography. It is because of the prevalent notion that the human behaviour is unpredictable. It is here that the revolution ran up against the notion of freewill and the unpredictability of human behaviour.

By 1970s, the revolution was over and a wide range of geographers, including Hartshorne, accepted that quantitative techniques could be usefully applied to the geographical problems. Thereafter, the best use of such techniques was made in the study of uniformities in locational patterns in geography. Among all the modern schools of thought, it was British geography that showed the earliest signs of change during quantitative revolution particularly through the works of Gregory, Hagget and Chorley. Initially, it was through the publication of, for instance, Prof. S. Gregory's book *"Statistical Methods and the Geographers"* (1963). This was followed in by *"Frontiers in Geographical Teaching"* by Hagget & Chorley (1965) and was the first of the whole series of books they have written or edited about the search for laws in geography. Also in 1965, Hagget's book *"Locational Analysis in Human Geography"* attempted to provide a theoretical framework for the study of regularities in human geography. In 1967 another book by Hagget and

Chorley: "*Models in Geography*" attempted to review the entire theoretical framework which geographers had used to study locational patterns. However, within two to three decades of the beginning of Quantitative Revolution, the overwhelming focus on statistical modelling itself eventually became responsible for the undoing of this revolution.

Nature and Objectives

The mathematical and statistical methods were introduced to attain a desired level of objectivity for the search of laws and theories in geography. All these efforts were devoted to the fundamental conception of geography as a spatial science. The Quantitative Revolution involved a mesh of techniques, including the use of, for instance, Descriptive statistics; Inferential statistics (e.g. correlation and regression); Basic mathematical equations; e.g. the Gravity Model[9]; Deterministic models (as those of Von Thunen's and Weber's location models), and Stochastic models, involving the concepts of probability, such as some of the work on spatial diffusion processes. The common factor, linking these techniques, was a preference for numbers over words, plus a belief that numerical work has a better scientific creditability. The "Quantitative Revolution" saw the first concerted attempt to apply quantitative methodologies within geography. The new approach aimed to make geography more scientific and was guided by three basic ideals of logical positivism, viz. that only one scientific method exists; that knowledge is neutral; and that the standards of accuracy and precision in the physical sciences offered the only genuinely explanatory framework for the generation of scientific knowledge[10].

The main objectives of this revolution in geography were first, to change the narrative character of the subject (geo + graphics) and make it a scientific discipline; the second objective was to explain and interpret the spatial patterns of geographic phenomena in a logical way; the third objective dealt with the use of mathematical and statistical techniques; fourthly, to make accurate statements (generalization) about location order; fifthly, to prepare estimates, principles and laws for testing estimates and estimates and forecasts; and lastly to provide a sound philosophical and theoretical base to geography.

Impact

The greatest impact of the quantitative revolution was not the revolution itself but the effects that came afterwards in the form of spread of positivist (post-positivist) thinking and counter-positivist responses.[11] The spread of positivist thinking led to the emergence of Spatial Science School; whereas, the counter-positivist response from geography was created in the form of introduction of anthropocentricism in the discipline later. Although, this movement is over now, it did lead to mathematization of much of our discipline, with an attendant emphasis on the construction and testing of theoretical models. In the words of Yeates[12], 'geography can be regarded as a science concerned with the rational development, and testing of theories that explain and predict the spatial distribution and location of various characteristics on the surface of the earth'. In order to achieve this objective and to obtain the real picture of a region, the geographers began to use and apply quantitative tools and techniques.

The quantitative revolution had enormous implications in shaping the discipline of geography into what it looks like today. The main claim for the quantitative revolution is that it led to a shift from a descriptive geography to an empirical law-making geography. It replaced the 'idiographic' approach based on areal differentiation by 'nomothetic' one, which had its roots in the search for models of spatial structure and phenomenon. The quantitative revolution led the basis of geography as a spatial science that dealt with the spatial analysis of phenomena that existed on the earth surface. In simpler words, it gave geography a scientific vision through the application of methodology rooted in statistical methods. The most obvious change has occurred due to the quantitative revolution that brought changes in the methods and techniques used to explain the geographical phenomenon in a spatial framework. The new method of inquiry led to the development of generalizations about spatial aspects in a wide range of natural and cultural settings. Generalizations took the form of tested hypotheses, models, or theories, turning geography into a nomothetic science. One of the most significant works to provide a legitimate theoretical and philosophical foundation for the reorientation of geography into a spatial science was David Harvey's book, *Explanation in Geography*, published in 1969. In this work, Harvey laid out two possible methodologies to explain geographical phenomena: an inductive route where generalizations are made from observation; and a deductive one where, through empirical observation, testable models and hypothesis are formulated and later verified to become scientific laws.[13]

Dissatisfaction with idiographic geography lies at the root of the quantitative revolution. The development of theoretical, model-building geography has been the major consequence of the quantitative revolution. The emergence of geography as an abstract, theoretical science appears to have been the most overriding development in geographic research during the 1960s. Attention, since then, has focused on the spatial organization of economic, social, physical, political and urban processes, and on how outcomes generated by these processes are evidenced at given times and in particular places. Special attention has been placed on the spatial aspects of social processes. Problems related to environmental perception and control have also been researched. Accompanying these developments has been a widespread acceptance of the need for more geographically oriented quantitative methods for use in model building. Ultimately, the Quantitative revolution has greatly influenced the practice of Physical and Human geographers alike, as well as vastly improving the scientific credibility of the discipline. Continued progress has been made in the observation and handling of data related to geographic problems. Eventually the quantitative revolution had its strongest repercussions on the fields of physical, urban and economic geography especially pertaining to its location theory. The statistical methods were found useful for bulk studies and were appreciated if used to study complex phenomenon and processes that constitute a large number of variables or indicators. In the case of climatology, there is little dispute about the use of quantification. This branch of geography whole-heartedly embraced these new statistical techniques to interpret various climatic phenomena. Examples can be cited from the works of Thornthwaite, Mather and Green, Bryson who have successfully implemented quantitative techniques to seek answers for climate problems; thus silencing their critics.

The true value of mathematical and statistical methods was discovered in the role that they play in the testing and verifying of ideas established, however tentatively, within a theoretical framework. According to Burton (1963) 'quantification is inextricably intertwined with theory'. The quantitative methods are especially used in presenting models or using models in geography. Prior to quantitative revolution, the geographers attempted to use their experiences to formulate geographical problems and then to solve them by the direct examination of empirical evidence. Now an increasing number of geographers began translating their problems into mathematical terms and then deriving the desired solutions. The capacity to convert problems into mathematical terms enabled geographers to tackle for more difficult problems, to formulate them more precisely, and to obtain, with the use of statistics, much more soundly based conclusions than was the case in pre-revolution phase. In addition, the link between scientific method and mathematics helped to explain why a focus on the search for theories and the posing of hypothesis in geography was accompanied by a sharp increase in the use of quantitative techniques.

The Quantitative Revolution pioneered statistical and mathematical methodologies to improve our understanding of complex processes. Human geographers have also used quantitative methodologies to study a multitude of topics including demographics, migration, housing and settlement patterns and ethnic segregation. Fotheringham *et al.* (2000) identifies quantitative geography as consisting of the analysis of numerical spatial data, the development of spatial theory or the construction and testing of mathematical models of spatial processes. However, by far the greatest struggle for the acceptance of quantitative methods has been in human and economic geography. It is here that the revolution runs up against notions of free-will and the unpredictability of human behaviour.[14]

Despite the criticisms, the quantitative revolution in geography was an essential development because it modernized a largely descriptive discipline (Holt-Jensen, 1999. Sayer, 1985). However, the lack of consideration of the philosophical underpinnings of the research of this period, made much of it susceptible to the strong criticisms of positivism (Unwin, 1992). As a result, quantitative methodologies experienced a downturn in popularity in the 1980s and 1990s, as geography experienced the `cultural turn' and split into various modes of thought (Robinson, 1998). The quantitative revolution was not restricted to geography; it also encapsulated many other social science subjects, such as economics. But, this mood changed by the late 1960s, with a significant number of geographers beginning to question the utility of this quantitative turn.

Criticism

Many geographers became increasingly concerned that the quantitative techniques simply put a highly sophisticated technical gloss on the approach to study that was barren of theory. Other critics argued that it removed the 'human dimension' from a discipline that always prided itself on studying the human and natural world alike. As the 1970s dawned, the Quantitative Revolution came under direct challenge. The mood changed with a significant number of geographers beginning to question the wisdom and even the morality of this quantitative turn. In spite of the fact that various criticism were raised against Quantitative Revolution, it could not be ignored that the quantification is valid if a correct use of methods is made.

NOTES

1. Gregory, Derek *et. al.* (2009).
2. Burton, I. (1963).
3. Kimble, G.H.T. (1951).
4. Livingstone, David N., (1993).

5. Ackerman, E.A. (1945), also cited in Johnston, Ron and Sideway James (2016).
6. Hartshorne, Richard (1954, 1955, 1958, 1959), also cited in Johnston, Ron and Sideway James (2016).
7. Gregory, Derek *et al.* (2009). *op. cit.*
8. Johnston, Ron; Sidaway, James (2016), pp. 64–76.
9. The Gravity Model: For decades, social scientists have been using a modified version of Isaac Newton's Law of Gravitation to predict movement of people, information, and commodities between cities and even continents. The gravity model, as social scientists refer to the modified law of gravitation, takes into account the population size of two places and their distance. Since larger places attract people, ideas, and commodities more than smaller places and places closer together have a greater attraction, the gravity model incorporates these two features. The relative strength of a bond between two places, for example, may be determined by multiplying the population of city A by the population of city B and then dividing the product by the distance between the two cities squared.
10. Schafer (1953).
11. Zipf, G. K. (1949).
12. Johnston, Ron; Sidaway, James (2016). *op. cit.*
13. Spate, O. H. K. (1956).
14. Gregory, Derek *et. al.* (2009). *op. cit.* p. 608.

REFERENCES

Ackerman, E.A. (1958). *Geography as a Fundamental Research Discipline*. University of Chicago, Department of Geography, Research paper no. 53.

Bansal, Taruna, "Quantitative Revolution in Geography", *E-text*.

Braithwaite, R. B. (1955). *Scientific Explanation*. Cambridge, Cambridge University Press.

Bunge, W. (1962). "Theoretical Geography", *Lund Stud. in Geog.*, C, 1, Dept. of Geog., Royal University of Lund. 53, 1963 pp. 331-33.

Burton, I. (1963). "The Quantitative Revolution and Theoretical Geography", *Canadian Geographer*, 7, 4, pp. 151-182.

Chorley, R.J. (1957). "Climate and Morphometry", *Journal of Geology*, 65, 628-638.

Cloke, P.; Philo, C. and Sadler, D. (1991). *Approaching Human Geography*. London: Chapman.

Davies, W.K.D. (1972). *Conceptual Revolution in Geography*, New Jersey: Rowman & Littlefield.

Dury, (1954). "Contribution to a General Theory of Meandering Valleys", *American Journal of Science*, 123, pp. 167-176.

Fortheringham, A.S., Brunsdon, C., and Charlton, M. (2000). *Quantitative Geography: Perspective on Spatial Data Analysis*. London: Sage.

Gregory, S. (1963). *Statistical Methods and the Geographers*, London: Longman.

Hagget, P. (1965). *Locational Analysis in Human Geography*, Edward Arnold, London.

Hagget, P. and R.J. Chorley (1965). *Frontiers in Geographical Teaching*, Mathuen.

Hagget, P. and R.J. Chorley (1965). *Models in Geography*, Routledge.

Hartshorne, R. (1939), *The Nature of Geography: A Critical Survey of Current Thought in the Light of the Past*, Association of American Geographers, Lancaster.

Hartshorne, Richard (1954). "Comment on 'Exceptionalism in geography'". *Annals of the Association of American Geographers*, Vol. 38, pp. 108–9.

Hartshorne, Richard (1955). "'Exceptionalism in Geography' re-examined". *Annals of the Association of American Geographers*, Vol. 45, pp. 205–44.

Hartshorne, Richard (1958). "The concept of geography as a science of space from Kant and Humboldt to Hettner". *Annals of the Association of American Geographers*, Vol. 48 : 97–108.

Hartshorne, Richard (1959). *Perspective on the Nature of Geography*. Chicago: Rand McNally.

Harvey, D. (1969). *Explanation in Geography*, New Delhi: Arnold Publishers.

Holt-Jenson, A. (1999). *Geography History and Concepts*. London: Sage.

Johnston, Ron and Sideway James (2004, 2016). *Geography and Geographers: Anglo-American Human Geography since 1945*. New York: Routledge.

Kimble, G.H.T. (1951). *The Inadequacy of the Regional Concept*, London Essays in Geography, edd. L.D. Stamp and S.W. Wooldridge, pp. 492-512.

Livingstone, David N., (1993). *The Geographical Tradition: Episodes in the History of a Contested Enterprise*, Blackwell Publishers ISBN 0-631-18586-0.

Livingstone, D N. (1992). Statistics don't bleed: quantification and its detractors. *The Geographical Tradition*, Ch. 9.

Mackay, (1963). *The Mackenzie Delta Area*, N.W.T., Ottawa: Queen's Printers.

Marshall, Alan, "A critique of the development of quantitative methodologies in human geography", *Internet*.

Peet, R. (1998). *Modern Geographical Thought*. Oxford: Blackwell.

Robinson, G. (1998). *Methods and Techniques in Human Geography*. London: Hodder.

Sayer, A. (1985). "Realism and Geography", in Johnston, R. (ed.) *The Future of Geography*, London: Methuen.

Schaefer, Fred K. (1953). "Exceptionalism in Geography: A Methodological Examination", *Annals of the Association of American Geographers*, Vol. 43 (3) : 226-249.

Spate, O.H.K. (1956). "Quantity and Quality in Geography". *Annals of Association of American Geographers*, 46, pp. 369-77.

Stamp, L. Dudley (1957). "Geographical Agenda: A Review of some Tasks Awaiting Geographical Attention". Pres. address, *Trans. and Papers, Inst. Brit. Geog.*, 23, 1-17.

Taylor, P.J. 1976. "An interpretation of the quantification debate in British geography." *Transactions of the Institute of British Geographers*, N.S. 1, pp. 129-142.

Unwin, T. (1992). *The Place of Geography*. Harlow: Longman.

Wolman, M.G. (1955). "The Natural Channel of Brandywine Creek, Pennsylvania", *Geological Survey Professional Paper*, 271.

19

Emergence of Spatial Science School (Paradigm Phase VII)

The Spatial Science School had taken a lead by 1970s and represented the Paradigm Phase VII in geography. It developed as a response to the perceived inadequacies of the prevailing regional paradigm during 1950-1970. Since then it remained a significant component of the disciplinary portfolio, and was termed as the discipline's 'quantitative and theoretical revolution'.[1] The traditional paradigm of geography, being largely classificatory, was under great stress. The new development within the subject not only represented a wider range of methods, but also demanded a fundamental paradigm shift. Model and theory building was set up as the aim of geographical investigation, a task to be performed with the aid of quantitative methods and the use of computer to handle the data. The 'new geography' provided the discipline with notable research projects that could serve as exemplars for new students as, for instance, 'Hagerstrand's Diffusion Model'.[2] The chorological viewpoint, emphasizing the character of and interrelationships within specific places or regions, was rejected in favour of a geography based on 'spatial analysis' which stressed the geometric arrangements and patterns of phenomena.

The Driving Force

The general acceptance of quantitative techniques, and the more complete training of a new generation in this regard made possible this development.[3]The former years of over-enthusiastic pressing of quantitative methods had given way to the subsequent phase in which mathematical methods were taken as significant tools for approaching geographic problems. Schaefer's paper undoubtedly had been instrumental behind this change.[4] Bunge[5] further extended the arguments of Schaefer to the effect that geography is the science of spatial relations and interactions. The major advances towards a unifying methodological and philosophical basis for the Spatial Science School were initially made by British geographers, notably Peter Hagget, Richard Chorley and David Harvey.[6]

However, about spatial science approach in geography, innovative pieces of work can be identified from 1940s and even earlier.[7] Such academic social movements emerged in both the U.S.A. and U.K. in the middle of Twentieth century, taking the discipline by storm and rapidly ensuring a major shift in its foundations. One influential work was that of Edward Ullman, who produced two pioneering essays on theoretical approaches to urban geography, one being co-authored with Chauncy Harris (Harris and Ullman, 1945[8], Ullman, 1941[9]). In the mid-1950s, many graduate students were attracted towards this new theoretically-based approach he was promoting. He applied this approach, particularly, focusing on the role of transportation as a major influence on human spatial behaviour and the resultant spatial patterns of areal differentiation[10]. As described in the previous chapter, a parallel introduction of similar approaches in Britain was promoted by Richard Chorley, Peter Haggett and Stanley Gregory. Chorley and Haggett introduced the teaching of quantitative methods at the University of Cambridge in 1960. Haggett later taught a course that became his pioneering text on Locational analysis in human geography[11]. Gregory, a London-trained climatologist, initiated the teaching of statistical methods at the University of Liverpool in the 1950s, and published the first textbook on the subject.[12] Their students promptly speeded the 'revolution' and inspired a further stream of students. Besides, a number of other contemporary writers including, for instance, Brian Berry, Ron Boyce, William Bunge, Michael Dacey, Art Getis, Duane Marble, Richard Morrill, John Nystuen, Forrest Pitts and Waldo Tobler also appeared in the scene to stimulate others to adopt 'new geography'.

From these origins, interest in quantitative and theoretical work rapidly proliferated, not only through those as mentioned above but also by many others, mostly English-speaking, such as Australia and New Zealand. By the early 1970s many pioneering works had been produced conveying the approach. Notable and highly influential among these were those by Haggett's (1965); Abler, Adams and Gould's (1970)[13] and Morrill's (1970)[14]. Besides, Harvey's *'Explanation in Geography'* (1969)[15] provided a detailed philosophical and methodological foundation for the practices the pioneers supported. Spatial science soon occupied a major role within the discipline, with compulsory courses in quantitative methods. This provided the foundation for continued substantive and technical developments over the next four decades.

Many of the stimuli for this fresh approach to human geography came from outside the discipline, notably economics in the first instance. The contributions of a number of economists provided the initial spurt that geographers followed up. Many of such works published in the early nineteenth century pertained to, for instance, Hoover, Hotelling, Lösch, Palander, Alfred Weber and von Thünen. They especially concentrated on the location of economic activities, particularly the manufacturing industries. They sought to account for observed distributions firstly by identifying the rationales for decisions through empirical case studies; and then by uncovering general principles underpinning all locational decisions through theoretical development followed by hypothesis testing.

The School

The Spatial Science School was about patterns and processes involving large numbers of decision-makers, seeking order in even the most complex situations (Haggett, 1990; Law, 2004). The idealizations of spatial structures and processes stimulated a large number of empirical studies seeking ordered patterns in the landscape. The emphasis shifted very much on pattern, as for example, point patterns of settlements and hierarchies, or line patterns of transportation routes, smooth (trend) surfaces of inter-nodal land use patterns and land values, diseases diffusing down through urban hierarchies and across the surfaces—as well as ordered patterns of behaviour, such as distance-decay patterns in various forms of interaction.

As Bunge (1966), Harvey (1969) and others argued, geometry was the language of this 'new geography'. To test the geometric hypotheses, the literature of statistical applications in a range of disciplines was explored for relevant procedures—such as nearest neighbour analysis for point patterns, graph theory for line patterns, the gravity model for migration flows, and trend surface analysis. The model became a key word in the vocabulary of the contemporary writers.

For these scholars, the basis principle in the theories and models they were attracted to was largely related to studies based urban transport, manufacturing, commercial activities and settlement patterns. The transport costs, for instance, reflected an assumed importance of distance (or space) in decision-making. Of the various costs involved in manufacturing products, it was the one that could be promoted as clearly geographical—as case made by Ullman (1956)[16] and adopted by his followers. Manufacturing industries, for example, have to bring raw materials and component parts for a product together—which involves costs (both time and money)—and then distribute them to their markets, involving further costs. Alongside industrial location and migration, a third area of work dominating much early spatial science was central place theory, addressing issues relating to settlement patterns. Many nineteenth and early twentieth century scholars had explored regularities in the distribution of central places—where consumers purchase goods and services from shops and offices. A German geographer, Walther Christaller (with a strong background in economics), first formalized this[17]. This theory and a more complex variant developed by Lösch[18], was the basis for much data collection and hypothesis-testing, seeking regular patterns with hexagonal foundations. Finally, among the most influential sets of ideas—though taken up a little later than some of the others—was work on spatial diffusion developed by a Swedish geographer, Torsten Hägerstrand (1968)[19].

Most of the above approaches adopted by so-called spatial scientists were summarized by Garrison (1959)[20] into six basic spatial problems: the transportation problem–searching for the most cost-efficient flow of goods through a network of supply and demand points; the spatial price-equilibrium problem—the resultant prices for goods and services from such flow patterns; the warehouse-location problem—looking for the best set of locations for a series of supply points, to serve a given geography of demand; the industrial location problem—i.e. the optimum location for an industry given the geography of supply for its inputs and that of demand for its outputs; the interdependencies problem—seeking the best pattern of linked plants so as to maximize profits jointly; and the boundary-drawing problem—exploring the most efficient set of boundaries for the hinterlands of places (such as schools), which minimizes transport costs.

The newly formed Spatial Science School emphasized the study of aggregate spatial patterns, including spatial behaviour, within clear theoretical frameworks, using quantitative methods to evaluate models and hypotheses. Geometry, as the measurement of space, became the language of geography. The theoretical approaches and quantitative methodologies evolved considerably over the past four decades were deployed. The relative position in space—distance measured in various ways—became the main explanatory factor. For, with the successful rise of the disciplines like geophysics, astronomy and geology, geography could no longer deal with the whole earth, but only the earth's surface and 'with the earthly things that fill its spaces'.[21] Harvey (1969) argued that the concept of geography as a chorological science of the individual case was not tenable because it was built on the assumption of Absolute Space.[22] Instead, the concept of Relative Space was advocated. Through this introduction, the spatial relations and distance measured in different ways were given more explanatory power. Distance, now, could be measured in terms of transport costs, travel time, mileage through transport network and even as a perceived distance. It has been observed that since distances in time, cost or even network mileage are partly artifacts of socio-economic demands and technological progress, these types of spaces are naturally dynamic and truly relative. This had led to a definition of 'Plastic Space'—a space that is continuously changing in size and form.[23] These conceptual and empirical considerations were intrinsic to the further development of the Spatial Science School.

The Spatial Science School, in general, deals with aggregate patterns in space and their generating processes. It is characterized by two major features: the description, explanation and prediction of general patterns of spatial behaviour and their reflection in the landscape; and the application of rigorous mathematical and statistical procedures in that task. However, the set of procedures that came to dominate the social sciences was the general linear model, as developed in statistics for testing hypotheses regarding associations and causal relationships. A great attention was also given to regression and associated measures of correlation. Regression modelling and associated techniques such as factor analysis remained at the heart of much spatial science. But over the half-century since they were introduced to human geography applications have become much more sophisticated, allowing a wider range of hypotheses to be tested and data sets of different types to be explored. Log-linear modelling, logit analysis, multi-level modelling and a range of other procedures are now widely applied: all are based on the general linear model and allow similar interpretations. The testing is undertaken in one of two ways: deductively and inductively. Deductively, the formal hypotheses regarding the size, strength and direction of relationships among variables are stated and those expectations compared against empirical data; and Inductively, the patterns are identified in data sets which suggest ordered behaviour. Both approaches use measures of statistical significance to assess the importance of their findings. Most researchers took the gravity model as the source for hypotheses about flow patterns. Wilson (1970)[24] restructured it as a way of representing an entire system, concerned, for example, not just with the decisions underpinning travel behaviour—as with commuting in a city—but also those regarding where homes and jobs might be located. Besides, the concept of a system attracted considerable attention among spatial scientists, largely as a rhetorical device rather than as a format for undertaking detailed research.

Spatial science became a substantial component of contemporary human geography. It was very closely associated with quantitative methods: without them and associated technologies for their application much spatial science would not be feasible. The approaches taken within this broad framework have changed considerably over the last half-century. There is also considerable interdisciplinary interaction with workers in other disciplines, such as economics, sociology, political science and archaeology, where the spatial approach—as in human geography—is one component of a broader disciplinary framework. And there are applied disciplines—notably city and regional planning—which build on those foundations.

Criticism

It has been claimed that "To the extent that is not widely recognized, the move to quantification took place within the basic framework of geography put forward by Hartshorne in his '*Nature of Geography* (1939)'.[25] There is one major aspect in which geography does differ from the other social sciences. The latter, as they mature, concentrate more and more on the discovery of process laws, that is, laws that are in one important aspect like the laws of Newtonian astronomy. Given the state of a system at a certain point in time, process laws allow for the prediction of the changes that will take place.[26] Geography is essentially morphological. Purely geographical laws contain no reference to time and change. Thus is not to deny that the spatial structures we explore are, like all structures anywhere, the result of processes. But the geographer, for most part, deals with them as he finds them, readymade. Technically, the morphological character of geography finds its expression in its own specific tool, maps and cartographic correlation. Mapping has been called the short-hand of Geography.

Many critiques of early spatial science argued that its models were too simplistic, reducing decision-making in a complex world to a single key influence—transport costs; distance was presented as a basic determinant of behaviour. The realization of the power of these arguments stimulated a major shift in the nature of much spatial science. Then, most research workers regarded advanced statistical methods as being useful in some branches of the discipline; other branches notably historical and cultural geography, felt less need for the new techniques. The models and theories that dominated its early years—such as central place theory—became much less important (indeed, some almost disappeared from the disciplinary portfolio), and were replaced by approaches that initially went under the name of behavioural geography. These now so predominate within the spatial science component of the discipline that the term is rarely used.

NOTES

1. Burton, 1963.
2. The purpose of this model is to understand the processes which create a "nebula-distribution," a common spatial pattern. This model explores the way that the diffusion of ideas across a social network might produce this pattern. The model attempts to recreate the chronologic and geographic patterns produced in a historical example, the diffusion of the adoption of government subsidies for improved pasture on farms in Sweden. Two related models were outlined and studied by the author. The first model, referred to as the isotropic model, begins with a uniform population number and a 5×5 grid. This model is designed to show how innovation might spread on a limited scale. The second model, the anisotropic model, initializes with different numbers of agents on each patch, a larger more complex spatial area and is intended to recreate a specific case study. The models rely on real-world data to create some of the models' parameters (e.g. probability distributions used in the model are calculated from actual migration data and telephone call data) and attempt to compare real world distributions of the spread of ideas with those predicted by the model.
3. Holt-Jenson, A. (2009), p. 121.
4. Schaefer, F.K. 1953. "Exceptionalism in geography: a methodological examination." *Annals of the Association of American Geographers,* 43: 226-249.
5. Bunge, W. (1962, 1966).
6. Holt-Jenson, *op. cit.* p. 120.
7. Johnston, 2006.
8. Harris, C.D. and Ullman, E.L. 1945. "The nature of cities", *Annals of the American Academy of Political and Social Science,* pp. 242: 7-17.
9. Ullman, E.L. (1941). "A theory of location for cities", *American Journal of Sociology,* 46, pp. 853-864.
10. Ullman, E.L. (1956). "The role of transportation and the bases for interaction". In W.L. Thomas, editor, *Man's role in changing the face of the earth,* Chicago: University of Chicago Press, 862-880.
11. Haggett, P. (1965). *Locational analysis in human geography*. London: Edward Arnold.
12. Gregory, S. (1963). *Statistical methods and the geographer*. London: Longman.

13. Abler, R.F., Adams, J.S. and Gould, P.R. (1971). *Spatial organization: the geographer's view of the world*, Englewood Cliffs NJ: Prentice-Hall.
14. Morrill, R.L. (1970). *The spatial organization of society*, Belmont CA: Wadworth.
15. Harvey, D. (1969). *Explanation in geography*, London: Edward Arnold.
16. Ullman (1956). *op. cit.*
17. Christaller, W. (1966). *Central places in southern Germany*, Englewood Cliffs NJ: Prentice-Hall.
18. The German economist **August Lösch** expanded on Christaller's work in his book *The Spatial Organization of the Economy* (1940).
19. Hägerstrand, T. (1968). *Innovation diffusion as a spatial process*, Chicago: University of Chicago Press.
20. Garrison (1959).
21. Schaefer, Fred K. (1953), p. 226.
22. Holt-Jenson, A. (2009), p. 116.
23. *Ibid.*
24. Wilson (1970).
25. Holt-Jenson, A. (2009), p. 119.
26. Schaefer, Fred K. (1953), pp. 243-245.

REFERENCES

Ackerman, E.A. (1958). *Geography as a fundamental research discipline*. Chicago: University of Chicago, Department of Geography, Research Paper 53.

Bunge, W. (1962, 1966). *Theoretical Geography*, Lund Studies in Geography, Series C1, Gleerup, Lund.

Burton, I. (1963). "The quantitative revolution and theoretical geography", *The Canadian Geographer* 7: 151–62.

Chorley, R.J. and Haggett, P. (Editors) (1967). *Frontiers in geographical teaching*. London: Methuen.

Chorley, R.J. and Haggett, P. (Editors) (1965). *Models in geography*. London: Methuen.

Forer, P. (1978). "A Place for Plastic Space", *Progress in Human Geography*, Vol. 2, pp. 230-67.

Fotheringham, A.S. (2006). "Quantification, evidence and positivism", in S. Aitken and G. Valentine, editors, *Approaches to human geography*. London: Sage, 237-250.

Fotheringham, A.S., Brunsdon, C. and Charlton, M. (2000). *Quantitative geography: perspectives on spatial data analysis*. London: Sage.

Garrison, W.L. (1959). "Spatial structure of the economy II". *Annals of the Association of American Geographers* 49: 471-482.

Haggett, P. (1990). *The geographer's art*. Oxford: Balckwell.

Haggett, P., Cliff, A.D. and Frey, A.E. (1977). *Locational analysis in human geography* (second edition). London: Edward Arnold.

Holt-Jenson, A. (2009). "Paradigms and Revolutions", *Geography: History and Concepts: A Systematic Guide*, Fourth Edition, London: Sage, pp. 100-124.

Johnston, R., "Spatial science School of Geographical Sciences", University of Bristol (Internet).

Johnston, R.J. (1978). *Multivariate Statistical analysis in geography: a primer on the general linear model*. Longman, London.

Johnston, R.J. and Sidaway, J.D. (2004). *Geography and geographers: Anglo American human geography since 1945* (sixth edition). London: Arnold.

Jones, E. (1956). "Cause and effect in human geography", *Annals of the Association of American Geographers* 46: 370-377.

Wilson, A.G. (1970). *Entropy in urban and regional modeling*, London: Pion.

Wilson, A.G. (2000). *Complex spatial systems: the modelling foundations of urban and regional analysis*, Chichester: John Wiley.

Wilson, A.G. and Bennett, R.J. (1985). *Mathematical methods in human geography and planning*, Chichester: John Wiley.

20
Critical Revolution (Crisis Phase VII)

Quantitative methodologies, undoubtedly, are a powerful research technique in any discipline that can provide valuable and accurate insights with an understanding of the limitations. However, by mid 1960s, the initial euphoria about quantitative revolution that Geography witnessed had receded. The geographers had begun to question its relevance to real life. According to critics, quantification creates a false sense of objectivity. During the 1970s and 1980s there was a downturn in the popularity of such methodologies. The backlash against the quantification trend led to 'critical revolution' under the humanistic and welfare approaches in geography, and geography saw another phase of transformation (Crisis Phase VII) since early 1970s. Precisely, the Critical Revolution is associated with opposition to quantification in geography. In the 1970s itself, a new type of criticism had been developing. Certain scholars felt that "the quantitative revolution has run its course, and diminishing marginal returns are apparently setting in; the paradigm is not coping well; it is ripe to overthrow."[1] The overwhelming focus on statistical modeling was, eventually, responsible for the undoing of the quantitative revolution.

Opposition

The debate on quantitative methodologies in human geography continued to be polarized between the supporters and opponents of such techniques (Fotheringham *et al.*, 2000). Opposing the so-called 'revolution' Dudley Stamp preferred to call it a 'civil war' and noted that quantification had many points common with a political ideology; it was more or less a religion to its followers: 'its golden calf is the computer'.[2] Many geographers became increasingly concerned that these techniques simply put a highly sophisticated technical gloss on an approach to study that was barren of fundamental theory. Other critics argued that it removed the 'human dimension' from a discipline that always prided itself on studying the human and natural world alike. As the 1970s dawned, the quantitative revolution came under direct challenge. The counter-positivist response came as geographers began to expose the inadequacy of quantitative methods to explain and address issues on humanistic grounds.[3]

The opposition to the Quantitative Revolution came in several ways[4]:

1. There were those who thought that the whole idea was a bad one and that quantification would mislead geography in a wrong and fruitless direction.

2. There were traditional scholars like Dudley Stamp who argued that geographers had spent too long perfecting their tools (maps, cartograms and other diagrammatic representations) and should get on improving these tools further rather going for something new like quantification.
3. A third kind of opposition holds that statistical techniques are suitable for some kinds of geography; but not all geography, because there are certain things that cannot be measured.
4. Another class of objection is that although quantitative techniques are suitable and their application to geographic problems is desirable, they are nevertheless being incorrectly applied; ends are confused with means.
5. A final kind of criticism to note is that quantification is all right, but the quantifiers are not.

Thus, Quantification was criticized for the reasons as discussed above.[5] The basic philosophy of the quantitative revolution was attacked on various grounds as, for example, scope of quantitative analysis, relevance of existing tools and their replacement, suitability of new techniques, their applicability and over-enthusiasm on the part of applicants. First, there were critics who believed that quantification would mislead geography towards a futile course. With time, such opponents have become weaker and vanishing. If such critics are still among us they have not made themselves heard for some time. Secondly, there were those like Stamp who argued that quantifiers were too busy in sharpening their instruments to the extent that they forgot the real purpose of the revolution. He asserted that geographers had spent too long perfecting their tools (maps, cartograms, and other diagrammatic representations) and should get on with some real building. Stamp was "a little alarmed by the view that the geographer must add to his training a considerable knowledge of statistics and statistical method, of theoretical economics."[6] Stamp's viewpoint led geography in the situation of another dichotomy. The notion that geographers either improve their tools or engage in research with available tools seemed false. Surely advances in technology are most likely to occur at the moment when we are wrestling with our toughest problems. Furthermore, to argue that geographers should not use statistical methods comes close to defining geography in terms of one research tool, namely the map. One weakness of this position has been can be demonstrated with the example of an Isopleth Map. The visual comparison of isopleth maps is not an adequate means of determining correlations between spatially distributed phenomena; it needs to be substituted by quantitative analysis for interpretation and understanding. Few opponents also commented on the suitability of statistical techniques for all kinds of geography. They opined that these techniques were appropriate for some branches and not the entire geography. Because there are certain things that cannot be measured. This may be true for some variables. However, even with qualitative characteristics, nominal observations can be made and there is an expanding body of literature on the analysis of qualitative data.[7] A variant of this argument is that the variables with which geography is concerned are too numerous and complex for statistical analysis. Quantifiers claim that it is precisely because of the number and complexity of the variables that statistical techniques are being employed. Another group condemned this revolution on a note that there was a confusion of ends and means. It argued that although quantitative techniques are suitable and their application to geographic problems is desirable, they are nevertheless being incorrectly applied. The quantitative analysis has failed on occasion to distinguish the significant from the trivial; the alleged discoveries of the quantifiers are not very novel; and so on. A final kind of criticism to note is that quantification is alright but quantifiers are not. They are vivacious; suffer from over-enthusiasm, vaulting ambition, or just plain arrogance. However, when one is involved in a revolution, it is difficult not to be a little overconfident or impudent.

That these criticisms have a grain of truth cannot be denied, but to the valid, correct use of quantitative methods (and this is surely what we are concerned with) they are merely irrelevant. Incorrect applications have been and no doubt will continue to be made, and in some cases for the wrong reason such as fashion, craze, or snobbery. More often, however, they are genuine and honest attempts to gain new knowledge and new understandings. The Quantitative Revolution, ultimately, led to a number of dichotomies within the discipline apart from the quantity-quality issue. Now, these included measurement by instruments *versus* direct sense-data; rational analysis *versus* intuitive perception; cold scientific constructs developed in the laboratories *versus* rich daily sensed—experience from the real world itself; constantly changing phenomena versus discrete cases; nomothetic *versus* idiographic, to mention a few. Another dichotomy lies in classifying and understanding; classification should never be mistaken for comprehension. Goodall's (1952) point is worth pondering where he states that quantitative methods or statistical techniques are only adjunct to elucidations or descriptions; they can neither provide explanations nor replace them. Therefore, these methods should be observed only as useful tools and not keys to universal knowledge (Spate, 1960).

The criticisms became more prominent as critical geographers started questioning the relevance and value of spatial science in the early 1970s. The quantitative geography was labelled as positivist and empiricist because it was based on the principles of scientific objectivity, value neutrality, and the search for universally applicable generalizations. (Kwan and Schwanen, 2009). These criticisms clearly point out that the quantitative methods have some severe limitations, especially when applied to the study of certain kinds of phenomena—for example when the purpose is to uncover the complex gendered, racialized or sexualized experiences of individuals or the socio-spatial construction of identities. The models and theories developed on the basis of empirical data exclude the normative questions like belief, taboos, emotions, desires, hopes, fears, likes and dislikes, prejudices and aesthetic values. This is mainly being done to make the study objective and scientific. In the real world, the normative questions have close bearing in the interrelationship between man and environment, and decision making process (Cloke *et al.*, 1991; Johnston and Sidaway, 2004). However, most of the criticisms mounted against the use of quantitative methodologies in geography were targeted at the positivist underpinnings of the approach and concerned the claim of objective research, the lack of consideration of agency and structure, the imposition of the natural sciences approach, the assumption that social systems could be considered closed and the notion that statistical relation implied causal relation.[8] The positivist claim that research should be value free was criticized by those who argued that this was not possible in social research. As researchers are part of society, their values, experiences and motives inevitably influence their research. Quantification was claimed to give a false sense of objectivity by artificially separating the observer from the quantitative techniques to appreciate the importance of observed (Cloke *et al.*, 1991). Another criticism was the failure of structure and agency. Quantitative researchers treated people as objects without any consideration of the values and meanings that make individuals human and the capabilities that they possess (Cloke *et al.*, 1991,

Smith, 1998). Concerns were raised that the complex economic, political and social structures that act on spatial patterns were not sufficiently taken into account by quantitative methodologies. A purely quantitative approach, it was argued, looked at how things seemed to be rather than how they might be under different social conditions (Cloke *et al.*, 1991). For some geographers the new quantitative approach seemed "socially and politically irrelevant"[9] The idea that a unity of scientific method existed was another positivist assumption that came to be challenged (Bryman, 2004). The imposition of the methods of the natural sciences were rejected by many geographers who felt that each discipline should have its own approach to reflect its unique focus. The positivist approach is suited to and often assumes a closed system and does not consider the difficulties of quantitative modelling of open systems (Cloke *et al.*, 1991). Sayer (1985) argues that there are two conditions that must both be satisfied for a closed system to exist. These are that there must be no change in the object possessing the causal powers and that the relationship between the causal mechanism and those of its external conditions must also be constant. From this definition it is clear that social science research involves open systems because humans have the capacity to change and human actions have the capacity to alter the configuration of systems (Sayer, 1985). In addition to the criticisms of the positivist underpinnings of quantitative geography other weaknesses emerged. The statistical techniques that were applied after the Quantitative revolution were largely imported from outside the discipline (Harvey, 1969). Some of these techniques were used in a `cookbook' fashion without consideration of the appropriateness of use for spatial data (Fotheringham *et al.*, 2004).

The major change that resulted from the criticisms of the quantitative revolution was the recognition that the philosophical basis and role of quantitative methodologies had to change. Harvey[10] recognized the danger of the inappropriate use of quantitative tools. The idea that quantification would lead to the generation of universal laws (as in the natural sciences) has been recognized as impossible due to the complexity and ephemeral nature of social systems. It has been acknowledged that ontological assumptions applied by those who used quantitative methodologies in the social sciences must differ from the natural sciences (Fotheringham *et al.*, 2000, Christensen, 1982). During the 1950s and 1960s quantitative analysis was used speculatively to develop theory on social processes. A consensus view developed that statistical analysis should not subsume theoretical development and should be accompanied by relational and contextual understanding of the social process (Robinson, 1998, Bennet, 1985). The research strategy has since matured from inductive and deductive approaches to a retroductive approach that recognizes the guiding role that social structures and mechanisms should play in social science (Blaikie, 1993). The speculative nature of much of the quantitative research in human geography led to false identification of causation between variables because researchers had not considered the mechanisms of the social process that they were studying (Blaikie, 2000). Sayer (1985) stresses the importance of qualitative techniques and causal knowledge to distinguish between those variables that are statistically related and those that are causally related. Whichever methodological approach that is used subjectivity is a difficult issue to address.

The philosophical changes described above led to a number of practical changes to the quantitative approach in human geography. One of these was the increase in use of quantitative and qualitative techniques with the thinking that there is "no fundamental clash between the purposes and capabilities of the two approaches[11]." Quantitative geographers have also shifted the focus of their research away from attempts to investigate similarities across space and the production of global, general laws to local analysis that tests for the presence of differences across space (Fotheringham *et al.*, 2000).

Outcome

The positivist assumptions upon which the approach was based were rightly attacked in the 1970s and 1980s as geography experienced its 'cultural turn' and split into a range of modes of thought (Peet, 1998). During the Critical Revolution, a 'welfare geography' working within the framework of the existing economic and social systems was promoted.[12] With these kind of developments, David Harvey, too, disregarded earlier works where he advocated for the quantitative revolution and adopted a Marxist theoretical framework.[13] Soon new subfields emerged in human geography to contribute a new vocabulary for addressing these issues. Critical geography is used as an umbrella term for Marxist, feminist, postmodern, post structural, queer, left-wing, and activist geography.[14] The rush to quantitative methodologies in human geography in the 1950s and 1960s did not include sufficient consideration of the philosophical underpinnings of such approaches (Harvey, 1969).

Thus, Critical Revolution of 1980s can be seen as a critique on positivism introduced by the Quantitative revolution. This proved as one of the major turning points in the history of geography. Major turning points that came before were, in chronological order, environmental determinism, regional geography and the quantitative revolution etc. Critical geography is theoretically informed geographical scholarship that seeks for social justice, liberation, and leftist politics.

NOTES

1. Holt-Jenson, A. (2009), p. 121.
2. Holt-Jenson, A. (2009).
3. Gregory, D. *et al.* (2009).
4. Davies, W.K.D. (1972).
5. Burton, 1963.
6. North, D.C. (1961).
7. Maxwella, E. (1961).
8. Peet 1998: 67.
9. *Ibid.*
10. Harvey (1969): 7.
11. Yeung, 1997: 64.
12. Holt-Jenson, A. (2009).
13. Harvey, D. (1973). *Social Justice and the City*. London: Edward Arnold; Harvey, D. (1972). "Revolutionary and counter revolutionary theory in geography and the problem of ghetto formation". Antipode. 4: 1–13.
14. Holt-Jenson, A. (2009).

REFERENCES

Bansal, Taruna, "Quantitative Revolution in Geography", *E-text*.

Bennet, R., (1985). "Quantification and Relevance". in: Johnston, R. (ed) *The Future of Geography*. London: Methuen.

Blaike, N. (1993). *Approaches to Social Enquiry*. Cambridge: Blackwell.
Blaikie, N. (2000). *Designing Social Research*. Cambridge: Blackwell.
Bryman, A. (2004). *Social Research Methods*. Oxford: Oxford University Press.
Burton, I. (1963). "The Quantitative Revolution and Theoretical Geography", *Canadian Geographer*, 7, 4, pp. 151-182.
Christensen. K, (1982). "Geography as a Human Science: a Philosophic Critique of the Positivist-humanist split", in Gould, P. and Olsson, G. *A Search for Common Ground*. London: Pion. pp. 37-57.
Cloke P, Philo C, Sadler D (1991). *Approaching Human Geography*. London: Chapman.
Davies, W.K.D. (1972). *Conceptual Revolution in Geography*, Rowman & Littlefield, New Jersey.
Fotheringham, A., Bunsden, C. and Charlton M. (2000). *Quantitative Geography*. London: Sage.
Gregory, Derek; Johnston, Ron; Pratt, Geraldine; Watts, Michael J.; Whatmore, Sarah (2009). *The Dictionary of Human Geography* (5th ed.). USA & UK: Willey-Blackwell. p. 612.
Harvey, D. (1969). *Explanation in Geography*. London: Arnold.
Harvey, D. (1973). *Social Justice and the City*, London: Arnold.
Holt-Jenson, A. (2009). "Paradigms and Revolutions", *Geography: History and Concepts: A Systematic Guide*, Fourth Edition, London: Sage, pp. 100-124.
Johnston R, Sidaway J, (2004). *Geography and Geographers—Anglo-American Geogrpahy since 1945*. London: Hodder.
Marshall, Alan, "A critique of the development of quantitative methodologies in human geography", *Internet*.
Maxwella, E. (1961). *Analyzing Qualitative Data*. London: Methuen and Co.
North, D.C. (1961). "Quantitative Research in American Economic History". *American Economic Review*, 53, pp. 128-30.
Parker, I. (1999). "Qualitative data and the Subjectivity of Objective Facts", in Simpson, L. and Dorling, D., *Statistics in Society*. London: Hodder.
Peet, R. (1998). *Modern Geographical Thought*. Oxford: Blackwell.
Poon, J., (2004). "Quantitative methods: past and present", *Progress in Human Geography*, Vol. 28 : 807-814.
Robinson, G. (1998). *Methods and Techniques in Human Geography*. London: Hodder.
Rose, G. (1997). "Situating Knowledge: positionality, reflexivities and other tactics', *Progress in Human Geography*, Vol. 21 : 305-320.
Sayer, A. (1985). "Realism and Geography", in Johnston, R. (ed.) *The Future of Geography*, London: Methuen.
Simpson, L. and Dorling, D. (2004). "Statistics of Racial Segregation: Measures, Evidence and Policy", *Urban Studies*, Vol. 41: 66, 1-681.
Smith, M. (1998). *Social Science in Question*. London: Sage.
Stamp, D. (1966). "Ten Years On", *Transactions, the Institute of British Geographers*, Vol. 40 : 11-20.
Yeung, M.W. (1997). "Critical Realism and realistic research in human geography: a method or a philosophy in search of a method?" *Progress in Human Geography*, Vol. 21 : 57-74.

21

Anthropocentric Tilt (Paradigm Phase VIII)

We have seen that the Quantitative Revolution was at its peak in Geography during mid-twentieth century. By 1970s, when the need for precision in research through increased use of statistical-mathematical techniques had been commonly accepted, the euphoria about quantitative revolution had also greatly receded. Now, many workers had begun to feel disillusioned with the quantitative models being developed and used by human geographers of the positivist orientation, since they failed to relate to real-life situations. Consequently, geography again witnessed a paradigmatic shift, bringing in great anthropocentricism in its studies. Critical geography that emerged thereafter led to three main schools of thought, viz. Behaviouralism, Radicalism and Humanism. Behaviouralism tried to challenge quantitative side of geography and provide a greater understanding of how people perceive space and place; Radicalism had a quest for social justice; and the last and most accepted ideology, Humanism, studies human awareness, human agency, human consciousness and human activity. Modern geography is a spatial science increasingly focusing on the study of spatial aspects of social and economic phenomena, and thus being pre-eminently anthropocentric. This humanistic tradition in geography reflected in various scholarly texts, as by David Stoddart (1986), David Livingstone (1992), Derek Gregory (1994), Richard Peet (1998) etc. Under Critical Revolution, the theoretical and practical lens of the field widened in subsequent years to open doors for various other anthropocentric approaches in Geography.[1] Subsequently, numerous schools of thought marched side by side in search of new paradigm. Geographers got divided themselves, for instance, in the category of positivists, pragmatists, phenomenologists, existentialists, idealists, realists, dialectical materialists, and so on. Critical geography is now situated at the very heart of the discipline of geography.

Behaviouralism

Behaviouralism is a movement that has spread into the social sciences from the behavioural sciences.[2] Dissatisfaction with the models and theories developed by the positivists, using the statistical techniques which were based on the 'economic rationality' of man led to the development of behavioural approach in geography. It was increasingly realized by the geographers that the models propounded and tested with the help of quantitative techniques, provided poor descriptions of geographic reality and man and environment relationship. Even the theories such as Central Place Theory, based on statistical and mathematical techniques, were found inadequate to explain the spatial organization of society. The economic rationality of decision-making was also criticized as it does not explain the behaviour of floodplain dweller, which does not leave his place despite the risk of flood. Many geographers began to feel that a deeper understanding of man-environment interaction could be achieved if geographers paid greater attention to the study of the mental images that the decision-makers have of the environment to which their decisions relate since it is the perceived reality that leads to action. The realization that decisions may be based on subjective or perceived reality than on objective reality as it exists in nature led to attempts towards derivation of theories of spatial behaviour. Behavioural geography sought to counter the perceived tendency of quantitative geography to deal with humanity as a statistical phenomenon. It was a psychological turn in human geography which emphasized the role of cognitive (subjective) and decision-making variables as mediating the relationship between environment and spatial behaviour.[3] During the cognitive turn the idea that only the environment determines our behaviour changed. Cognition became an important element in explaining behaviour.

The behavioural movement of the post-quantitative revolution phase had, particularly, been triggered by the writings of Julian Wolpert[4]. It was a paper based on his Ph.D. Thesis, published in the Annals of Association of American Geographers in 1964[5]. The paper was built around the results of a field study of agricultural land use in a part of Sweden. The author compared the actual labour productivity of farmers with what could have been achieved under optimizing decision-making by the farmers as rational economic persons, and found that there was up to 40 per cent less than the expected level of productivity. The lower than expected level of productivity, was explained as a result of the farmers' lower aspiration level which was itself a reflection of their economic and social environment, the context which motivated them to act differently from the theoretical economic man. Wolepert's various papers (1964, 1965, 1967, 1970) on the satisficing model of decision-making behaviour had contributed to heralding the arrival of a new behavioural movement in geography. The axiom of 'economic person' who always tries to maximize his profit was challenged by Wolpert. In his work, Wolpert showed that, for a sample of Swedish farmers, optimal farming practices were not attainable. He concluded that the farmers were not optimizers but satisficers. By the second half of the 1960s, a new branch of geography, termed as 'Behavioural Geography' had begun to be identified. As a consequence there was a behaviouralist turn in human geography.

This behaviouralist turn certainly represented a significant milestone in geographical thought (Gold, 1980). Following Gold (1980), the major achievements of behavioural movement in geography were recognized as the assertion of the role of an individual in geography; a thorough reappraisal of the methodological approaches to man-environment relations; a revived interest in the study of some older themes, such as cultural landscape and idiographic analysis of places; opening up new channels of dialogue and debate with other disciplines, and thereby contributing to a trend towards cross-fertilization of ideas; and lastly, promoting interest in problems of social concern.

Major objectives of behavioural approach have been identified as to develop models for humanity which were alternative to the spatial location theories developed through quantitative revolution; to define the cognitive (subjective) environment that determines the decision-making process of man; to unfold the spatial dimensions of psychological, social and other theories of human decision-making; to change in emphasis from aggregate populations to the disaggregate scale of individuals and small groups; to search for methods other than the mathematical and statistical that could uncover the latent structure in data and decision-making; to emphasize on procession rather than structural explanations of human activity and physical environment; to generate primary data about human behaviour and not to rely heavily on the published data; and to adopt an interdisciplinary approach for theory-building and problem-solving. The objective of behaviouralism was to replace the simplistic and mechanistic conceptions of man-environment relationships with a new theory that views human actors in space as thinking beings.

The basic philosophy of behaviouralism is that man and environment are dynamically interrelated. Man is viewed as a motivated social being, whose actions are mediated by the mental processes, by means of which he acquires, organizes and uses knowledge. His decisions are arbitrated by his cognition of the spatial environment and creation of an environmental imagery.[6] As an approach, it looks at man-environment relationships with a new and sharp outlook. In between the environmental influences and the man's actions there are mental processes of various kinds, which govern man's behaviour in a given set of environmental elements. The behaviouralist rejected the assumption that decisions are always based upon a sequence of reasoned thoughts involving systematic evaluation of all available options.[7] He is interested in the way in which people come to terms with their physical and social milieu. He is also interested in the factors that influence the interrelationships of thought and action of man. Seen in this light, Behavioural Geography has been regarded as the most common name for that part of geography that adopts the behaviouralist approach to man-environment relationships, where explanation of spatial patterns of behaviour is sought primarily in cognitive process. Behavioural geographers focus on the cognitive processes underlying *spatial reasoning, decision making* and *behaviour.* The fundamental arguments put forward for behavioural geography are that people have environmental images; those images can be identified accurately by researchers; and there is a strong relationship between environmental images and actual behaviour (Fig. 21.1).

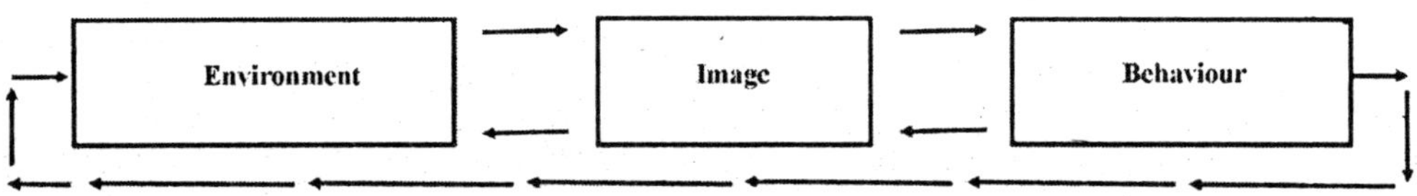

Fig. 21.1: A Conventional Model of Man-Environment Relationship (after Boulding, 1953)

In the Behavioural Model, as shown in Fig. 12.1, man has been depicted as a thinking individual whose transactions with the environment are intervened by mental processes and cognitive representation of external environment. In geographical circles, this concept is derived primarily from the work of Boulding (1956) who suggested that over time individuals' developmental impressions of the world (images) are formed through their everyday contacts with the environment and that these images act as the basis of their behaviour.

The approach to man-environment relationships that many geographers adopted has several distinguishing features. The behavioural geographers argued that environmental cognition (perception) upon which people act may differ markedly from the true nature of the environment of the real world; the *Space,* here, is said to have dual character, i.e. the objective environment, i.e. real and the behavioural environment, the world of mind; the meaning of environment includes both the physical and social environment; the behavioural geographers give more weight to an individual rather than to groups, or organizations or society; and the behavioural geography is multidisciplinary in its outlook. The behaviouralists focus on the study of individual, not the group or community. They assert that research must recognize the fact that the individual shapes and responds to his physical and social environment. The approach adopted in behavioural geography is closely related to that of psychology; but, it also draws on research findings from a multitude of other practitioners including economists, psychologists, philosophers, historians, sociologists, anthropologists, ethnologists and planners.

'Environmental Perception' and 'Mental Maps' are recognized as the basic concepts of Behavioural Geography. Environmental Perception is the total perception an individual has of his surroundings, which creates mental maps; whereas, a mental map is an individual's own internal map of their known world. Different kinds of environments around an individual may be delimited, as the 'Local' or 'Behavioural', Perceptual, Operational and Geographical. The environment immediately next to an individual is the 'Local' or 'Behavioural' Environment. Usually these environments are the places of dwelling, e.g. the urban areas. In local environment an individual performs most of his daily functions. The 'Perceptual Environment' is immediately next to the Local Environment, e.g. the recreational centres. It is the portion of the Operational Environment within which man operates. It is created, e.g. by the people working in neighbouring towns or cities, but dwelling in rural areas. The 'operational environment' is the environment human beings live in, affecting them in some way or the other, whether or not they are conscious of it. Then, ultimately, is the 'Geographical Environment', i.e. the one whose only a fraction or a part has been so far perceived by the individual.[8] The most important of all these environments remains the behavioural environment or the personal space, which every individual is very conscious of. Thus, within the perceptual environment one can focus more narrowly on the portion that draws out a behavioural response or towards which behaviour is directed. This behavioural environment receives much attention in current research in social science disciplines, including geography. It is in this behavioural environment that the man acts and interacts most closely. It is in this behavioural environment that physical features acquire values and potentialities that attract or repel human actions. On the other hand, a person's perception of the world is known as a mental map, defined as an internal representation of a place based on memory. Of its two types, the first is that formed from man's personal experiences and deals more with hisperception of the attributes of a place. For example a trip to 'Disneyworld' in Florida might conjure up images of Mickey Mouse, fun and sun provided that the trip was a pleasant experience. The second type of Mental Maps is the one formed from references to maps we have seen. This can sometimes be dangerous if maps are inaccurate. A concept a person has of the shape of the United States is an example of a mental map based on the printed map. If it were not for maps we would have no idea of the shape of our country. Mental maps not only provide a better visual conception of our surroundings but also shape our attitudes about people and places. The aspect of behavioural analysis that has been most enthusiastically adopted by geographers is the concept of mental maps. The concept has been referred to as the 'geographies of human mind'. The seminal influence in this regard came from a paper by Peter Gould (1966)[9]. Through his work, he argued that since locational decisions are guided by how the decision maker perceives the quality of his environment, it is necessary for the geographer to map out the mental image of the environment on the basis of which a decision was taken. The theme was fully developed in the book entitled: "Mental Maps" by Gould and White (1974). Among various concepts used as surrogates for the cognitive representations underlying the work on mental maps, the one most commonly employed was 'the concept of image'. Over a period of time, individuals develop impressions (or images) of the world through their every day contact with the environment. These images later guide the individual or group behaviour, as in the case of, for instance, response to natural hazards, spatial behaviour in urban areas and the role of neighbourhood preferences in Intra-Urban migration.

As the behavioural perspective in geography became generally accepted as a useful analytical device, the scope of behavioural geography widened. Various areas of behavioural research in human geography began employing this approach, the dominant ones being, for instance, the Studies of decision-making and choice of behaviour (e.g. locational choice, route selection and patronage patterns), analysis of information flows (in relation to innovation–diffusion); models of learning (derived from theories in psychology); perception-research focusing on hazard perception, image formation and mental maps and studies in voting behaviour (geography of elections).

Radicalism

Besides Behaviouralism, the Radicalism also grew as a major outcome from the criticism of quantitative geography, positivism and traditional regional geography. Radical geography is an approach to geographic research that seeks to understand social and spatial problems, and advocate solutions. Radical geographers are interested in everyday lives, the lived experiences of members of society. Radical geographers tried to transform the scope of the discipline of geography by responding to the great issues of the time as the civil rights, environmental pollution, and war. This emphasis reflected a desire for geographers to do research that is relevant to, and useful for, society.[10] Radical geography originated, by and large, as a negative reaction to the established discipline during 1970-1980 in America (Peet, 1977). According to radicalists, inequality is inherent in the capitalist mode of production. Redistribution of income through taxation policies will not solve the problems of poverty. Alternative environment designs, with removal of central bureaucracies and their replacement by anarchistic models of community control are needed, and geographers should work towards their creation.

The radical approach developed as a reaction to 'quantitative revolution' and positivism which tried to make geography as a spatial science, with great emphasis on locational analysis. Events of the late 1960s, such as burning of large cities in the Western World, student unrest, work-uprising in Paris in 1968, massive anti-Vietnam war protest actions and radical cultural reformation exposed the social and political irrelevance of geography as a spatial science and proved the hollowness of locational analysis. It was in this background that the traditional approaches in geography were challenged. The young radical geographers published papers dealing with, e.g. urban poverty, discrimination against women, coloured people and minority groups, and unequal access to social amenities, crime, deprivation, permissiveness and sexism. They also published articles on underdevelopment, poverty, malnutrition, and unemployment and resource misuse in the Third World countries. Thus, the radicalists took the side of the oppressed, advocating their causes and pressing for fundamental social change. In brief, the radical geography was a quest for social relevance of the discipline of Geography at a time of contradiction and crisis in the capitalistic society of the West. There were three prominent issues of international concern behind the movement, viz. the Vietnam war; Civil rights (especially of the American Blacks); and the pervasive poverty and inequality suffered by the residents of urban ghettos and deprived rural areas all of which were generating increased social unrest and tension. The radicalists put emphasis on the need for a revolution in both theory and practice of geography.

The followers of radical approach in geography mainly concentrated on the issues of great social relevance. The radical geographers introduced the study of topics that were previously ignored by the geographers. The radical approach is value based, especially the theory of labour value, as against the supposedly value-free approaches. Radicalists stress that, with the changing production techniques, the symbiotic relationship between human beings and the environment also changes accordingly, a relationship from which the vital elements of the composition of society arise. Radicalism believes in economic classes and the subsequent class struggle as the cornerstone of historical materialism. Most of the radicalists have a strong Marxist base and take a holistic view of economics, society and polity. Radical geographers reject the spatial determinism implicit in spatial science; instead they seek to place questions of geography within broader social and political contexts. They emphasize context and relationships between places and people and broaden the geographical research agenda by considering previously neglected issues including poverty, hunger, urban decay, and social inequality.[11]

The Salient Features of Radical Geography include, for instance, to expose the issues of inequality, deprivation, discrimination, health, exploitation, crime and environmental degradation in the capitalist countries; to highlight the weaknesses of the positivism and quantitative revolution; to bring a cultural revolution to eradicate permissiveness, sexism and discrimination against females; to remove regional inequalities; to oppose political centralization and economic concentration; to oppose imperialism, nationalism, national chauvinism and racism; to oppose the idea of superiority of the White and the West; and to develop a more just, equal, tension-free and peaceful society.

However, certain criticisms have also been raised against radicalism as, for example, the radicalists are victims of Maxist orthodoxy as they stress more on time over space; the redicalists lack flexibility in a fast-changing world of knowledge and therefore the radical interpretation suffers from an undue authoritative analysis of 'space'; radical geographers had no empirical study on the erstwhile Communist countries; the theoretical base of the radical paradigm was weak and the radical geographers attempted to borrow theoretically sophisticated ideas from other disciplines without much success; radical geography was radical in topics (imperialism, anti-war, discrimination, poverty, crimes etc.) and politics (anti-capitalist) but not in theory or method of analysis; the radicalists could not develop an appropriate model about population resource in an integrated dynamic way to remove inter-regional and intra-regional inequalities; the radicalists gave priority to time over space. Thus, despite Marxian leaning the radicalists could not liberate human beings, especially the oppressed class from the natural and social restraints. In the process of creation of space, and man and environment relationship, the followers of radicalism in geography ultimately realized that eradication of social injustice and removal of regional inequalities from both the capitalistic and socialistic societies is a gigantic task which demands deep thinking and more empirical research.

Humanism

Humanism is a tradition of thought which forms the basis to humanistic geography[12]. The emergence of humanism in geography in the 1970s, as that of behvioural and radical geography, owed much to a deep dissatisfaction with the more mechanistic model developed during the 'quantitative revolution'. In recent years, the geographers have begun to give greater attention to environmental issues, the plight of the world's indigent, and other global humanitarian concerns, strengthening the connections between the humanities and geography. Geographers became increasingly active in applying their skills of spatial analysis to finding equitable solutions to the many problems that now confront humankind. Early positivist science or spatial science was based on statistical values, where human subjects react to their environment more or less in the same way using reason. While spatial science had little regard for individuals, who were nothing more than dots and lines on a chart, the humanist epistemology sees that studying the human world is very different from studying the physical, biological or material world. The contemporary 'humanist' approach in geography has received increasing attention from geographers and has paralleled the growth of humanist perspective in other human sciences.[13] The humanist approach in geography is best illustrated in the works of geographers such as Tuan (1976), Buttimer (1976), Relph (1981) etc. An amalgamation of the quantitative and qualitative methodologies seems to be necessary for a reliable interpretation of man, space, place and environment relationship. This kind of thinking led to Humanism. Humanism also grew as a criticism against positivism and quantification in geography.

Humanist tradition is often traced back to the intellectual shift in Europe during "Renaissance": The shift was about replacing the medieval belief with God as a centre of cosmic order by a world-view in which belief in schooling, cultural achievement and sciences gave a more central place to human beings and their intellectual capabilities. This led to the invention of the notion of the "human subject" as something active and basic to human action and thought[14]. The cultural and historical geographers attacked the positivism from the early 1970s. In fact, it was a rejection of the geometric determinism in which men and women were made to respond automatically to the dictates of universal spatial structures and abstract laws. The followers of spatial science (positivists) treated people as dots on a map, data on a graph and number in an equation. It was at the same time a claim for a human geography was made with the human being at its very centre, a people's geography, about the real people and for the people to develop human being for all.

One of the first geographers with his advocacy for humanistic approach was William Kirk (1951)[15]. But, it was Yi-Fu-Tuan (1976) who used the term 'humanistic geography' for the first time and argued for it[16]. For Tuan, Humanistic geography was the perspective that disclosed the complexity and ambiguity of relations between people and place (man and environment). During this time, human geography came to be viewed as 'people's geography', i.e. human beings are placed at the centre of the discipline. So, social construction of space, place and landscape became a dominant tradition during the 1970s. As a result, various humanistic approaches (as Marxism, Existentialism, Phenomenology, Functionalism, Pragmatism, feminism etc.) came into existence.

Deep in antiquity, humanism in geography is traced back to the works of French geographers, especially Febvre, Blache and Emmanuel Kant. Its real beginning is attributed to the Kantian philosophy, followed by French and American Schools of Thought. Kant asserted that: "History differs from geography only in consideration of time and space. The former is a report of phenomena that follow one another and has reference to time, the latter is report of phenomena beside each other in space. History is a narrative, geography a description. Geography and history fill up the entire circumference of our perception: geography that of space, history that of time. The French school of Possibilism advocated the view that the physical environment provides the opportunity for a range of possible human responses and that people have considerable opportunity to choose between them. The possibilists emphasized that 'nature is never more than an adviser' and that milieu interne revealed the human being as 'at once both active and passive'. Vidal de la Blache's writings, however, bear many of the hallmarks of functionalism and pragmatism. Carl Sauer, an American scholar, wrote of the phenomenology of landscape in 1925[17]. In 1936, Wooldrige claimed that historical geography must seek to view the countryside through the eyes of the farmer[18]. In 1947, John Wright introduced the term 'geosophy' as part of his contention that geographical knowledge is part of the mental stock of all humans[19]. In 1939, Hartshorne pleaded the cause of humanistic geography in his book: 'Nature of Geography'[20]. He accepted that geography's basic task was essentially Kantian. Subsequently, it was Kirk (1951) and Tuan (1976) who laid the foundation of humanism in geography.

Humanistic geography studies human awareness and human agency, human consciousness and human activity. It, therefore, deals with the meaning, value and human significance of life events. Contemporary humanism in geography emphasizes the study of meanings, values, goals and purposes. Within this humanistic perspective concepts of traditional significance in geography are given new meanings. For example, place is defined as a focus of human emotional attachment. Humanistic geographers argue that their approach deserves the appellation 'humanistic' in that they study the aspects of man which are distinctively 'human'. Humanists are not in favour of reducing space to mere geometrical concepts of surface as viewed by the positivists. From a humanistic point of view, space is inseparable from consciousness of humans who inhabit it.[21] Humanism is a rejection of the 'geometric determinism' which views human beings as always dictated by 'universal spatial structures and abstract spatial logics'. The basic objection of humanists against quantitative revolution is that its tools and assumptions do not adequately explain human world and human issues, especially relating to social institutions, attitudes, morals, customs, traditions and aesthetics.

Precisly, Humanistic geography is characterised by four subjects: Human awareness, human agency, human consciousness and human creativity. It is an attempt at understanding meaning, value and the human significance of life events. In the humanistic strand, the intent has been to understand and recognize the dignity and humanity of the individual. The focus of humanistic geography is on people and their condition. Humanists explain and interpret man and space relationship with historical approach. Humanism does not treat humans as machines. It is a subjective approach which aims at an understanding of man in his environment. Humanism is a conviction that men and women can best improve the circumstances of their by thinking and acting for themselves, and especially by exercising their capacity for reason (Relph, 1981). Humanist geographers propose that reasoning in humanistic geography should conserve contact with the world of everyday experience and recognize, if not celebrate, the human potential for creativity. The followers of this approach consider geography as "the study of the earth as home of man." Humanistic geography, as such, is not an earth science in its ultimate aim. It belongs to the humanities and to the social sciences to the extent that they all share the hope of providing an accurate picture of the human world. In fact, in humanities knowledge of human world is acquired by examining social institutions. These institutions can be viewed both as example of human inventiveness and as forces limiting the free activity of individuals. Humanistic geography achieves an understanding of the human world by studying people's relation with nature, their geographical behaviour as well as their feelings and ideas in regard to space and place. Humanists reject the reduction of space and place to geometrical concepts of surface and point as perceived and preached by the positivists through the methodology of quantitative techniques.

Place (landscape, region) is a key concept in humanistic geography. Much humanistic writing is devoted to illustrating and clarifying space. From a humanistic perspective, the meaning of a place (landscape, region) is inseparable from the consciousness of those (people) who inhibit it. The scope of place as concept varies according to the extension of the thoughts, feelings and experiences that make the consciousness of the inhabitants. In humanistic geography, central importance is given to the actor's (man's) definition and behaviour for examining the social world. The researcher needs to discover the actor's definition of the situation, namely, his or her perception and interpretation of reality and how these relate to behaviour. In other words, the researcher must be able to see the world as the actor sees it.

The proponent of humanistic geography (Tuan 1976), explored five themes of general interest of geographers, namely: geographical knowledge (personal geographies), territory and place, crowding and privacy, livelihood and economies, and religion. Man is the superior form of life and has special capacity for thought and reflection. The primary task of humanistic geographers, therefore, is the study of articulated ideas (geographical knowledge). In general, broadly conceived knowledge of geography is necessary to biological survival. All animals must have it, and even the migratory birds have a mental map. All living beings, including humans, occupy and use space, and have a sense of territory. Much more than animals, man develops an emotional attachment to place as he satisfies his biological needs (drinks, eat and rest). How mere space becomes an intensely human place is a task of geographer to study. Due to this reason, during the initial years, both behavioural geography as well as humanistic geography were inseparable in their approach. But humanistic geography soon established itself as a distinct identity because of its emphasis on the 'essential subjectivity of both the investigator and the investigated'. As another theme, Crowding of a place leads to physical and psychological stress. It has been observed that the human behaviour becomes abnormal at a crowded place. Culture, social institutions and infrastructures, however, help in reducing these stresses. For example, people in crowded Hong Kong are no more prone to crime than people living in relatively spacious American, European and Australian cities. Contrary to this, in the Kalahari Desert, the Bushman are crowded by choice, and biological indicators of stress are absent despite of high density of places where water is available. Similarly, privacy and solitude also affect the thinking process and decision-making of a person regarding space. In solitude, a person creates his own world. All people need privacy; the degree and kind may vary. Then, man sustains himself by doing some economic and social activities. All human activities appear to be economic and functional in the sense that they support the social system outside of which people cannot live. While working for his livelihood, man differentiates between life-sustaining and life-destroying activities. Production of armaments, for example, is an economic activity that provides a livelihood for many workers, but its contribution to the survival of the species is in doubt. All people and professional planners plan their economic activities according to their knowledge and technology. 'To what extent do planners make use of economic theory and facts in reaching the decision' or 'how good are the such questions need to be asked by the humanistic geographers. Lastly, religion is present at varying degrees in all cultures. It appears to be a universal trait that distinguishes man from other life forms. A humanistic approach to religion would require that we should be aware of the differences in the human desire for coherence, and not how these are manifest in the organization of space and time in attitude to nature or physical environment.

This approach has been criticized on several grounds. It has been claimed (Latour, 1993) that there are phenomena which cannot explained by the humanistic approach. For instance global warming, genetic engineering and deforestation cannot be understood properly within the confines of humanist orientations. Accordingly, this approach is hopelessly asymmetrical and one-sided in giving value, importance and causal power exclusively. Further, the humanistic approach ignores the importance of networks and relations (Aitken & Valentine, 2006, p. 34). A general criticism of human geography is that the investigator can never know for sure whether one has actually succeeded in providing true explanation. Undoubtedly, one can never know with certainty that a humanistic explanation is true. On methodological grounds it separates physical geography from human geography. In physical geography, the scientific techniques can be applied for theory and model building and hypothesis testing as it mainly deals with non-living objects. Contrary to this, in human geography, such quantitative techniques may not give the authentic and reliable results as the behaviour of man varies in space and time. The dichotomy of physical and human geography is thus harmful for the growth and development of the discipline. In humanistic geography which is largely based on participant observation, it is difficult to develop theory, abstraction, generalization and spatial geometry. Thus, it has no sound and valid methodological base as it involves more subjective than objective research. There is insignificant emphasis on the applied research. For example, it does not give emphasis on applied research or policy related to the location of industry, locational analysis of land use and crop intensity. The indifference towards applied research may destroy the base of the subject. Humanistic geography does not offer a viable alternative to, nor a pre-supposition less basis for, scientific geography. Rather, the humanist approach is best understood as a form of criticism (Entrikin, 1976). Ultimately, the Humanistic approach has been considered 'methodologically obscure'. The goal of understanding man's meaningful experience seems to lead to a situation in which any method is acceptable. It is not a practical philosophical as it involves thinking rather than practical activity. Its methodology is eclectic and sources of interpretation are numerous and therefore it becomes difficult to ascertain the reality.

Most of the criticisms of humanistic geography are, however, ill-founded. Nonetheless, six common themes of critical geography behavioural, radical and humanistic) have been identified[22] encompassing: a commitment to theory and a rejection of empiricism, a commitment to reveal the processes that produce inequalities, an emphasis on representation as a means of domination and resistance, an optimistic faith in the power of critical scholarship, a commitment to progressive practices, a claim to be united with social movements and activists with commitments to social justice and an understanding of space as a critical tool. Critical geographers consciously deploy theories of some form, but they draw from a variety of theoretical wells, such as political economy, governmentality, feminism, anti-racism, and anti-imperialism. Critical geographers seek to unveil power, uncover inequality, expose resistance, and cultivate liberating politics and social changes. A common focus of critical geography is to study how representations of space sustain power; or on the contrary, how representations of space can be used to challenge power. Critical geographers believe that scholarship can be used to resist the dominant representations, and scholar can undo power domination and free the oppressed. There exists an implicit confidence in the power of critical scholarship to reach the ignorant, and in the capacities of people to defeat alienation by means of reflexive self-education. Critical geographers pay special attention to how spatial arrangements and representations can be used to produce oppression and inequality. Critical geographers identify knowledge based upon intuition and description of 'direct' experience.

NOTES

1. Agger, Ben (1988).
2. Aitken, S. and G. Valentine (Ed.) (2006), p. 338.
3. Cloke, P., Philo, Ch. and Sadler, D. (1991), pp. 66-67.
4. Julian Wolpert was a professor of Geography, Public Affairs and Urban Planning at the Woodrow Wilson School of Public and International Affairs, Princeton University, U.S.A.
5. Rana, L. (2008), pp. 276-290.
6. Rana, L. (2014), pp. 284-299.
7. Hussain, M. (1995), p. 308.
8. Gold, (1980), p. 4.
9. Gould, P.R. (1966), pp. 182-220.
10. Jenny Pickerill, (2017).
11. Olson, E. and Sayer, A. (2009).
12. Cloke, P., Philo, C., Sadler, D. (1999), p. 59.
13. Entrikin, J. Nicholas (1976), p. 615.
14. Ibid., p. 61.
15. Kirk, W. (1951). "Historical geography and the concept of behavioural environment," pp. 152-160 in Silver Jubilee Souvenir and N. Subrahmanyam Memorial Volume. Madras: Indian Geographical Society.
16. Tuan, Yi-Fu (1976). "Humanistic Geography", *Annals of the Association of American Geographers*, **66**. pp. 266–276.
17. Sauer, C.O. (1925). *The Morphology of Landscape*. University of California Publications.
18. Wooldrige, S.W. and W.G. East (1951). *The Spirit and Purpose of Geography, Hutchinson*, London.
19. Lowenthal, D. and Bowden, M.J. (Editors) (1976). *Geographies of the Mind: Essays in Historical Geosophy In Honour of John Kirtland Wright.*
20. Hartshorne, R. (1939). *The Nature of Geography: A Critical Survey of Current Thought in the Light of the Past*, Association of American Geographers, Lancaster.
21. Tiwari, P., "Principles of Humanism and Themes in Humanistic Geography", *Wikipedia*.
22. Holt-Jenson, A. (2009), p. 121.

REFERENCES

Agger, Ben (1988). *Critical Social Theories: An Introduction*. Avalon Publishing. ISBN 9780813321745.

Aitken, S. and Valentine, G. (Eds.,) (2006, 2009). *Approaches to Human Geography*. Sage: California, London, New Delhi, Singapore.

"Behaviouralism: It's Objectives, Salient Features and Historical Perspective | Geography", *Wikipedia, the free encyclopedia.*

Best, U., "Critical Geography" in Kitchin, Rob, Thrift, Nigel, (eds.) (2009). *International Encyclopedia of Human Geography.* Vol. 2, pp. 345-347, Oxford: Elsevier. ISBN 9780080449104.

Blomley, Nicholas (2006). "Uncritical critical geography?". *Progress in Human Geography.* 30 (1): 87–94.

Boulding, K.E. (1956). *The Image: Knowledge in Life and Society,* University of Michigan Press.

Buttimer, Anne (1976). "Grasping the Dyanism of Life World", *Annals of the Association of American Geographers,* 66. pp. 277–292.

Cloke, P., Philo, Ch. and Sadler, D. (1991). *Approaching Human Geography.* Chapman, London.

Entrikin, J. Nicholas (1976). "Contemporary Humanism in Geography", *Annals of the Association of American Geographers,* Vol. 66, No. 4, (December, 1976), pp. 615-632.

Fuller, Duncan; Kitchin, Rob (2004). 'Radical Theory/critical Praxis: Making a Difference Beyond the Academy?'. *Praxis (e) Press.* ISBN 9780973456103.

Fay, Brian (1987). 'Critical social science: liberation and its limits'. *Cornell University Press.* ISBN 9780801494581.

Gold, J.R. (1980). *An Introduction to Behavioural Geography,* Oxford University Press.

Golledge, G., and Timmermans, H. (1990). 'Applications of behavioural research on spatial problems'. *Human Geography,* Vol. 14.

Gould, P.R. (1966). "On Mental Maps", Michigan Inter-University Community of Mathematical Geographers, Discussion Paper 9, reprinted in Downs, R.M. and D. Stea (Eds.) (1973), *Image and Environment,* Edward Arnold, London.

Gould, P. and White, R. (1974). *Mental maps,* Harmondsworth: Penguin Books.

Gregory, Derek (1994). *Geographical Imaginations.* Blackwell. ISBN 0631183310.

Hussain, M. (1995). "Paradigm in Geography", *Evolution of Geographic Thought,* New Delhi: Rawat Publications, pp. 300-340.

Harald, B., Engel-Di Mauro (2008). *Critical Geographies: A Collection of Readings,* Praxis (e) Press, Critical Topographies Series.

Harvey, D. (1973). *Social Justice and the City,* London: Arnold.

Harvey, David (2000). *Spaces of Hope.* University of California Press. ISBN 9780520225787.

Holt-Jenson, A. (2009). "Paradigms and Revolutions", *Geography: History and Concepts: A Systematic Guide,* Fourth Edition, London: Sage, pp. 100-124.

Kirk, W. (1951). "Historical geography and the concept of behavioural environment," pp. 152-160 in Silver Jubilee Souvenir and N. Subrahmanyam Memorial Volume. Madras: Indian Geographical Society.

Latour, B. (1993). *We Have Never Been Modern,* Trans. C. Porter, Cambridge: Harvard University Press.

Livingstone, David (1992). *The Geographical Tradition: Episodes in the History of a Contested Enterprise,* Blackwell.

Norton, W. (1997). "Human geography and behaviour analysis: An qapplication of behaviour analysis to the evolution of human landscapes." *The Psychological Record,* 47, 439–460.

Olson, E. and Sayer, A. (2009). *Radical Geography and its Critical Standpoints: Embracing the Normative.* Antipode, 41: 180-198.

Peet, J. Richard, (Ed.) (1977). *Radical Geography: Alternative Viewpoints on Contemporary Social Issues.* Maaroufa Press.

Peet, Richard (1998). *Modern Geographical Thought,* Blackwell.

Peet, Richard (2000). "Celebrating Thirty Years of Radical Geography", *Environment and Planning* A. 32 (6): 951–953.

Pickerill, Jenny (2017). 'Radical Geography', Published Online: 6 March, 2017, DOI: 10.1002/9781118786352.wbieg0506.

Rana, L. (2008). *Geographical Thought: A Systematic Record of Evolution,* New Delhi: Concept Publishing Company.

Rana, L. (2014). *Geographical Thought: Classical to Contemporary,* New Delhi: Concept Publishing Company.

Relph, E. (1981). *Rational Landscapes and Humanistic Geography.* New York: Barnes and Noble.

Saraswati Raju, Alex Jeffrey (2017). 'Critical Geography', *The International Encypedia of Geography,* Published Online: 6 March, 2017.

Stamp, D. (1966). "Ten Years On", *Transactions, the Institute of British Geographers,* Vol. 40 : 11-20.

Stoddart, D.R. (1986). *On geography and its history*. New York: Blackwell.

Tiwari, P., "Principles of Humanism and Themes in Humanistic Geography", *Wikipedia.*

Tuan, Yi-Fu (1976). "Humanistic Geography", *Annals of the Association of American Geographers,* 66. pp. 266–276.

22
Digital Revolution (Crisis Phase VIII)

With the widespread introduction of computer technology, geography has entered another revolutionary phase (Crisis Phase VIII) in the twenty first century. Such a change had been brought about by the Quantitative Revolution in 1950s, and now it is Digital Revolution. In fact, the latter rejuvenated the spirit of quantification among geographers about half a century after the Quqntitative Revolution. However, this change had come in the West in late 1950s itself, but the revolution becomes universal in the present times. The adoption and proliferation of digital computers and digital record keeping have become more widespread now. The modern techniques have opened up new avenues for geographers, and with their help it is possible to collect and interpret a variety of information about the earth's surface. Since geographers are directly concerned with analyzing the variable data on the earth's surface, this has unwrapped vast areas for analysis and research.

The Revolution

The Digital Revolution refers to the advancement of technology from analogue to the digital. Implicitly, the term refers to the sweeping changes brought about by digital computing and communication technology during (and after) the latter half of the twentieth century. The Digital Revolution marks the beginning of the Information Age. Central to this revolution is the mass production and widespread use of digital logic circuits, and its derived technologies, including the computer, digital cellular phone and the Internet. These technological innovations have transformed traditional practice of data collection, presentation and analysis.

The development and advancement of digital technologies started with one fundamental idea, i.e. the Internet. The Digital communication became economical for widespread adoption after invention of the personal computer. The government, military and other organizations began making use of computer systems during 1950s and 1960s itself. This research eventually led to the creation of the World Wide Web[1]. As digital technology proliferated, the switch from analogue to digital record keeping became the new standard in various fields, including business and academics alike. During 1980s, the computer became a familiar machine and by the end of the decade, being able to use one became a necessity. By 1992, the World Wide Web had been introduced, and by 1996 the Internet became a normal part of most business and academic operations. By the late 1990s, the Internet became a part of everyday life for a large number of people, especially in the developed world. The digital revolution became truly global after revolutionizing society in the developed world in the 1990s. By 2000s the Digital Revolution had begun to spread all over the developing world as well as the mobile phones were commonly seen and the number of Internet users continued to grow. The connection between Internet websites and mobile gadgets has now become a standard in communication. In the decade 2010, nearly 70 per cent of the world's population owned a mobile phone. Thereafter, the innovation of tablet computers has far surpassed personal computers with the use of the Internet and the promise of cloud computing services. By 2012, over 2 billion people used the Internet, twice the number using it in 2007. Along with tablet computers and smart phones, Cloud computing had entered the mainstream by the early 2010s[2]. In the late 1980s, less than 1 per cent of the world's technologically stored information was in digital format, while it was 94 per cent in 2007, with more than 99 per cent by 2014. The year 2002 is estimated to be the year when human kind was able to store more information in digital, than in analogue format, the "beginning of the digital age".[3] It is assessed that the world's capacity to store information has increased from 2.6 (optimally compressed) exabytes in 1986, to some 5,000 exabytes in 2014 (5 zettabytes). Today we live in an age of easy global communications, instantaneous information access, and the ability to store, search and manipulate large amounts of information. The Information and Communications Technologies (ICT) sector represents a variety of devices, technologies and services built on scientific breakthroughs in computers, software design, photo-optics, circuit switching, and satellites, among others. The most visible representation of the ITC Revolution is the Internet, which integrates telecommunications and computing. The Internet allows anyone with a small capital investment in a computer and access to a server to connect to other computers all over the world in an interconnected web of machines, data and people. The number of messages sent *via* email is currently about 6.5 billion per day worldwide and growth will be exponential as more users come on-line[4]. In addition, the technology is going wireless; further decreasing the costs of being connected and potentially bringing more people into the digital world.

The most recent development, in this direction, is geo-computation (Longley *et al.*, 1998). Geo-computation is the process of applying computing technology to geographical problems, i.e. the ways in which new computational tools and methods are used in the depiction of spatial process[5].

Impact

History has witnessed the search for ways to transmit information over distance greater than range of the human voice, at lower cost and with improved accuracy. Beginning with the advent of the written word and continuing with the invention of the printing press, information and communication technologies (ICT) have changed in rapid succession, from the telegraph, radio, telephone, fax and most recently the Internet. While previously individuals had to be face to face to communicate, technology makes it possible to exchange information at great distance, and enables the transmission, storage and access of more advanced and complex context. The result is a decrease in the cost of communication and correspondingly, economic relationships become easier to negotiate across geographic space. Geographically disparate activities may more easily be connected. Geographic separation no longer implies information deprivation.

By way of conversion of technology from analogue into a digital format, the information may be more easily collected, stored, processed, accessed and communicated. Digital information is compact, transportable, and, therefore, more efficient to use. It became possible to make copies that were identical to the original. In digital communications, for example, repeating hardware was able to amplify the digital signal and pass it on with no loss of information in the signal. Of equal importance to the revolution was the ability to easily move the digital information between media, and to access or distribute it remotely. Positive aspects include greater interconnectedness, easier communication, and the exposure of information. As a result of the widespread impact of Digital Revolution, various analogue technologies, as e.g. Mail (parcel), Telegram, Typewriter, Fax, Landline phone etc. are fast declining or disappearing.

There have been huge benefits to society from the digital revolution, especially in terms of the accessibility of information. Expanded powers of communication and information sharing, increased capabilities for existing technologies, and the advent of new technology brought with it many potential opportunities for exploitation. The digital revolution helped usher in a new age of mass surveillance, generating a range of new civil and human rights issues. The digital revolution made it possible to store and track facts, articles, statistics, as well as minutiae hitherto unfeasible.

The economic impact of the digital revolution has, particularly, been large. Without the World Wide Web (WWW), for example, globalization and outsourcing would not be nearly as feasible as they are today. The digital revolution radically changed the way individuals and companies interact. Evidence is mounting that digital technologies have significantly increased the productivity and performance of businesses.

Relevance in Geographical Studies

The computer has drastically transformed both the world of geography as an academic discipline and the geography of the world in which we live. From being a tool for geographers to collect, analyze, map and visualize data since 1950s, the computers have increasingly become an integral part of the geographers' study by the end of twentieth century. Computers are crucial in the history of geography and technology. They have enriched the discipline of geography with the development of automated geography, GIS, and the virtual geography department.[6] Computers and the computer-led information revolution have changed both scientific practices and many facets of human society in fundamental ways, as evidenced by the increasing emphasis of mapping and visualization as a discovery tool (Hall, 1993; Chen, 2002), and the maturing information society led by the digital economy (Malecki, 2009). The computers now penetrate almost every aspect of human lives. It is under this broad scientific and societal milieu that geographers entered the twenty first century.

The traditional geography has been radically changed by the past fifty years of the computer age, and especially by the last twenty years. Computers have fundamentally changed not only how geographers study the world, but also what we study, as spaces and places; the natural and human landscapes are constantly being refigured by computers. The increasing use of computers has allowed geographers faster processing of larger sets of data. The enhanced speed and capacity of computers allow and encourage the use of datasets that were simply unmanageable and tedious only twenty years ago. Especially in physical geography (e.g. huge satellite imagery datasets, such as Landsat 1972) but also in human geography (very large census files), far more comprehensive datasets can be used and far more complex models can be run (MacDougal, 1976). Although coding of qualitative data (recorded interviews, answers to open-ended survey questions) remains slow, computers have vastly eased through direct computer answering of surveys, data entry and analysis, and textual search.

The increasing use of computers has also contributed to the creation of vastly more data with better spatial and temporal resolutions. The digital processing of satellite imagery began as early as 1965 in the military, but has exploded in the last decade with the availability of high-quality, frequently updated images of most parts of the earth's surface. These images permit the analysis of change and processes, both short-term (weather, traffic) and long-term (physical and human land use), and also allow recognition of complex and previously hidden patterns revealing geological and historic processes. By the late 1990s, satellite images with 1-metre resolution were commercially available for civilian applications. Also access to the vast amounts of spatial and non-spatial data has become unbelievably easy and convenient. With faster processing of more data comes the capacity to estimate ever more complex and distinct relationships and structures in data and to design and test far more comprehensive models of behaviour. It is now possible to execute astoundingly intricate and useful simulations of the evolution of human landscapes, both past reconstructions and future scenario development (Batty and Longley, 1994). The geo-computation, especially in data mining and knowledge discovery techniques (Miller and Han, 2001), have pushed the computing capabilities for spatial data to a new height.

Computers have made writing, editing and publishing much faster and easier. Computers have extended geographers' eyes (*via* remote sensing and air photos), their hands (*via* computer-assisted cartography), their mouth (*via* the World Wide Web and telephones), and their minds (*via* data processing and artificial intelligence). These extensions have greatly facilitated the acquisition, visualization, processing and communication of unprecedented amounts of geographic data. In addition, rapid innovations have fostered the development of specialized instruments and software tools that can greatly facilitate geographic research. For example, GPS, laptops, portable digital assistants (PDA), handheld computers, digital cameras, portable scanners and electronic data loggers all raise the efficiency of fieldwork as well as library retrieval.

Computers have not simply automated the manual processes geographers were working on for years; they have also stimulated new lines of inquiry, accelerated the quest for laws and theories, and addressed serious spatial issues in statistics, including the modifiable areal unit problem (Openshaw, 1994), spatial autocorrelation (Cliff, 1970), and development of new spatial statistical techniques (Getis and Ord, 1992; Fotheringham *et al.*, 2000). Computers have also encouraged the growth of new branches of geography, the most visible of all being the birth of quantitative geography (Barnes, 2003), computer-assisted cartography, analytical cartography, remote sensing, and GIS. With the maturing of GIS technology and the further convergence among remote sensing, global positioning systems (GPS), and computer cartography, visualizing spatial data and mapmaking have been made much easier. The proliferation of GIS and the increasing use of geo-referenced data have made the broader social science community recognize the key role that space plays in human society.

In reviewing the application history of computers in geography, it becomes apparent that computers not only helped make geography a more respected discipline among social scientists, but also that geographers, as they enter the twenty first century, are beginning to play the leadership role among social scientists in addressing important social issues that have a spatial dimension. So dependent have we become so quickly on the web for information, and for data files, maps, and photos to download for our research and teaching, that we can scarcely remember that this has been possible for only seven or eight years. Hundreds of maps, created by scores of agencies or organizations can be downloaded and printed. Geographers and geography departments across the world, especially those in the U.S., have been quick to tap into the vast potential of this new telecommunication medium, culminating in the ambitious project to build a Virtual Geography Department (VGD). With the growth in routine use of the Internet (such as e-mail and web surfing) among geographers, and with more and more courses, journals, books, and data available on-line, a new geography department emerged around 1997, the Virtual Geography Department (VGD). The goal of the VGD is to produce and disseminate geographic knowledge and geography course materials.

This new world, different from the physical, tangible world geographers studied for thousands of years, is virtual, digital, and ephemeral. In this virtual world, everything geographers study has become bits of information, which can be transmitted across the globe instantaneously (Brunn, 1998). Each one of us is not only an information processor, but also information processed. In other words, we have literally become digital individuals: our identity is more and more equated by digital information such as Social Security and credit card numbers, multiple ID and PIN numbers. Indeed, even the entire earth is now becoming digital as embodied in the concept and initiative of digital earth (www.digitalearth.gov). This new digital virtual world itself provides geographers with a fascinating new subject of study (Kotkin, 2001; Kellerman, 2000). Geographers have tried to visualize this basically invisible world of cyberspace using various mapping techniques (Dodge and Kitchin, 2000); explored the utility of classic location theory in analyzing distributed computing facilities and e-commerce (Goodchild, 2000); presented new results on how economic geography as defined by the Internet's infrastructure contributes to the urban competitiveness among cities (Malecki, 2009); explored Environmental impacts of the emerging digital technology (Wilsdon, 2001); tackled issues of accessibility in the age of Internet (Janelle and Hodge, 1999); studied the overall impacts of the digital revolution on the landscape (Kotkin, 2001) and the spatial distribution of economic activities (Brunn and Leinbach, 2001).

From these empirical studies and theoretical explanation, the computers have had and will continue to have enormous impact on society, on human behaviour and interaction, on human settlement and the character of places, on the meaning of distance and of regions, and on the degree of spatial variation and inequality across the landscape (Wilson, 2001). As a consequence, the content of geography has changed, for example, *via* the very concept of globalization, space-time compression, and space-time distanciation (Giddens, 1984). The rapid, cheap worldwide transfer of messages, documents, data, and images has brought about yet another shrinking of distance and abetting of globalization in all its economic and cultural forms.

The Digital Earth

The most ambitious concept to exemplify the potential of computers both as tools and as reality is the concept of "Digital Earth." Digital Earth is a virtual representation of the Earth that is geo-referenced and connected to the world's digital knowledge archives[7]. It enables to access vast amounts of scientific and cultural information to help them understand the Earth and its human activities.[8] According to the 6th International Symposium on Digital Earth (held in September 2009 at Beijing), the Digital Earth is an integral part of other advanced technologies including: earth observation, geo-information systems, global positioning systems, communication networks, sensor webs, electromagnetic identifiers, virtual reality, grid computation etc. It is seen as a global strategic contributor to scientific and technological developments, and as a catalyst in finding solutions to international scientific and societal issues.[9] Digital Earth is believed to play a strategic and sustainable role in addressing such challenges to human society as natural resource depletion, food and water insecurity, energy shortages, environmental degradation, natural disasters response, population explosion, and, in particular, global climate change.

Several key elements of the Digital Earth have been suggested[10], viz.:

- The Digital Earth is not one, but it is multiply connected globes/infrastructures addressing the needs of different audiences: citizens, communities, policymakers, scientists, educationalists;
- It is problem oriented, e.g. environment, health, societal benefit areas, and is transparent on the impacts of technologies on the environment;
- It allows search through time and space to find similar/analogue situations with real time data from both sensors and humans (different from what existing GIS can do, and different from adding analytical functions to a virtual globe);
- It asks questions about change, identification of anomalies in space in both human and environmental domains;
- It enables access to data, information, services, and models as well as scenarios and forecasts: from simple queries to complex analyses across the environmental and social domains;
- It supporting the visualization of abstract concepts and data types (e.g. low income, poor health, and semantics);
- Based on open access, and participation across multiple technological platforms, and media (e.g. text, voice and multi-media); and
- It is engaging, interactive, exploratory, and a laboratory for learning and for multidisciplinary education and science.

The digital earth can be envisioned as "a multi-resolution, three-dimensional representation of the planet, into which we can embed vast quantities of geo-referenced data". Digital Earth is inherently an intriguing concept for geographers. First, it has some of the properties of a moonshot, or a vision that can motivate a wide range of interdisciplinary research and development activities. Digital Earth challenges our state of knowledge about the planet, not only in terms of raw data, but also in terms of data access and the ability to communicate data through visualization. Moreover, it challenges our understanding of process in the invitation to model, simulate, and predict, since the concept is not limited to static portrayal. Second, Digital Earth is interesting because of its implications for the organization of information. The prevailing metaphor of user interface design is the office or desktop, with its filing cabinets and clipboards. Many prototype digital libraries employ the library metaphor, with its stacks and card catalogs. But Digital Earth suggests a much more powerful and compelling metaphor for the organization of geographic information, by portraying its existence on a rendering of the surface of the

Earth. This idea can be seen in limited form in many current products and services, including Microsoft's Encarta Atlas. Third, at a much deeper level, Digital Earth may be regarded as a materialization of what has been called the 'Noosphere', dominated by flows of digital information.[11] Together with geographers' focus on the atmosphere, hydrosphere, and biosphere, the study on the geography of Noosphere could be a new challenge for geographers in the twenty first century.

The empirical studies have shown that ICT (Information and Communication Technology), particularly Internet, has affected the locational patterns of economic activity. The Internet has facilitated information exchange and resulted in management efficiency and greater consumer choice, particularly in economic sector. The productivity impacts have been profound and promise to continue. The tendency for innovative activity to cluster spatially is well known and identified with places such as Silicon Valley, Research Triangle Park, Route 128 in Boston in the United States, Wireless Valley in Finland, Oxbridge in England, and Singapore, among others. Economists have been studying this phenomenon since Marshall (1890) and more recently, have considered the reasons why high rent and high wage places such as New York, San Francisco, London, Paris and other prominent cities continue to grow.

Joel Kotkin[12], renowned economic and social-trend forecaster, discusses how the digital revolution is changing where and how we live and work in the bricks-and-mortar world. Historically unprecedented forces are at work buffeting cities, suburbs and towns across the country. In his *The New Geography*, Kotkin focuses on the digital revolution's surprising impact on cities: their traditional role as the centres of creativity and the crossroads for trade and culture is becoming ever more essential in a globalized information-age economy. He also identifies two new species of communities: *Nerdistans*: high-end, self-contained, office park-oriented suburbs, built to be attractive to a certain class of techie, and *Valhallas*, wealthy rural enclaves for information-age plutocrats. *The New Geography* is a brilliant beachhead onto a subject that affects us all.

The spatial analysis through digital mapping can transform education and society through better decision-making using the geographic perspective (Kerski, 2015)[13]. This may be best accomplished in the various areas of geographic interest, as Rural, Urban, more effectively Economic, Land-Use, and Population Issues. Such concepts as settlement, age, birthrate, growth rate, and human-environment interaction in Population Dynamics; zoning and land-use practices in Land Use; historical and current development of cities, site *versus* situation, and challenges facing cities in Urban Development; industrialization, energy, employment, and measures of development in Economic Geography; and watersheds, landform dynamics, and natural hazards in physical geography may be more effectively understood and studied with the help of new technology. Embedded in studying these issues are core geography themes such as considerations of scale in patterns and processes; interpreting maps and analyzing geospatial data; understanding and explaining the implications of associations, networks, and interconnections among phenomena in places; defining regions and the regionalization process; establishing a sense of place; and understanding the nature and limitations of geographic data.

The spatial perspective is critical to understanding the geographic content and processes. Being competent and confident in the application of the spatial perspective to geographic understanding enhances and strengthens geographers' ability to handle the discipline. A number of powerful web mapping and related tools such as ArcGIS Online (arcgis.com) and story maps (storymaps.arcgis.com) are now available for the geography instructor.

A geospatially literate population is the goal of geographers in the twenty first century. Such a population is better equipped to recognize, understand, and resolve those critical issues, whether local or global, that will confront us today and in the future. New technologies have always been important in advancing geographic understanding, but never have they been so thoroughly and rapidly transformative of the discipline as at this stage in geography's evolution. Just as new technologies have profoundly expanded both research possibilities and the knowledge base of other disciplines, such as biology, physics or medicine, so too are the revolutionary new geographic technologies developed during the past few decades extending frontiers in geographic research, education and applications. They are also creating new and resurgent roles for geography in both society and in the university. This trend is still accelerating, as the integration of geographic technologies, such as the global positioning system and geographic information systems (GPS/GIS), is creating an explosion of new "real-time, real-world" applications and research capabilities. The resultant dynamic space/time interactive research and management environments created by interactive GPS/GIS, among other technologies, places geography squarely at the forefront of advanced multidisciplinary research and modelling programmes, and has created core organization management tools (geographic management systems) which will dramatically change the way governments and businesses work in the decades ahead. As obvious, this crisis phase with revolution will lead to new paradigm phase.

NOTES

1. The World Wide Web (WWW), also called the Web, is an information space where documents and other web resources are identified by Uniform Resource Locators (URLs), interlinked by hypertext links, and accessible *via* the Internet. English scientist Tim Berners-Lee invented the World Wide Web in 1989.
2. "World Internet Users Statistics and 2014 World Population Stats". Retrieved 17 April, 2015.
3. Martin Hilbert (11 June 2011). "World_info_capacity_animation"–via *YouTube.*
4. Lange, L. (1999).
5. Simply defined, geocomputation is the process of applying computing technology to geographical problems. It just means the universe of computational techniques applicable to 'spatial problem'. However, GeoComputation involves combination of four leading edge technologies: GIS, which creates the data; Artificial intelligence (AI) and computational intelligence (CI), which provide smart tools; High performance computing, which provides the power; and Science, which provides the philosophy.
6. Litan, Robert, E., and Alice M. Rivlin (2001), pp. 313-317.
7. It is the name given to the concept by former US Vice President Al Gore in 1998; "The Digital Earth-Al Gore". *digitalearth-isde.org. 1998-01-31.*
8. Foresman, T.W. (2008). "Evolution and implementation of the Digital Earth vision, technology and society". *International Journal of Digital Earth.* **1** (1): 4–16.
9. *"The 6th International Symposium on Digital Earth-Digital Earth in Action".* 12-09-2009, Beijing.
10. Craglia, M. *et al.*, (2008). "Next-Generation Digital Earth", *International Journal of Spatial Data Infrastructures Research,* (3):146-167.
11. Noosphere is a postulated sphere or stage of evolutionary development dominated by consciousness, the mind, and interpersonal relationships. It is the sphere of human consciousness and mental activity especially in regard to its influence on the biosphere and in relation to evolution.
12. Kotkin, Joel (2002). *The New Geography: How the Digital Revolution Is Reshaping the American Landscape.*

13. Joseph Kerski is a geographer who serves on the ESRI education team and is active in GIS communication and outreach, creates GIS-based curriculum, conducts research in the effectiveness of GIS in education, teaches online and face-to-face courses on spatial thinking and analysis, and fosters partnerships to support GIS in formal and informal education at all levels, internationally. He is the co-author of Spatial Mathematics, The Essentials of the Environment, The GIS Guide to Public Domain Data, and other books.

REFERENCES

Armstrong, M. (2000). "Geography and Computational Science", *Annals of the Association ofAmerican Geographers,* 90: 146-56.

Balchin, W. and Coleman, A. (1967). "Cartography and Computers", *The Cartographer* 4: 120-27.

Barnes, T.J. (2003). "The Place of Locational Analysis: A Selective and Interpretive History", *Progress in Human Geography* 27: 69–95.

Brunn, S.D. (1998). "The Internet as the New World of and for Geography: Space, Structures, Volumes, Humility and Civility", *GeoJournal* 45: 5-15.

Brunn, S.D. and Dodge, M. (2001). "Mapping the Worlds of the World Wide Web: (Re) Structuring Global Commerce Through Hyperlinks", *American Behavioural Scientist* 44: 1716-39.

Brunn, S.D. and Leinbach, T.R. (1991). *Collapsing Space and Time: Geographic Aspects of Communication and Information*. London: Harper Collins Academic.

Burton, I. (1963). 'The Quantitative Revolution and Theoretical Geography", *Canadian Geographer* 7: 151-62.

Chen, C. (2002). *Mapping Scientific Frontiers: The Quest for Knowledge Visualization.* London: Springer.

Cliff, A. (1970). "Computing the Spatial Correspondence Between Geographic Patterns", *Transactions of the IBG* 50: 143-54.

Curry, M.R. (1998). *Digital Places: Living with Geographic Information Technologies*. London and New York: Routledge.

Daniel Sui, D. and Richard Morrill (2004). "Computers and Geography: From Automated Geography to Digital Earth", in S.D. Brunn, S.L. Cutter, and J.W. Harrington (eds.), *Geography and Technology*, pp. 81-108, Wikipedia, the free encyclopedia.

"Digital Revolution", *Wikipedia.*

Dobson, J.E. (1983). "Automated Geography", *The Professional Geographer* 35: 135-143.

Dobson, J.E. (1993). "The Geographic Revolution: A Retrospective on the Age of Automated Geography", *The Professional Geographer* 45: 431-39.

Dodge, M. and Kitchin, R. (2000). *Mapping Cyberspace*. London: Routledge.

Feldman, Maryann P., "The Internet Revolution and the Geography of Innovation", *Internet.*

Feldman, M.P. (2000). "Location and innovation: the new economic geography of innovation, spillovers, and agglomeration" in G. Clark, M. Feldman and M. Gertler, editors. Oxford Handbook of Economic Geography, Oxford: Oxford University Press.

Fortheringham, A.S., Brunsdon, C., and Charlton, M. (2000). *Quantitative Geography: Perspective on Spatial Data Analysis*. London: Sage.

Gaits, G. (1969). "Thematic Mapping by Computer", *The Cartographic Journal* 6: 50-68.

Getis, A. and Ord, J.K. (1992). "The Analysis of Spatial Association by Use of Distance Statistics, *Geographical Analysis* 24: 189-206.

Gilbert, D. (1995). "Between Two Cultures: Geography, Computing, and the Humanities", *Ecumene* 2: 1-13.

Goodchild, M.F. (2000). "Communicating Geographic Information in a Digital Age", *Annals of the Association of American Geographers* 90: 344-55.

Gould, P. (1970). "Computers and Spatial Analysis: Extensions of Geographic Research", *Geoforum* 1: 53-69.

Gould, P. (1985). *The Geographer at Work,* Boston: Routledge and K. Paul.

Hall, S.S. (1993). *Mapping the Next Millennium: How Computer-Driven Cartography is Revolutionizing the Face of Science*. New York: Vintage Books.

Hagerstrand, T. (1967). "The Computer and the Geographer", *Transactions of the Institute of British Geographers* 42: 1-20.

Haggett, P. (1969). "On Geographical Research in a Computer Environment", *Geographical Journal* 135: 497-505.

Harvey, F. (1997). "From Geographic Holism to Geographic Information System", *The Professional Geographer* 49: 77-85.

Irena, Bojanova (2014). "The Digital Revolution: What's on the Horizon?" *IT Professional (January-February. 2014). **16** (1):* pp. 8–12.

Janelle, D. and Hodge, D. (1999). *Information, Place, and Cyberspace: Issues in Accessibility.*, Berlin: Springer-Verlag.

Kellerman, A. (2002a). *The Internet on Earth: A Geography of Information*. New York: John Wiley.

Kerski, Joseph (2015). "Teaching Geography in the Twenty-First Century", *Internet,* January 26, 2015.

Kotkin, J. (2001). *The New Geography: How the Digital Revolution is Reshaping the American Landscape*. New York: Random House.

Lange, L. (1999). "The Internet." *IEEE Spectrum,* 36 (1), January: 35-40.

Longley, P., Brooks, S.M., McDonnell, R., and Macmillan, B. (1998). *Geocomputation: A Primer*. Chichester, John Wiley and Sons.

Litan, Robert, E., and Alice M. Rivlin, (2001). "Projecting the Economic Impact of the Internet." *American Economic Review,* 91 (2): 313-317.

MacDougal, E.B. (1976). *Computer Programming for Spatial Problems,* London: Edward Arnold.

Malecki, Edward J. (2009). *The Economic Geography of the Internet's Infrastructure,* Wiley.

Miller, H.J. and Han, J.W. (2001). *Geographic Data Mining and Knowledge Discovery*. London: Taylor and Francis.

Openshaw, S. (1994). "Computational Human Geography: Towards a Research Agenda", *Environment and Planning* A 26: 499-505.

Perez, C., (2004). "Technological Revolutions, Paradigm Shifts and Socio-Institutional Change" in E. Reinert, *Globalization, Economic Development and Inequality: An alternative Perspective,* Cheltenham: Edward Elgar, pp. 217–242.

Perez, C. (1983)."Structural change and assimilation of new technologies in the economic and social systems", *Futures,* Vol. 15, pp. 357–375.

Pitts, F. (1967). "Chorology Revisited", *The Professional Geographer* 14: 8-11.

Robinson, A. (1954). "Geographical Cartography", In James, P.E. and Jones, C.F. (Eds.) *American Geography: Inventory and Prospect*, 553-77. Syracuse, NY: Syracuse University Press.

Slocum, T.A. (1990). "The Use of Quantitative Methods in Major Geographical Journals, 1956-1986", *The Professional Geographer* 42: 84-94.

Tarrant, J.R. (1968). *Computers in Geography,* IBG Newsletter 6: 11-25.

Techopedia: "Digital Revolution".

Wilsdon, J.P. (2001). *Digital Futures: Living in a Dot-Com World,* London; Sterling, VA: Earthscan Publications.

Wilson, M.I. (2001). "Location, Location, Location: The Geography of the Dot Com Problem", *Environment and Planning* B: Planning and Design 28: 59-71.

Wright, J.K. (1947). "Terrae Incognitae: The Place of the Imagination in Geography", *Annals of the Association of American Geographers* 37: 1-15.

23
Post-Modern Tradition

Post-modernism brought with it the questioning of previous approaches of knowing, and introduction of new methods of study. Geography witnessed the maximum number of critical phases with revolution one after the other since 1950s, resulting in faster paradigmatic shifts in comparison to earlier phases of development. The nomothetic twist, emergence of spatial (locational) perspective and anthropocentricism have been instrumental in bringing about the paradigmatic change in geography in this period. The crises in the discipline, on the other hand, have been brought about by Schaefer-Hartshorne Debate, Quantitative Revolution, Critical Revolution and Digital Revolution. Interestingly, the Post-Modern period in Geography began with a Crisis Phase and the discipline is still undergoing another one.

Post-modernism in geography initiated with the criticism to chorological view and installation of more systematic and nomothetic perspective. The philosophy and methodology of geography were apparently debated most earnestly in the writings of two American scholars, namely Richard Hartshorne and Fred K. Schaefer during mid twentieth century. Geography's identity and methodology were called into question in this debate. Following Hartshorne, Regional geography was the major philosophical school during 1940s and 1950s. All of these events had presented a great threat to geography's position as an academic subject and thus geographers began seeking new methods to counter critique. Schaefer advocated the use of spatial laws as a basis for geographic explanation. Hartshorne's work, on the other hand, advocated for chorological (regional) approach in geography. However, the dissatisfaction developed with chorological perspective and the new generation of geographers was increasingly drawn to the systematic studies. Main reason was that the chorological viewpoint was promoting the concept of uniqueness of areas, neglecting theory and generalizations. The paradigm had become sterile in geography and was not yielding laws and theories. Schaefer criticized the claim made for geography as a chorological science, and put forward a strong case for the discipline to adopt the philosophy and methodology of scientific positivism. The essential difference between the positions taken by Hartshorne and Schaefer was that Hartshorne's was a positive view of geography, i.e. geography is what geographers have made it. Schaefer's view, on the other hand, was a normative one, i.e. what geography should be irrespective of what it had been. Geographers in increasing numbers had, by the end of 1950s, turned to Schaefer's view of geography as a spatial (locational) science. For this they used methods of other systematic sciences and were increasingly concerned with quantification and development of theory. The net outcome of Schaefer-Hartshorne Debate was that geography, by the end of the 1950s, had come increasingly to be viewed as a science requiring the use of the 'scientific method' like other sciences. This brought about a distinctive shift in emphasis from 'regional' to 'systematic' studies. This meant that geography thereafter began increasingly to be viewed from a nomothetic standpoint, i.e. it required developing the habit of seeking the general in the particular. This also involved a shift from 'areal' to 'locational' studies; from 'absolute' to 'relative' locations; and from 'areal differentiation' to 'spatial interaction'. The mechanistic approach of environmentalism, which dominated much of the nineteenth century, was soon replaced by a more nomothetic and scientific approach.

The dualism developed between Nomothetic and Idiographic got resolved after Schaefer-Hartshorne Debate. Nomothetic method advocated for the aspect of science that can be expressed in universals and Idiographic that which is concerned with the individual attributes of objects. It was agreed upon that geography 'like other sciences' can develop 'the knowledge of principles, laws, and general truths'—and thus lay claim to the name of science. The idiographic view was very strong in the earlier part of the twentieth century, but from 1950s onwards various objections were raised against this method of study. In the light of vast geographical knowledge accumulated by the end of Age of Discovery, the idiographic method proved inefficient to handle this information. The Nomothetic method was, therefore, was proposed as the key to the problem.

The Quantitative Revolution began under the banner of the scientific method. Apparently new perspective was opened up under the impact of this Revolution during mid twentieth century. It sought to develop a more rigorous and systematic methodology for the discipline that later lead to a switch over to a new paradigm. It came as a response to the inadequacy of regional geography to explain general spatial dynamics. The quantitative revolution, that occurred during the 1950s and 1960s, marked a rapid change in the method behind geographical research. The revolution was geography's attempt to redefine itself as a science. The revolutionaries declared that the purpose of geography was to test general laws about the spatial arrangement of phenomena. Adopting the philosophy of 'positivism' from natural sciences, they turned to 'mathematics'—especially 'statistics'—as a way of proving solutions to their problems. The mathematical and statistical methods were introduced to attain a desired level of objectivity for the search of laws and theories in geography. All these efforts were devoted to the fundamental conception of geography as a spatial science. Although, this movement is over now, it did lead to mathematization of much of our discipline, with an attendant emphasis on the construction and testing of theoretical models. The main claim for the Quantitative Revolution was that it led to a shift from a descriptive (Idiographic) geography to an empirical law making (Nomothetic) geography. Then, during the Second World War technology became very influential in our society. This culminated in the Quantitative Revolution, during which science and mathematics enjoyed unrivalled importance. Nomothetic aspects of geography heightened in importance and quantitative methods were applied to virtually all investigation and experimentation. The changes introduced during and this phase initiated the ever increasing use of computer-based practices that improved precision, and theory-based practices to conceptualize location and space in geographical research. By the mid-1960s the quantitative revolution had successfully displaced regional geography from its dominant position and the paradigm shift was evident by the myriad of publications in academic journals and textbooks of geography. However, the greatest impact of the quantitative revolution was not the revolution itself but the effects that came afterwards in the form of spread of post-positivist thinking and counter-positivist responses. The counter-positivist response was created in the form of introduction of anthropocentricism in the discipline later.

The quantitative revolution had enormous implications in shaping the discipline of geography into what it looks like today. It replaced the idiographic approach based on areal-differentiation by nomothetic one, which had its roots in the search for models of spatial structure and phenomenon. The revolution led the basis of geography as a spatial science that dealt with the spatial analysis of phenomena that existed on the earth surface. In simpler words, it gave geography a scientific vision through the application of methodology rooted in statistical methods. The most obvious change has occurred due to the quantitative revolution that brought changes in the methods and techniques used to explain the geographical phenomenon in a spatial framework. The new method of inquiry led to the development of generalizations about spatial aspects in a wide range of natural and cultural settings. Attention, since then, has focused on the spatial organization of economic, social, physical, political and urban processes, and on how outcomes generated by these processes are evidenced at given times and in particular places. Special concern has been placed on the spatial aspects of social processes. The theoretical approaches and quantitative methodologies evolved in mid twentieth century lead to the evolution of Spatial Science School of Geography by 1970s. The School emphasized the study of aggregate spatial patterns, including spatial behaviour, within clear theoretical frameworks, using quantitative methods to evaluate models and hypotheses. The general acceptance of quantitative techniques and the more complete mathematical training of a new generation made possible this development. The Spatial science approach dealt with aggregate patterns in space and their generating processes. However, within two to three decades of the beginning of Quantitative Revolution, the overwhelming focus on statistical modeling itself eventually became responsible for the undoing of this revolution.

Critical Revolution came as a critique on positivism introduced by the *Quantitative revolution*. This again proved as one of the major turning points in the history of geography. The discipline of Geography saw another phase of transformation since early 1970s under this Revolution. The Critical Revolution, in general, is associated with opposition to quantification in geography. By mid 1960s, the initial euphoria about quantitative revolution had receded. Many geographers became increasingly concerned that these techniques simply put a highly sophisticated technical gloss on the approach to study that was barren of theory. Their relevance to real life was widely questioned. As the 1970s dawned, the Quantitative Revolution came under direct challenge. The mood changed with a significant number of geographers beginning to challenge the wisdom and even the morality of this quantitative turn. The backlash against the quantification trend led to Revolution under the humanistic and welfare approaches in geography. Quantitative methodologies, undoubtedly, are a powerful research technique in human geography that can provide valuable and accurate insights if used appropriately and with an understanding of the limitations. However, during the 1980s and 1990s there was a downturn in the popularity of such methodologies. The overwhelming focus on statistical modelling was, eventually, responsible for the undoing of the quantitative revolution. The critics argued that it removed the human dimension from a discipline that always prided itself on studying the human and natural world alike. As the 1970s dawned, the quantitative revolution came under direct challenge. The counter-positivist response came as geographers began to expose the inadequacy of quantitative methods to explain and address issues regarding race, gender, class and war. The positivist assumptions upon which the approach was based were attacked in the 1970s and 1980s as geography experienced its `anthropocentric turn' splitting into a range of modes of thought.

The Critical geography of this phase lead to three main schools of thought, viz. Behaviouralism, Radicalism and Humanism. Behaviouralism tried to challenge quantitative side of geography and provide a greater understanding of how people perceive space and place; Radicalism questioned the social relevance of the discipline and argued that quantitative research is not useful when there are neither solutions nor alternatives to human problems; and the last and most accepted ideology of Humanism, studies human awareness, agency, consciousness and activity. Modern geography is a spatial science increasingly focusing on the study of spatial aspects of social and economic phenomena, and thus being pre-eminently anthropocentric.

During 1970s, when the need for precision in research through increased use of statistical-mathematical techniques had been commonly accepted, and the controversy regarding systematic versus regional geography resolved, with the scale tilting in favour of the former, but both being accepted as independent and equally useful components of the discipline, the euphoria about quantitative revolution had also greatly receded. Now, many workers had begun to feel disillusioned with the quantitative models being developed and used by human geographers of the positivist orientation, since they failed to relate to real-life situations. Consequently, geography again witnessed a paradigmatic shift, as many geographers began to feel that a deeper understanding of man-environment interaction could be achieved if geographers paid greater attention to the study of the mental images that the decision-makers have of the environment to which their decisions relate since it is the perceived reality that leads to action. This shift in focus inevitably led to a shift to the behavioural sciences. The objective of behaviouralism was to replace the simplistic and mechanistic conceptions of man-environment relationships with a new theory that views human actors in space as thinking beings whose actions are mediated by cognitive processes, i.e. the mental processes by means of which people acquire, organize and use knowledge. The focus was on studying how people come to terms with their physical and social milieu. The behaviouralist rejected the assumption that decisions are always based upon a sequence of reasoned thoughts involving systematic evaluation of all available options. The behaviouralist was concerned with the full spectrum from the sub-conscious impulse to the self-conscious deliberation in the process of decision-making. The basic philosophy of behaviouralism is that man and environment are dynamically interrelated. Man is viewed as a motivated social being, whose decisions and actions are mediated by his cognition of the spatial environment and creation of an environmental imagery. As an approach, it looks at man-environment relationships with a new and sharp outlook. It aimed at replacing the simplistic and mechanistic conception that previously characterized man-environment theory with new version that explicitly recognizes the complexities of human behaviour within a given set of environment; and the degree of their impact on space. The behaviouralist viewed the human, subjects of his studies, as thinking beings whose actions are intervened by mental processes. In between the environmental influences and the man's actions there are mental processes of various kinds, which govern his behaviour in a given set of environmental elements. The behaviouralist is interested in the way in which people come to terms with their physical and social milieu. He is also interested in the factors that influence the interrelationships of thought and action of man. The 'environmental perception' and 'mental maps' are recognized as the basic concepts of Behavioural Geography. Environmental Perception is the total perception an individual has of his surroundings, which creates mental maps; whereas, a mental map is an individual's own internal map of their known world.

Radical geography is suggested as an approach to geographic research that seeks to understand social and spatial problems, and advocate solutions. Radical geographers are interested in the issues of relevance to everyday social life: the lived experiences of members of society. Radical geographers tried to transform the scope of the discipline of geography by responding to the great issues of the time-civil rights, environmental pollution, and war. This emphasis reflected a desire for geographers to do research that is relevant to, and useful for, society. A quest for social justice permeates much of radical geography. The radicalists put emphasis on the need for a revolution in both theory and practice of geography. The radical approach is value based, especially the theory of labour value, as against the supposedly value-free approaches. Most of the radicalists have a strong Marxist base and take a holistic view of economics, society and polity. Radical geographers reject the spatial determinism implicit in spatial science; instead they seek to place questions of geography within broader social and political contexts. They emphasize context and relationships between places and people and broaden the geographical research agenda by considering previously neglected issues including poverty, hunger, urban decay, and social inequality.

Humanism in geography also developed due to a deep dissatisfaction with the mechanistic models of spatial science that had developed during the quantitative revolution. It evolved as a criticism against positivism and quantitative revolution. The basic objection of humanists against quantitative revolution is that its tools and assumptions do not adequately explain human world and human issues, especially relating to social institutions, attitudes, morals, customs, traditions and aesthetics. Humanistic geography studies human awareness and human agency, human consciousness and human activity. It, therefore, deals with the meaning, value and human significance of life events. The Humanists are not in favour of reducing space to mere geometrical concepts of surface as viewed by the positivists. From a humanistic point of view, space is inseparable from consciousness of humans who inhabit it. Humanism is a rejection of the geometric determinism which views human beings as always dictated by universal spatial structures and abstract spatial logics. Geographers became increasingly active in applying their skills of spatial analysis to finding equitable solutions to the many problems that now confront humankind. The humanist epistemology sees that studying the human world is very different from studying the physical, biological or material world. Within this humanistic perspective concepts of traditional significance in geography are given new meanings. The focus of humanistic geography is on people and their condition. Humanists explain and interpret man and space relationship with historical approach. Humanism does not treat humans as machines. The followers of this approach consider geography as 'the study of the earth as home of man.' Humanistic geography, as such, is not an earth science in its ultimate aim. It belongs to the humanities and to the social sciences to the extent that they all share the hope of providing an accurate picture of the human world. Humanistic geography achieves an understanding of the human world by studying people's relation with nature, their geographical behaviour as well as their feelings and ideas in regard to space and place.

Geography, in the twenty first century, is experiencing another revolutionary phase. With the widespread introduction of computer technology, geography has entered a new methodological revolution. In 1950s, such a change has been brought by the quantitative revolution, and now, i.e. after five decades or so, it is Digital Revolution. Computers, remote sensing, GIS and related fields are opening up new avenues for geographers. With the use of these modern techniques there may be collected and interpreted a variety of information about the earth's surface. Since geographers are directly concerned with analyzing the variable data on the earth's surface, this is going to open up vast areas for analysis and research. The Digital Revolution refers to the advancement of technology from traditional mechanical devices to the digital technology available today. The Revolution also marks the beginning of the Information Era. The Digital Revolution is sometimes also called the Third Industrial Revolution. The Revolution is the shift from mechanical and analogue electronic technology to digital electronics. However, this change had come in the West in late 1950s, the revolution became universal in the present times. The adoption and proliferation of digital computers and digital record keeping have become more widespread now. Implicitly, the term refers to the sweeping changes brought about by digital computing and communication technology during and after the latter half of the twentieth century. Central to this revolution is the mass production and widespread use of digital logic circuits, and its derived technologies, including the computer, digital cellular phone, and the Internet. These technological innovations have transformed traditional practice of data collection, presentation and analysis. The most visible representation of the Revolution is the Internet, which integrates telecommunications and computing. The information may be more easily collected, stored, processed, accessed and communicated after the conversion of technology from analogue to digital format. Digital information is compact, transportable, and, therefore, more efficient to use. There have been huge benefits to society from the digital revolution, especially in terms of the accessibility of information. Expanded powers of communication and information sharing, increased capabilities for existing technologies, and the advent of new technology brought with it many potential opportunities for exploitation. The digital revolution helped usher in a new age of mass surveillance, generating a range of new civil and human rights issues. The digital revolution made it possible to store and track facts, articles, statistics, as well as minutiae hitherto unfeasible.

The computer has drastically transformed both the world of geography as an academic discipline and the geography of the world in which we live. From being a tool for geographers to collect, analyze, map and visualize data since 1950s, the computers have increasingly become an integral part of the geographers' study by the end of twentieth century. Computers have enriched the discipline of geography with the development of automated geography, GIS, and the virtual geography department. No other technological innovation in human history has affected the practice of geography in such a profound way as the computer. The traditional geography has been radically changed by the past fifty years of the computer age, and especially by the last twenty years. Computers have fundamentally changed not only how geographers study the world, but also what we study, as spaces and places, the natural and human landscapes, are constantly being refigured by computers. The increasing use of computers has allowed geographers faster processing of larger sets of data. The increasing speed and capacity of computers allow and encourage the use of datasets that were simply unmanageable and tedious only twenty years ago. Not only in physical geography but also in human geography, far more comprehensive datasets can be used and far more complex models can be run. The computers have vastly eased the coding of qualitative data, i.e. recorded interviews, answers to open-ended survey questions, through direct computer answering of surveys, data entry and analysis, and textual search. The Computers have also stimulated the growth of new branches of geography, as quantitative geography, computer-assisted cartography, analytical cartography, remote sensing, and GIS; the process culminating in building of the Virtual Geography Departments the world over. In this virtual world, everything geographers study has become bits of information, which can be transmitted across the globe instantaneously. Each one of us is not only an information processor, but also information processed.

Indeed, even the entire earth is now becoming digital. Digital Earth is inherently an intriguing concept for modern geographers. The concept of Digital Earth is the most ambitious concept to exemplify the potential of computers both as tools and as reality. The digital earth is envisioned as "a multi-resolution, three-dimensional representation of the planet, into which we can embed vast quantities of georeferenced data." Digital Earth challenges our state of knowledge about the planet, not only in terms of raw data, but also in terms of data access and the ability to communicate data through visualization. Moreover, it challenges our understanding of process in the invitation to model, simulate, and predict, since the concept should is not limited to static portrayal. Second, Digital Earth is interesting because of its implications for the organization of information. As a consequence, the content of geography has changed, for example, *via* the very concept of globalization space-time compression, and space-time distanciation. The rapid, cheap worldwide transfer of messages, documents, data, and images have brought about yet another shrinking of distance and abetting of globalization in all its economic and cultural forms. Digital Earth is regarded as a materialization of what has been called the Noosphere, dominated by flows of digital information. Together with geographers' focus on the atmosphere, hydrosphere, and biosphere, the study on the geography of Noosphere is a new challenge for geographers in the twenty first century.

A geospatially literate population is the goal of geographers in the twenty first century. Such a population is better equipped to recognize, understand, and resolve those critical issues, whether local or global, that will confront us today and in the future. New technologies have always been important in advancing geographic understanding, but never have they been so thoroughly and rapidly transformative of the discipline as at this stage in geography's evolution. Just as new technologies have profoundly expanded both research possibilities and the knowledge base of other disciplines, such as biology, physics or medicine, so too are the revolutionary new geographic technologies developed during the past few decades extending frontiers in geographic research, education and applications. They are also creating new and resurgent roles for geography in both society and in the university. This trend is still accelerating, as the integration of geographic technologies, such as the global positioning system and geographic information systems (GPS/GIS), is creating an explosion of new "real-time, real-world" applications and research capabilities. This would, undoubtedly lead to a new paradigm shift in Geography.

V
CONCLUSION

24

Geographical Thought: A Methodological Examination

Geographical thought, at any point in time, represents the prevailing philosophical viewpoints and major methodological approaches in vogue. The methodology of a discipline is related to its content, approaches to study, and methods of collecting and organizing information. These are continuously either refined or discarded in a progressive field. Methodology is the logic of this process and it thrives on change and evolution. The methodological debate, therefore, is a sign of health in advancing disciplines. In Geography, as well, there has been extreme diversity of viewpoints with constant extension and also a shift in focus of study. The methodology of geography came under debate for the first time during the middle of twentieth century with the publication of a powerful paper by F.K. Schaefer that brought both the qualitative and quantitative revolution in the field. Shortly after this publication were discussed the paradigmatic shifts in a discipline by Thomas S. Kuhn through his model, 'Paradigm of Science'. Since then, the concept of a paradigm shift is being used in numerous scientific contexts. Paradigm Shift is a change from one way of thinking to another.

The knowledge of geographical nature has infiltrated to its present in the form of various fields of learning, philosophies, terminologies, concepts and techniques, as bequeathed to Geography from its various Schools of Thought that refined its methodology from time to time. However, the history of development of geographical thought has not been an affair of continuous and smooth progress. Two major revolutions have changed the course of its development, viz. the Industrial Revolution in 1750s and the Methodological Revolution in 1950s. Accordingly, the phases of geographic development, in general, may be identified as Modern, pre-Modern and post-Modern. Hence, Pre-modernism, Modernism, and Post-modernism may be considered as distinct epistemological systems, each of these being defined by the dominant philosophies and methodologies of the time.

The Pre-Modern Trad`ition is reflected in the developments during Ancient and Medieval phases. The Ancient schools of geographical thought represented the Western, Indian and Chinese traditions, whereas the Arabian civilization generated the medieval tradition. During this entire period, the quest for knowledge was seen primarily through religious source in both, the European and Oriental worlds alike. This long phase of evolution recorded a gradual transition of geographical tradition from cosmological to scientific. The ancient philosophy largely revolved around the questions pertaining to fundamental causes and principles of the Universe and their understanding. The reasoning and argumentation were established as crucial philosophical methods. The Modern period saw the shift in power away from religion, with the politics and universities taking over as the primary sources of authority. Two new approaches to knowing became dominant, viz. empiricism and logic. Along with these, the nature of geographical query also became more explanatory from descriptive. In an attempt to answer each set of new questions, geography gained a different and more meaningful definition every time. With the development of natural sciences in the eighteenth and nineteenth centuries it became apparent that mere description would not do. Geography came to be conceived as the science concerned with the formulation by the laws governing the spatial distribution of features on the surface of earth. This implied that geographers must describe and explain the manner in which things exist in association with others in an area; the combinations differing from area to area. The viewpoint was called chorological. This chorological perspective presented geography with a problem of great methodological controversy, the major one being regional versus systematic. The anti-scientific sprit of historicism was another major intellectual force of this phase. It penetrated geographical thought and powerfully affected its course. Schaefer called it Exceptionalism and traced its roots to Kant, to whom history and geography find themselves in an exceptionalist position different from that of the so-called systematic sciences. Schaefer's challenge to chorological view in mid twentieth century brought another turning point in the history of geographical development. Post-modernism brought with it the questioning of previous approaches of knowing, and introduction of new methods of study. Instead of relying on one approach to knowing, it advocated for an epistemological pluralism. Along with the revelation (pre-modern way) and science and reason (modern ways), Post-modernism also includes many other ways of knowing such as intuition, relational and spiritual. Post-modern approaches seek to deconstruct previous authoritatative sources and power. However, the Post-modern period reflects more methodological concerns than conceptual. During this phase, Geography witnessed a number of critical phases with revolution one after the other, resulting in faster paradigmatic shifts in comparison to earlier phases of development. The nomothetic twist, emergence of spatial (locational) perspective and anthropocentricism (behaviouralism, radicalism and humanistic trends) have been instrumental in leading to the paradigmatic shifts in geography in its Post-modern period. The crises in the discipline, on the other hand, have been brought about by Schaefer-Hartshorne Debate, Quantitative Revolution, Critical Revolution and Digital Revolution.

Kuhn's model gives a scientific explanation of the growth phases of scientific knowledge. According to him, the scientific development consists of a series of phases in which the sequence of crisis, revolution and paradigm continues throughout the history of science and helps in its advancement. It is argued that geography is paradigmatic science and has undergone the metamorphosis. Any idea, concept, method or technique that is capable of generating scholarly concern in a discipline, propelling it to 'normal science' stage, may be called a 'paradigm. As far as geography is concerned, this cause of scholarly concern has been, for instance, the philosophy largely related to cosmology and theology during Pre-modern period; environmentalism, chorology, logic and reasoning during Modern period; and anthropocentricism in Post-modern period. These philosophies have generated several paradigmatic shifts in the discipline in the past. In the light of Kuhn's Model one can discuss the evolution of geography through eight Paradigmatic Phases interrupted by eight Crisis Phases during its entire course of evolution. Geography, during its Pre-modern period, had already passed through two paradigmatic and two crisis phases. It experienced other three paradigmatic shifts intervened by two crisis phases during its Modern period of development. The paradigmatic shifts in the discipline have been relatively faster in the Post-modern period than earlier. More recently, the discipline has encountered four critical phases intervened by three paradigmatic shifts, and it is still undergoing a methodological revolution.

The Pre-modern period witnessed the Pre-paradigm phase of Geography followed by Professionalization and two initial Paradigmatic shifts. These paradigmatic shifts in the discipline are linked to the emergence of Graeco-Roman School in ancient times and Arab School in medieval times. The Dark Age and Age of Discovery brought crises in the discipline during this period. Emergence of Chorological perspective, Videlian Tradition and Man-Environment based studies were the major developments during Modern period. However, the major methodological concerns of modern period have been the issues related to systematic and regional approach to geography. The German, French, American and British Schools have largely been responsible for these developments. The prevalence of numerous dualistic tendencies has been crucial in bringing about crises in the discipline during this period. The Radicalism, Humanistic trends and Social Relevance of Geography have been instrumental in bringing about the paradigmatic shifts in Geography in its Post-modern period. The methodology of geography has been under discussion more elaborately now than earlier. Therefore, since 1950s, the discipline has undergone four crises phases.

Throughout Pre-modern phase of its development, geography was considered as a descriptive field of study. In general, the religion was the primary source of knowledge in pre-modern times in the European, Oriental and Muslim worlds alike. The scholarly observations of the Western and Eastern civilizations during ancient period mark the Pre-Paradigm phase of geographic development. Towards West the earliest developments took place in Western Asia, Egypt and Europe. Various cultural groups were involved besides Greeks and Romans. The Graeco-Roman people had a deep sharing of culture, knowledge, science and technology with many of the preceding civilizations that emerged successively in Mesopotamian region, viz. the Sumerians, Babylonians, Assyrians, Egyptians and Phoenicians. These people invented the first language script, the cuneiform; and made large advances in the techniques of agriculture and science. In fact, the origins of philosophy and reasoning are traced back to the early Mesopotamian wisdom. An extremely large body of texts has survived on cuneiform tablets here. The pre-Greek civilizations bequeathed Greeks, particularly, the sexagesimal system of numerals, the earliest known form of writing (cuneiform) and a system of alphabet, besides a number of valuable mathematical and astronomical contributions. The main feature of Greek thought, as obvious, represents the mathematical and astronomical knowledge of the Mesopotamians.

The philosophical developments in ancient Graeco-Roman world are associated with the Ionian people, and are represented by four periods, viz. the period of Naturalism, Metaphysical, Ethical and the Religious. Beginning with the Milesian School and culminating in the Neo-Platonic School, there emerged numerous schools of thought in between. The period of Naturalism, for instance, was dominated by the Milesian, Pythagorean, Ephesian, Eleatic, Pluralists' and Atomists' streams of thought. The Metaphysical period of Greek philosophy includes the philosophies, especially those of Sophists, Socrates, Plato, and Aristotle. The Ethical Period saw the displacement of metaphysics by ethics and the philosophy followed its ethical development through three streams of thought, viz. Stoicism, Epicureanism and Skepticism. During the Religious or the last period of development, the Greek philosophy sought refuge in religion, with an intermingling of Greek, Egyptian, Chaldean, and Jewish religions; the period being represented by the Judaic-Alexandrian, the Neo-Pythagorean the Neo-Platonic thought.

The Milesians, in general, focused on cosmological concerns, particularly the underlying substance of the universe. Pythagoreanism, revolving around the philosophy of Pythagoras, cultivated the mathematical sciences. The Ephesian school believed everything that exists is in a continuous process of passage from one state to another. The Eleatic School was based on the doctrine of Permanence, changelessness; the basic tenet of its philosophy revealing the fundamental approach to the problem of Being and Non-Being. The Pluralists attempted reconciliation between the Eleatic and Ephesian factions, and tried to overcome the opposition between 'Being' and 'Non-Being'. Accordingly, they stated that the world as we know it is composed of four basic material particles (earth, air, fire, and water) calling them "the roots of everything" that make all the structures in the world. These four elements are combined with each other in different proportions. It is in the aggregation and segregation of elements thus arising, that the atomists found the real process which corresponds to what is popularly termed growth, increase or decrease. The Atomists believed that the matter in Universe is composed of an infinite number molecules (atoms) differing in size and form but endowed with perpetual motion which is derived from their essence. Along with motion, the School also admits a second primordial element, the void. The differences in gravity cause the atoms to whirl into motion, thus giving origin to the formation of things. Accordingly, from the primitive void have come the stars and the earth and all beings, including man. Motion of individual distinct atoms has produced an orderly cosmos. During Metaphysical period, the Sophists were the first philosophers to introduce the study of man and held that knowledge is essentially empirical and relative to man. Socrates, on the other hand, concentrated on the universal aspects of man, as expressed in his doctrines of Concepts and Ethics. To arrive at understanding of these, Socrates used the inductive method of dialogue, popular as the Socratic Method. The greatest of the Socratic schools, after his death, was the Academy of Plato, which stayed closer to the original intent of the teachings of Socrates. Plato's metaphysics divides the world into two distinct aspects: the intelligible world of "forms", and the perceptual world we see around us. Aristotle, on the contrary, placed much more value on knowledge gained from the senses, and correspondingly earned the label of empiricist. Aristotle's method is largely inductive, while Plato's is essentially deductive. The unavoidable consequence of the times resulting from political instability in the region, developing due to the endless wars between the successors of Alexander and the Roman Empire, led the displacement of metaphysics by ethics. During Ethical period, the solutions to the existing problems of life were found in three possible means: to appeal to reason; to appeal to pleasures; and to deceive oneself in the belief that every appeal is useless, respectively constituting the three streams of thought which philosophy followed in its ethical development, viz. Stoicism, Epicureanism and Skepticism (the renunciation of every solution). During the Religious period of Greek philosophy, the Judaic-Alexandrian School attempted to coordinate Oriental religion with Greek speculation. The Neo-Pythagorean School attempted to construct a world religion upon Pythagorean doctrines, the teaching of Pythagoras aiming at ethico-religious reform. Neo-Platonism, in turn, conceived God as the source and goal of everything. The School tried to resolve this question by overcoming dualism with monism. While, on the one hand, the natural, metaphysical, ethical and religious philosophies inspired the thinking of Western scholars in ancient period, the interest in Natural Sciences also continued to survive. Along with philosophy, the theoretical interest of the time also concentrated on the physical sciences during Greek Age. The knowledge of new lands, new phenomena, and new facts made possible through the expeditions of Alexander also fostered the rise of science. The mathematics, physics, astronomy, zoology and botony were the principal sciences developed during this epoch. Geography, actually, began to be cultivated largely in its astronomo-mathematical aspect during Graeco-Roman period in ancient times in the West.

Towards Orient, the Indian contribution to geography came largely from the astronomers and astro-physicists, and the information contained in the religious records and travellers' accounts. The different systems of Indian philosophy represent the special perception of Indian thought (Samkhya, Yoga, Nyaya, Vaisheshika, Purva Mimamsa and Vedanta). Samkhya, the oldest of all, is a dualistic philosophy that postulates everything in reality stems from '*purusha*', i.e. 'self, *atma* or soul' and '*prakriti*', i.e. 'matter, creative agency or energy'. The philosophical system of yoga is closely allied with the Samkhya School. Nyaya is the closest Indian equivalent to analytic philosophy. Just as Aristotelian logic, its methodology is based on a system of logic that has subsequently been adopted by the majority of the Indian schools. The Vaisesika School postulates an atomic pluralism, as proposed by the ancient Greek School of Atomists. The main objective of the Purva Mimamsa School was to establish the authority of Vedas. Lastly, the Vedanta or Later Mimamsa School concentrates on the philosophical teachings of the Upanishads. Ancient Indians possessed great knowledge in all areas of the science, technology and human development. In fact, several inventions and discoveries, that are believed to have originated from the Western world have, actually been carried out by ancient Indians. Besides the observations regarding the sphericity of Earth and the gravitational pull, their discoveries and inventions ranged from aviation technology, to medicine and botanical science, atomic theory, plastic surgery, yoga etc. The knowledge contained in ancient scriptures covers the views of ancient Indian scholars on the study of the Universe, the Earth and the Indian Subcontinent. The ancient Indian literature deals with many problems pertaining to cosmology, cosmogony and cosmography. The principle of gravity and the concept of 'zero' is believed to be the Indian contribution to the world. Observations about the earth are related to its origin, sphericity, eclipses, size and dimensions, latitudes-longitudes and local time, cardinal points, earthquakes and volcanoes, atmosphere and seasons, and physical divisions of this planet. *Bhugol*, the term used for Geography, appropriately suggests that the ancient Indians endorsed the earth being a sphere. Besides proving Earth's sphericity, the Indian astronomers also formulated ideas about gravity and gravitation. The ancient Indian scholars were aware of the causes of occurrences of eclipses. The facts related to the size and dimension of the earth were quite near to accuracy. Puranas have a reference of the imaginary lines of latitudes and longitudes. They have used three latitudinal divisions of the globe into Equatorial, Northern Polar and Southern Polar regions. The South Pole was truly considered as the antipode of the North Pole. However, the world was not believed to exist beyond Equator, as the region here was compared to hell of the earth. The Eastern part, on the other hand, was believed to be 'the land of Gods'; the thinking is in consonance with that of the Europeans in the Early Medieval period. The ancient Indian scholars have also drawn Prime Meridian. These imaginary lines, the position of Sun and various stars have helped them to determine local time at various places. The ancient Indian scholars have identified the vacuum between the earth and the heaven as *Antariksha* and believed in the occurrence of various weather phenomena here.

Based largely on the studies in India, six seasons have been identified in *Valmiki Ramayana*; they are Spring, Summer, Rainy, Autumn, and Severe Winter. The known world during the Puranic period was divided into seven *Dwipas* or 'regions'. These divisions exclude the parts of world discovered later i.e. towards the close of medieval period, as American Continents, Greenland, England and Antarctica. These seven regions were known as Jambu, Krauncha, Kusha, Plaksha, Pushkara, Shaka and Shalmali. In Vedic and Puranic literature, the entire country from Himalayas to Kanyakumari has been referred to as Bharatvarsha by Aryans. Geographical knowledge about the Indian Subcontinent is associated with its physical identification and description of its people, culture, relief and drainage. However, the religious flavour is very strong in these descriptions. Although, the subsequent invaders ruthlessly destroyed the architectural achievements of this country and imposed their autocratic rule, they also transmitted Indian sciences to the outside world.

The ancient Chinese scholars and travellers also possessed a rich heritage of geographical knowledge. China's history has been largely characterized by repeated and violent imperial dynastic change. For centuries, Imperial China was one of the world's most technologically advanced civilizations. As in Egypt, Mesopotamia, and along the Indus River, Chinese civilization began within a major river valley, the Yellow River Valley. China, with a huge geographical expanse, contained an almost infinite number of ethnic groups and languages. Most Chinese dynasties emerged in the historical heartland of China, the China Proper. These earliest dynasties were based on agriculture, augmented by hunting and animal husbandry. They were also in use of a writing system and bronze metallurgy as revealed in inscriptions found on oracle bones. They initially embraced the philosophy of Confucianism and later the Buddhism. The major contributions of these dynasties include, for instance, the concept of a centralized government, the unification of the legal code, written language, measurement, and currency. The Chinese script is the only logographic script still used in the world. Besides, they also promoted extensive trade with distant foreign countries, through Silk Road. The Shang was the first confirmed historical dynasty. The Han Dynasty was the first one to embrace the philosophy of Confucianism, which became the ideological underpinning of all regimes until the end of imperial China. Under the Han Dynasty, China made great advances in many areas of the arts and sciences. The first opening of trading connections between China and the West, the 'Silk Road' was also enabled during this regime. The Chinese Thought has been mainly represented by Confucianism, Taoism and Buddhism. A number of Chinese philosophers, travellers and scholars contributed to geographical knowledge since ancient period. Their contribution goes to the fields of Astronomy and Astrology, Human and Regional Geography, Physical Geography, Mathematical and Practical Geography, and Accounts of Travels and Explorations. The Chinese mathematics had developed the Decimal System, which was introduced to the Arab world later. The Chinese were expert in map-making on the latitudinal-longitudinal grid, and this art is believed to have developed much earlier than in Greece or Rome. Also commendable are their observations on weather and climate. The earliest extant maps found in archeological sites of China date back to the fourth century B.C. The ancient Chinese started the trend of the gazetteer in China; and these gazetteers featured a wealth of geographic information. The Chinese were also concerned with documenting geographical information of foreign regions far outside of China.

The professionalization in the discipline had started becoming evident in the attempts of the European, Indian and Chinese philosophers, scholars, mathematicians, administrators and explorers in antiquity as mentioned above. In spite of physical isolation, fascinating parallels were exhibited in the concepts and methods of study among them. Geography, as a discipline of scholarship, has filtered through several fields of learning of their study as Philosophy, Cosmology and Cosmogony, Astrology and Astronomy, Physical Studies, Mathematical Sciences, Cartography, Humanistic Studies, Sciences, Technology and Travels and Explorations. Among all these, the fields of philosophy, astronomy, mathematics and science have been commonly contributed by Graeco-Romans, Indian and Chinese; whereas history and cartography have especially interested the Greeks and Chinese, as also the travels and exploration.

The philosophical developments in Geography are rooted in its Metaphysical, Logical, Epistemological and Ethical thinking. However, the earliest Greek speculators were philosophers of nature. Most of the schools of Indian philosophy were largely metaphysical, whereas the ancient Chinese were more ethical in their thought. The Universe and its origin remained a point of speculation among all the ancient civilizations. The traditional theories generally depicted an original unity which begins to become separated off into a series of opposites. The early philosophers sought to explain the cosmos on the basis of their observation of the regular behaviour of materials that make up the cosmos and all its phenomena, without reference to extra-natural causes or the personified gods as by traditional religions. The early Greek philosophers, for instance, sought to explain the cosmos and all its phenomena by appealing to regularities within the cosmic system itself. The Cosmological theory of Anaxagoras, for instance, suggests the origin of the world caused through the continuous process of motion and separation from a unified mixture. Likewise, Anaximenes also gave an account of how the world formed out of previously existing matter. Philolaus' ideas revolved around a hypothetical astronomical object, the Central Fire, and the planetary bodies moving around it. According to Empedocles, on the other hand, the world is in a constant state of organic evolution due to the combined action of two cosmic forces which work upon the elements in both creative and destructive ways, personified as 'Love'—a force of attraction and combination, and 'Strife'—a force of repulsion and separation. In consonance with the Greek thinking, the Universe (or *Brahamand*) is conceived as very immense and wide in the ancient Indian literature. Four types of theories have been advanced about the origin of Universe. Of these, the theory of Artistic Origin compares the Universe to a house, constructed by a number of Gods by weaving various materials into a pattern, and shaping it by blasting and smelting; the theory of Mechanical Origin suggests the dismemberment of a primeval body (*Adi-Purusha*), and the creation of Universe a part of its sacrifice ceremony; the theory of Philosophical Origin believed in the origin of Universe out of a vacuum; and lastly, the theory of Instrumental Origin talks about the occurrence of two parent bodies, viz. the the Earth (*Dharti*) and the Heaven (*Swarga*), from which the Universe was created, the union of these two producing the Sun, the creator of the Universe. By combining the basic concepts of Taoism and Confucianism, the ancient Chinese also created a traditionally consistent and uniquely harmonious view of the universe. They advocated primarily three models of celestial orientation: '*Gai Tian*' was the theory of the dome shaped heaven that placed what we today call Ursa Major at the centre of the celestial dome and China at the centre of the Earth; the 'School of Hun Tian' envisioned a spherical heaven shaped much like a hens egg where the earth is like the yolk; and finally, the theory of Xuan Ye suggested the unprecedented thought of an infinite universe with suspended celestial bodies. These theories lead to many technological advances in astronomy such as armillary rings and spheres.

The first references to astronomy in ancient world are interwoven with astrology. The astronomical observations were made with the help of some simple instruments as the gnomon, scaphion etc. The Greeks made many important and lasting contributions to the field in synthesis with Babylonian astronomy. Of them, the most intriguing is Hipparchus' work which also served the basis of Ptolemy's attempts later who compiled the astronomical knowledge of ancient world in his *Almagest*, one of the most influential books of classical antiquity. The Ptolemaic explanation of the motion of planets remained the accepted wisdom until the Polish scholar Copernicus proposed a heliocentric view in 1543. Besides, Philolaus' astronomical system is also popular that first proposed the heavenly bodies being arranged in concentric circles around a central fire. Anaximander's merits lie in the field of speculative astronomy, and his vision implied depth in the universe, i.e. the idea that celestial bodies lie behind one another in increasing order of brightness. Indian astronomers also propounded the theory that the earth is a sphere. The concept of Prithvi, the Earth, has been profusely used in the Vedas and Puranas. The term *'Bhugol'* as used for Geography clearly suggests that the ancient Indians endorsed the earth being a sphere. Through *'Aryabhattiya'*, the work of Aryabhatta, the Indians were the earliest to discover that the orbits of the planets around the Sun are ellipses, to describe numerical and geometric rules for eclipse calculations, to discuss about the locations of the planets in relation to its movement around the sun, to refer to the light of planets as reflections from the sun, and to explain eclipses in terms of shadows cast by and falling on earth. His working was later adopted by the Greeks and then the Arabs. Aryabhatta had described a geocentric model of the solar system. Subsequent Indian astronomers improved on his calculations, but his methods provided the core. Another Indian astronomer, Brahmagupta, calculated the circumference of the earth more or less accurately as known today. Besides, Bhaskara's work is based on a model of the solar system which is heliocentric and whose movements are determined by gravitation. Ancient Chinese astronomy, on the other hand, was primarily a government activity, and the astronomer's role was to serve the ruling emperor. The ancient Chinese invented the first planetarium and sophisticated observatory buildings. Based on the theories of his predecessors, Zhang Heng systematically developed the celestial sphere theory. An armillary constructed on the basis of his hypotheses bears a remarkable similarity to Ptolemy's earth-centred theory. The celestial model Zhang Heng constructed was almost a physical model of Ptolemy's earth-centred theory. Shen Kuo, another scholar of medieval astronomy, improved the designs of several astronomical instruments as the gnomon, armillary sphere and clepsydra clock. Shen is also known for his cosmological hypotheses in explaining the variations of planetary motions, including retrogradation.

The Earth studies of earlier scholars, particularly of Indians, were very elaborate and dealt with its origin, cardinal points, meteorology, geomorphology, seismology and physical divisions. As far as the origin of the earth is concerned, many of the facts as put forward by the ancient Indian scholars were more or less accurately known. The ancient Indian scholars believed in the solidification of earth from gaseous matter, its crust being made of hard rocks, clayey material and sandy material. They mention the earth to be apparently floating on the water like a sailing boat on the river. They were also aware of the fact that there is more land surface in the Northern Hemisphere. The ancient Indian scholars also used the concept of imaginary lines of *'Akshansh'* the latitudes and *'Deshantar'* the longitudes. On the basis of these, they attempted to divide the Earth into three global realms, viz. Equatorial, Northern Polar and Southern Polar with the identification of North

Pole and South Pole as the antipode of each other. They have also drawn Prime Meridian. These imaginary lines, the position of Sun and various stars have helped them to determine local time at various places. However, the world was not believed to exist beyond Equator, as the region here was compared to hell of the earth. The Eastern part, on the other hand, was believed to be 'the land of Gods'. This thinking is in consonance with that of the Europeans in the Early Medieval period, when the Dark Ages prevailed and the East in 'T-in-O' Maps was assumed to be the place of Adam and Eve. The Puranic literature formulated the idea of ten directions. Besides the main directions of East, West, North and South, there were included Zenith and Nadir, and added four more, viz. Southeast, Southwest, North-west and Northeast. The designation of these directions is significant in the sense that it bears the concept of the ruling deities dominating in each of these directions. The studies on Meteorology stretch back millennia. The ancient Indian scholars were aware of the vacuum between the earth and the heaven as *Antariksha*, its vast extent or thickness and the occurrence of various weather phenomena here. As far as the knowledge about the seasons is concerned, six seasons have been identified in *Valmiki Ramayan*, viz. Spring, Summer, Rainy, Autumn, Winter and Severe Winter. The observations of Chinese scholars on weather and climate are also commendable. They are known to have prepared a number of weather reports on the basis of empirically based facts. Besides, the philosopher Wang Chong accurately described the process of the water cycle and Shen wrote vivid descriptions of tornadoes. There are evidences of the interest of earlier Chinese scholars in the field of Paleoclimatology as well. Contributing to Seismological studies, the ancient Greeks, Indian and Chinese all provided the rational explanations scientifically alike. Thales and Democritus, for instance, found in water the cause of earthquakes. The deities of Air, Fire and Water cause the earthquakes or Bhukamp, as per Indian version. The Chinese kept accurate records of not only celestial events, but of earthquakes as well. They also discounted the superstitions related to the cause of earthquakes. In the ancient period, the knowledge about various parts of the world was limited due to the poor means of communication. Even then, the attempts were made to divide the world into several regions on the basis of available information. The term *Dwipa* has been used to designate various regions (continents) of the earth in India. Accordingly, the known world during the Puranic period was divided into seven *Dwipas* or 'regions', viz. Jambu, Krauncha, Kusha Plaksha, Pushkara, Shaka and Shalmali. These divisions exclude the American Continents, Greenland, England and Antarctica, since they were discovered only during the Age of Discovery in the late medieval period. The regions seem to have derived their names from the existing popular trees or grasses here.

Early contributions were made to the mathematical *sciences* including mathematics, Arithmetic, Algebra, Geometry, Trigonometry, Algorithm, Calculus and concept of Zero. The Vedic literature in India is replete with these concepts that were transmitted to other parts of the world and led to further developments laying foundations of the discipline. Among Greeks, Pythagoras excelled in mathematics. Known as "the father of numbers", he thought numbers were divine and believed that everything was related to mathematics; and through mathematics everything could be predicted and measured, just as in music, in rhythmic patterns or cycles. With the roots in Vedic literature, various treatises were authored by Indian mathematicians. Bhaskaracharya's book *Lilavati*, for instance, is regarded as the first book on modern arithmetic. The Indian mathematics recognized geometry as *Rekha Ganita* (Line Computation). The brick work of Harappa and Mohenjo-daro excavations show that people of ancient India (2500 BCE) possessed knowledge of

Geometry. However, it has originated in Egypt and was introduced into Greece. It was put into an axiomatic form by Euclid, whose treatment—Euclidean geometry—set a standard for many centuries to follow. The Arab scholar Mohammed Ibn Jubair al Battani studied and introduced various Indian geometrical concepts among the Arab scholars. Algebra is the Arabic name for *Bijaganitam* (*Bija* means 'another' or 'second' and *Ganitam* means mathematics). It was recognized as a parallel system of computation. In fact, the term 'algebra' is a corruption of the term *'Al jabr'* which the Arabs gave to *Bijaganitam*. Whatever the origin of algebra, it is certain that this technique of computation originated in India and was current around 1500 years back. Trigonometry is believed to have evolved in the Hellenistic world. Hipparchus made an early contribution to trigonometry producing a table of chords, an early example of a trigonometric table. He also introduced the division of a circle into 360 degrees into Greece. Trigonometry was introduced into ancient India through the translation of Greek works and was further advanced here. Later, its Arabic and Latin translations were very influential in Europe and the Middle East. Aryabhatta's contribution in this field of mathematics is unparalleled. Aryabhatt formulated the rules for finding the area of a 'triangle', which led to further advancement of Trignometry. The Siddhanta Shiromani of Bhaskara II is an astronomical treatise and contains many theories related to the preliminary concepts of infinitesimal calculus and mathematical analysis. Evidence suggests Bhaskara was deeply acquainted with 'differential calculus'. The Kerala School mathematicians expanded on Bhaskara's work and further advanced the development of calculus in India. Also the technique of calculation, called algorithm, which is today widely used in designing software programmes (instructions) for computers was derived from Indian mathematics. This technique which was named by westerners after Al Khwarazmi as "Algorismi" gave us the modern term Algorithm. Al Khwarazmi was the chief exponent of this Indo-Arab amalgam in mathematics who evolved this technique of calculation from Indian sources. Just like other mathematical inventions, the decimal number system in use today was first recorded in Indian mathematics. It was then transmitted to the Islamic world, and eventually to Europe. The earliest surviving evidence of decimal place value numerals in India and southeast Asia is from the middle of the first millennium CE. Decimal numerals recording the years 683 CE have also been found in stone inscriptions in Indonesia and Cambodia, where Indian cultural influence was substantial.

As far as Cartography is concerned, the Greek scholar, Ptolemy, was most influential among all. His contribution in the field of cartography and map-making is unparalleled. He designed a graticule and plotted the world using the concept of conical projection. On the Chinese realm, Pei Xiu and Zhang Heng were the earliest ones to promote and use cartographic techniques. Zhang Heng was the first to establish the grid reference system. However, the earliest known reference to the application of a geometric grid and mathematically graduated scale to a map was contained in the writings of the cartographer Pei Xiu who outlined and analyzed the advancements of cartography, surveying, and mathematics up until his time. The Chinese were familiar with the use of scale and Pei Xiu outlined six principles that should be observed when creating a map. The first three principles outlined the use of scale, direction, and road distance, and the last three used to properly calculate distances on uneven terrain as represented on a flat, two dimensional map.

The human aspect was intelligently dealt in various Historical, Regional, Anthropological, Psychological and Linguistic studies by the scholars of antiquity. The

Greeks like Herodotus, Polybius and Posidonius, and the Chinese like Ban Gu, and Sima Qian were the pioneers in the field of history. Although the writings of Herodotus cover only a short period in the history of Greek civilization, his account of the history of the wars is the first example of historical writing in a systematic manner in the Western World. Polybius was another historian to attempt to present history as a sequence of causes and effects. Polybius provided a unified view of history rather than a chronology. Following Polybius, Posidonius, in his 'Histories', gave an account of the rise and expansion of Roman dominance. It was not concerned with isolated political history of peoples and individuals, but included discussions of all geographical forces and factors which let humans act and be a part of their environment. The Chinese historians regard Sima Qian's work as their model, which stands as the "official format" of the history of China. Sima initiated a new writing style by presenting history in a series of biographies. Ban Gu's written work, on the other hand, set the trend for the establishment of geographical sections of history texts, and sparked the trend of the gazetteer in ancient China. Regarding the study of human beings, the explanations offered suggest that human culture developed as a response to necessity and the hardships of our environment. The study such phenomena as perception, cognition, emotion, personality, behaviour and interpersonal relationships had also been a concern in the past. Philolaus, for instance, recognized four distinct psychic faculties, viz. head, heart, navel, genitals. This hierarchy of four faculties is strikingly associated with the hierarchy of living beings (human, animal, plant) as well as a hierarchy of parts of the body. The core text for linguistics in ancient times, on the other hand, has been discovered in the '*Aṣṭādhyāyī*' of Pāṇini, the *sine qua non* of learning. Pāṇini's grammar can be considered to be the world's first formal system.

The sciences like physics, chemistry, botany, biology, Pharmacology, Medicine etc. are rooted in the works of ancient Greeks and Indians. Among Greeks, the foundation of physics lies in the writings of Empedocles and Posidonius. Empedocles had asserted that the matter is created by four 'elements', viz. earth, air, fire and water, by being 'mixed' in different combinations and proportions. The four elements correspond, at macroscopic level, with the traditional quadripartite division of the cosmos into earth, sea, air, and the fiery ether of the heavenly bodies. The knowledge of Indian scholars in this regard was more extensive. The concepts of atom (*Anu, Parmanu*) and relativity (*Sapekshavada*) were explicitly stated by them nearly 600 years before Christ in an elaborate manner. From the Vedic times, Indians not only classified the material world into four elements, viz. Earth (*Prithvi*), fire (*Agni*), air (*Maya*) and water (*Apa*); but, also added a fifth element, viz. ether or *Akasha* to these. These five elements, or *Pancha Mahabhootas*, were identified with various human senses of perception; earth with smell, air with feeling, fire with vision, water with taste and ether with sound. Indian philosophers also believed that these elements were physically palpable and hence comprised miniscule particles of matter. The Indian philosopher, Pakudha Katyayana, had also propounded ideas about the atomic constitution of the material world. The Atom was considered to be indestructible and hence eternal. Parallel to the development of the concepts of atom and atomic combinations in physics there also was a similar development of ideas in the area of biology, botany and chemistry. The science of living organisms, biology, reflected only in the works of Greeks in earlier times; the most successful scientific writings in this field were those of Aristotle. His ideas led to the development of a rational theory of evolution, as put forward by Darwin later. Again, it was a Greek philosopher, Theophrastus, a pupil of Aristotle, who is regarded as the founder of botany. He covered most aspects of botany from descriptions of plants, classification, plant distribution, propagation, germination to cultivation. Theophrastus classified plants into trees, shrubs, under-shrubs, and herbs. He described and discussed several species and varieties of plants from lands bordering the Atlantic and Mediterranean. For pharmacology, the ancient Chinese Scholars, as Shen Kuo and Su Song, held great concern. As far as Medicine is concerned, the *Charaka Samhita* of Acharya Charak is among the earliest and the most authoritative surviving medical manuals. The work reveals the facts on human anatomy, embryology, pharmacology, blood circulation and various diseases, and a description on the medicinal qualities and functions of about on lakh herbal plants. It represented a major advance over the superstitious ways of treating medical problems. Instead of appeasing deities and making sacrifices, practitioners were now looking at clinical problems and deciding how to treat them based on the specific disease.

The ancient civilizations had acquired proficiency in various technological fields as, for instance, metallurgy, Aviation, Civil engineering and architecture. Particularly, the Ancient Chinese are known for not only their travels and trade, but also for their many inventions and technological advancements. However, in the field of Metallurgy, the Indians excelled and there are many such examples that bear testimony to the excellence in smelting metals achieved in India in ancient times. Indian steel and iron were reportedly being used by the Romans for manufacturing armour as well as cutlery. Apart from this, Ashoka pillar at Mehrauli, New Delhi another iron pillar in Karnataka, and the copper statue of Gautama Buddha found in Bihar stand proof of India's metallurgical heritage. The discovery of urban settlements of Mohenjodaro and Harappa indicate existence of civil engineering and architecture, which blossomed to a highly precise science of civil engineering and architecture and found expression in innumerable monuments of ancient India. Through his innovative discoveries, Acharya Bharadwaj has been recognized as the pioneer of aviation technology.

Pinhole Camera, Water clock, armillary sphere, seismograph, Clock Tower, Magnetic needle compass, Movable type printing, Odometer and South Pointing Chariot were among several inventions in the field of Scientific Instruments and Appliances by ancient people. Shen Kuo was the first to experiment with the pinhole camera. The Water clock or Clepsydra, a timekeeping device, was in use in China as long ago as the Shang Dynasty. The Greek astronomer, Eratosthenes, invented the first armillary sphere, an astronomical instrument representing the celestial sphere. Su Song's armillary sphere was the first to be provided with a mechanical clock drive. However, Zhang Heng is the first person known to have applied hydraulic motive power, i.e. by employing a waterwheel and clepsydra, to rotate an armillary sphere. Zhang Heng's water-powered armillary sphere had profound effects on Chinese astronomy and mechanical engineering in later generations. Zhang Heng also presented the first seismometer. Shen Kuo's writing on magnetic compasses has proved invaluable for understanding China's earliest use of the compass for seafaring navigation. An obscure commoner and artisan known as Bi Sheng invented ceramic movable type-printing. Bi Sheng's methodical innovation was revolutionary for his time. Zhang Heng is often credited with inventing the first odometer. Similar devices were used by the Roman and Han-Chinese empires at about the same period. The South Pointing Chariot was another mechanical device credited to Zhang Heng.

The travels and explorations fascinated the Europeans and Chinese alike. Hanno the Navigator, Pytheas and Alexander the Great among Greeks, explored Africa, north-west Europe, British Isles and much of what was then the civilized world. Travels beyond

Chinese horizons were undertaken by orders of an emperor or by missionaries or traders which reveal the story of high adventures. Travelling beyond their original cultural hearth area, the discovery of Mediterranean civilization, for instance, is credited to Chinese traveller, Chang Chien. Among the other Chinese travellers most distinguished ones included the Buddhist monk, Huan Tsang who crossed the high windswept plateau of Tibet and the world's highest mountain on the way to India; Chou Ta Kuan who visited Combodia and wrote a detailed account of the strange customs of the Combodians; and Cheng Ho, a Chinese admiral who sailed towards Java, Malaya and even as far as India. These journeys had also promoted a great variety of economic and cultural exchanges among regions, and opened regular trade routes between the West and the East.

With the efforts of these European, Indian and Chinese scholars, a large amount of geographical knowledge was accumulated. New researches were thereby attracted and the research made progress. With these three parallel schools of thought, geography witnessed a great degree of professionalization, With time, however, the West became the centre of continuously increasing scholarly research; and the Graeco-Roman (European) school started becoming dominant as it developed new methods that come to be regarded as more interesting or significant. In other words, geography witnessed its first paradigmatic phase of development. The major contribution of Greeks and Romans was a body of philosophical and cosmological ideas. The Paradigm Phase-I of geography saw the emergence of various sub-fields of its learning as physical, human, historical, theoretical, political, regional, mathematical and practical covering almost entirely the gamut of geographical knowledge. Geography as a discipline got enriched by a number of concepts, viewpoints, techniques and terminologies. Besides, coining the term geography, a number of others are also credited to the ancient Greeks, as for instance, cartography, chorography, longitude, latitude, torrid, frigid, temperate etc. that still constitute the basic vocabulary of geography. The ancient Greeks were the originators of literary and mathematical traditions in geography. Thales and Anaximander, particularly, are known as the beginners of mathematical tradition. Academically, the ancient Greek and Roman scholars formed strong ties of geography with history and this inter-disciplinary bond could not be broken for several centuries to come, and has been intensely represented in the writings of geographers till the modern times. The roots of the concept of 'Areal differentiation' as adopted by Hartshorne in the modern period lie in the writings of Strabo, the first century B.C. Roman scholar. Ptolemy's geocentric view of the universe remained an accepted doctrine until the Age of Discovery. The regional descriptions reveal that the existing world of the times was believed to have two major regional divisions, namely, Europe and Asia; besides, parts of northern Africa were also known. One of the major contributions of the Greek scholars was towards suggesting the methods of theory building and hypothesis testing. Plato, a well known Greek scholar, favoured theory building by intuition and reasoned from general to particular, the deductive approach. On the contrary, Aristotle suggested the method of reasoning from particular to general, the inductive approach. These methods of deductive and inductive reasoning were introduced to geography as early as fourth and fifth centuries B.C. The Greeks are also credited for the introduction of Gnomon, one of the first scientific instruments ever made. The Graeco-Romans School, thus, led to concentrated research in geography, an activity described as Normal Science. It was the Greeks who first handled the information gathered by the middle-Easters scientifically.

The first phase of Crisis with Revolution in geography, popularly known as Dark Age in Europe, was witnessed in the Early Medieval Period. This Phase saw geographic horizons closing in due to theocratic slant in human thinking and writings. This kind of thinking had an adverse impact on the art of map-making, which got distorted. There emerged maps, which were purely fancy and unscientific. The earlier concepts of sphericity and the nearly correct distances of places, latitudes and longitudes were ignored. Christianity influenced the social, economic, literary and philosophical developments of the time. The Christian era was marked by the loss of ancient scientific concepts about the world and their replacements by unscientific, uncritical cosmogonies based largely on scriptures. The missionary travels were the only source of knowledge for the regional account of different places during this period. Consequently, Geography made a little progress as the organized body of knowledge. The quest for scientific learning stopped; instead, there was regression of the already acquired scientific knowledge. The image of the world was molded to fit the Biblical references, which were considered more valid. The earth was considered as a flat disc with Jerusalem at the centre. The ideal map of this period was 'Orbis-Terrarum' or 'T-in-O' Map which originally included the Mediterranean Sea and its surrounding territories. The three divisions, namely Europe, Asia and Africa were accepted as standard. Along with the inhabited world, there were inserted the mythical places, beasts and dragons in all these maps. These maps were made very beautiful and extremely decorative, but did not have any relation with actual facts, as they were far from reality with no scientific base. Also, there was vagueness in the direction. The sphericity and the nearly correct distances of places, latitudes and longitudes were ignored. The descriptions made were also purely aesthetic and without logic. Thus, most of the classical correct concepts were forgotten, and the old errors appeared again about the world maps and habitable portions of the earth. Moreover, the involvement of Christians and Muslims (Arabs) in crusade wars further aggravated the problem. When Europe was plunged into Dark Ages, the Arab School of Thought was rising to prominence in medieval times. The rise of Arab school marks the end of the recession in development of geographic literature and return to more scholarly research. This lead geography to enter in a new paradigm phase of development.

Medieval philosophers were the historical successors of the philosophers of antiquity. Medieval Thought in geographic development represents two major philosophical trends, viz. European and Middle-Eastern. However, the Paradigm Phase-II is associated with the Rise of Arab School of Thought during High Medieval Phase. Medieval philosophy in the West represents the philosophy of Western Europe during the Fall of Rome and the Renaissance. It was shaped by two pieces of the cultural legacy of late antiquity that survived the collapse of Roman civilization. The First of these was the 'Latin language', which provided medieval thinkers with access to some important ancient resources that gave early medieval thinkers an acquaintance with classical ideas. The Second was 'Christianity', stimulating the philosophical activity. Its ideas and doctrines constituted a rich source of philosophical subject matter. Medieval philosophy, here, took root in an intellectual world sustained by the Church and permeated with Christianity's texts and ideas. This, in turn, lead to, First the influx into the West of a large and previously unknown body of philosophical material newly translated into Latin from Greek sources, and Second, the emergence and growth of the great medieval universities. There was rise of Islam in Arab in seventh Century and the new zest of learning spreaded through the world of Islam. The vast expanse of the Muslim Empire facilitated the coming together of intellectual and scientific traditions of various civilizations, particularly Greek and Indian. The Arabian conquest of Palestine, Syria, Persia and Bactria, where Greek culture still flourished, gave to the Muslims the rich fruit of Greek learning. Baghdad was the major centre of intellectual world. Islamic geography was primarily a continuation of the Greek and Roman scholarship which had been lost in Christian Europe. The Arab scholars and travellers presented a new outlook of the world, which later led to the onset of the Age of Discovery. With the rise of Arab School, geography witnessed another paradigmatic shift in development. Many Arab scholars contributed to it, their outstanding contributions going to various fields of learning as Mathematical and Practical Geography, Regional and Historical Geography, Physical Geography and Climatology and Human Geography.

Since Greek science was the starting point in Arab School, it is not surprising, therefore, that mathematical geography received the maximum attention. Besides, the factors like semi-desertic conditions, open atmosphere, clear skies and vast distances to scan that facilitated the astronomical observations also helped in the promotion of studies in mathematical and practical geography. The early Arabs considered the earth as the centre of the Universe, around which revolved the planets. There were several attempts to ascertain the size of the earth. Sea-Charts and maps based on cylindrical projection were made. Borrowing from and improving over the works of Greeks, the Arabs divided the globe of earth into Torrid, Frigid and Temperate zones on the basis of temperature conditions. Besides, another greatest contribution of the Arabs to the advancement of mathematical knowledge was the introduction of the Hindu invention of the zero in arithmetic (*Hindsa*). Before this, the Roman numerals were in use and their range was limited. A wealth of regional descriptions exists in the works of Arab scholars, covering both the physical and human aspects. The writings of Ibn Batuta described the ecumene areas of the world. Accordingly, the limit of the habitable world was identified as up to the end of Mediterranean sea in the west, Japan in the east, Siberia in the north and about 10 degrees beyond equator in the south. The travels of Ibn Batuta brought empirical proof that Aristotle, the ancient Greek scholar, had been wrong in asserting that the Torrid Zone was too hot and unfavourable for human habitation. Batuta, on the contrary, confirmed that numerous tribes occupied it. The physical geography finds a special place in the writings of the Arab scholars. Their important observations are related to the study of the processes shaping landforms, alluvial deposits and phenomenon of tides. Al Biruni in his book entitled '*Kitab-al-Hind*' describes the fluvial processes and alluvial deposits of Himalayan region. Ibn Sina keenly observed the processes shaping landforms, viz. weathering and denudation. He formulated the idea that mountains were being constantly worn down by rivers and that the highest peaks occurred where the rocks were especially resistant to erosion. He also noted the presence of fossils in the rocks of high mountains. The observations related to the phenomena of tides and proved that they are caused because of the gravitational pull of the sun and the moon. Al Masudi, for instance, recorded the spring and neap tides while navigating the Caspian Sea. Besides, Al Idrisi corrected the erroneous ideas about the Indian Ocean and the Caspian Sea. The Indian Ocean was supposed to be enclosed, whereas the Caspian Sea was shown as an arm of the World Ocean in 'T-in-O' maps of Early Medieval period. However, the most notable under 'physical' were the contributions made to climatology. The important contributions in the field of Climatic Geography have come in the form of coining of the term '*mausim*' (Monsoon), preparation of the first Climatic Atlas, and division of the world into Climatic Regions. As far as the human studies of Arab scholars are concerned, modern geographers are particularly impressed with the cultural interpretation of the physical environment by Ibn-Khaldun and his analysis of the role of the city in regional economy. He tried to correlate man and environmental relationships in a scientific way. Ultimately, this Paradigm phase is known for laying roots for the philosophy of environmental (climatic) determinism, generating a voluminous literature on the ecumene and non-ecumene areas of the known world of their times and presenting a new outlook of the world that later led to the onset of the Age of Discovery, another crisis phase in geography.

Largely inspired by the writings of Arab scholars, the geographic horizons were pushed back during post-Arab period. The discovery of new lands, improvements in cartography and correction of old maps, and various scientific and technological inventions characterized this age. With the improvements in the art of navigation and the use of magnetic compass in the travels became very popular. Great explorers emerged in the scene and discovered the world that was unexplored earlier. As a result of the developments like the establishment of the heliocentric concept of the universe, numerous astronomical observations by Nicholus Copernicus, the presentation of Laws of Gravitation by Isaac Newton, and the invention of Printing Machine in Europe, there came a scientific revolution not only in materialistic terms but also in the ideas and thinking of man. Geography was enriched by an immense collection of facts and information about the world. Ptolemy's maps were restudied and the missing details were filled in and the discrepancies removed. There emerged lot of scope for scientific investigations and generalizations began in geography. Besides the discovery of new lands the Age of Exploration was rooted in new technologies and ideas growing out of the Renaissance and these included advances in cartography, navigation, and shipbuilding.

Although the travels had become very popular in Arab period itself, a number of other developments gave further impetus to the explorations; and they included, e.g. colonization, use of magnetic compass and improvement in the art of navigation. Great explorers emerged in the scene. Numerous journeys of geographical exploration were commissioned by a variety of nation states in Europe. Most of these voyages were financed because of the potential commercial returns from resource exploitation. The voyages also provided an opportunity for scientific investigation and discovery. These voyages also made significant contributions to geographic knowledge. Among these explorers most were of Italian, Portuguese and Dutch origin, the others being Spanish, Russian and English. The Italian discoverers explored the lands across the Atlantic Ocean. John Cabbot was the first European to search for the North-west Passage; he discovered the island of Newfoundland. Christopher Columbus is commonly credited as the first European to see the Americas and his voyage marked the beginning of European exploration and colonization of the Americas. However, this new continent was named as "America" after Amerigo Vespucci's first name, another Italian voyager. Among the Portuguese, Pedro Álvares Cabral is generally regarded as first European discoverer of Brazil; Vasco da Gama discovered an ocean route from Portugal to the East and was the first person to sail directly from Europe to India, through Cape of Good Hope in Africa; and Ferdinand Magellan led the first successful attempt to circumnavigate the Earth and became the first person to lead an expedition sailing westward from Europe to Asia and to cross the Pacific Ocean. The Dutch navigators explored the parts of Europe and Oceania. Among these, Willem Barents was a leader of early expeditions to the far north, after whom the Barents Sea is named Willem Janszoon was the first European known to have seen the coast of Australia; and Abel Janszoon Tasman is best known for his voyages to reach the islands of Tasmania, New

Zealand Fiji islands. Among other explorers, Juan Ponce de León (Spanish) was the first European to set foot in Florida, establishing the oldest European settlement in Puerto Rico and discovering the Gulf Stream, a current in the Atlantic Ocean; Yermak Timofeyevich (Russian) was the explorer of Siberia; and Captain James Cook made several voyages to the Pacific Ocean, accurately charting many areas and recording several islands and coastlines on European maps for the first time. His sailing around the Pacific Ocean contributed to European knowledge of the area. Several islands such as Easter Island and the Sandwich Islands (Hawaii) were encountered for the first time by Europeans. With the discovery of new lands across the Atlantic, Pacific and Indian Oceans, the Age of Exploration is generally said to have ended in mid eighteenth century. Almost all the expeditionary leaders relied on Ptolemy's calculations about latitudes and circumference of the earth. As a result of these voyages huge knowledge was accumulated. There was a huge mass of literature that had to be tested, to be made accessible to others.

Mainly the invention of printing machine and the works of the scholars like Copernicus, Kapler, Galileo and Newton brought the scientific and technological developments in geography during the Age of Discovery. The invention of 'Printing Machine' in Europe aided the scientific developments by publishing the accounts of explorations. Copernicus' numerous experiments on celestial mechanics explained planetary motions as elliptical and provided the first modern formulation of a heliocentric theory of the solar system. Galileo's achievements included invention of the telescope, a variety of astronomical observations, the laws of motion, and effective support for Copernicanism. Isaac Newton, the English physicist and astronomer, presented the Laws of Gravitation. The unifying and deterministic power of his laws was integral to the scientific revolution and the advancement of heliocentrism.

The field of geography that was the most benefited one during Age of Discovery was Cartography. Ptolemy's maps were unearthed and restudied. His calculations about latitudes, longitudes and circumference of the earth were corrected. The shape of the earth, i.e. elliptical, was finally ascertained. With the filling up of missing details and the removal of discrepancies, the world map became almost complete. On this basis the first accurate globe of the world was made by Martin Behaim, a Portuguese navigator, depicting the Earth in its true three-dimensional form. Besides, Mercator's Projection, based on the concept of the line of constant bearing, was devised, especially, for navigation. With all these developments, there came a scientific revolution not only in materialistic terms but also in the ideas and thinking of man. The quality of geographic literature, as such, improved. The impact of the developments during Age of Discovery on Geography was tremendous.

The positive gains of this period may be summarized as: the Discovery of new lands and completion of world-map; the revival of interest in geography from a scientific perspective; Academic Freedom of geography, i.e. identification as an independent field of study; accumulation of huge amount of data of geographical nature; split of geography in various specialized fields of learning as physical, human etc.; and advancements in Cartographic technique of data representation and map-making. Bringing all these together, Bernard Varenius, the German Scholar, presented the first modern literature of geography in the form of *Geographia Generalis*. Geography, as such, witnessed the beginning of scientific investigations and generalizations. This lead to the emergence of various modern schools of thought, as German, French, American and British. Especially, through the work of Varenius, Geography acquired a new definition and methodology and in his attempts, the foundations of Modern period of geographical thought were already laid before the Medieval period came to a close after the Age of Discovery.

The Modern period witnessed three major and rapid paradigmatic shifts (III, IV and V) in geography. It was as a result of the rise and dominance of German School inducting the studies with Systematic and Regional approach; origin of Determinism focusing the role of environment in human affairs; and emergence of Videlian tradition as an antithesis to Determinism, with a thrust on man in man-environment study. Major paradigms of this phase included Determinism, Possibilism and Chorological (Human and Regional) view. The intervening Crisis phases developed various dualistic tendencies addressed to these doctrines in geography. Particularly, three modern schools of thought have been instrumental behind this development, viz. German, French and American.

The Paradigm Phase III represents developments during the next 300 years after the Age of Discovery when geography was largely a German science. Lot of research was carried on towards defining geography and deciding upon its methodology. Geography was no more than encyclopedic description of places in late medieval and early modern period. The principles behind spatial arrangement of phenomena, their causal connections and inter-relationships that could give geography a true scientific status were yet to be established. The credit for this goes to the German scholars particularly to the writings of Varenius, Kant, Humboldt and Ritter. Undoubtedly, the initial impetus to German School had come from the works of Bernard Varenius. Varenius brought together contemporary knowledge of astronomy and cartography and subjected the different theories of his day to sound critical analysis. Through this he tried to provide a scientific methodology to geography by advocating two approaches for its study, viz. 'General' (systematic) and 'Special' (regional). General Geography consists of description of three forms of earth features, viz. Absolute, Relative and Comparative; similarly, Regional Geography also comprises descriptions of three properties of the constituent areal units, viz. Celestial, Terrestrial and Human. The geography of this period included both general studies of particular kinds of phenomena of the earth's surface, and descriptions of many kinds of phenomena found in particular areas. The terms that Varenius used: 'General' and 'Special' geography became standard in the discipline. Among later geographers, Humboldt and Ritter, popularly known as the fathers of modern geography, drew attention to Varenius' genius. They did it respectively in the form of a natural scientist and a human geographer. In the meanwhile, writings of Emmanuel Kant guided the research in Germany. For Kant, Geography was a chorographic and mainly a descriptive science, distinct from the systematic sciences. He considerably modified the scientific structure of geography as proposed by Varenius. Kant agreed on the two ways of acquiring knowledge, viz. through 'sense perceptions' (inductive) and through 'pure reasoning' (deductive). However, Kant's approach to geographical studies was purely deductive. Kant also pointed out that there are basically two ways of classifying knowledge for the various phenomena on the surface of earth. They are: 'Logical Classification', i.e. grouping according to the nature of the inherent characteristics of the phenomena; and 'Physical Classification', i.e. grouping according to the position or occurrence of phenomena in time or space. The former deals with organizing knowledge according to the type of objects studied, and accordingly covers a series of systematic or pure sciences that are devoted to the study of particular kinds; the latter, on the other hand, deals with organization of knowledge in its temporal or spatial dimensions and, therefore, includes historical and spatial sciences. Considering both history and geography as descriptive in nature, Kant also regarded both as complementary to each other. Besides, Kant laid the greatest emphasis on the study of Physical Geography. Along with it, he also identified some other branches of geography, namely: Mathematical Geography, Political Geography, Moral Geography, Mercantile Geography and Theological Geography. Kant has been labelled as 'the father of exceptionalism' in geography by F.K. Schaefer, a twentieth century American scholar, for laying an exceptionalist claim not only for geography, but also for history. However, in spite of all the criticisms put forward against Kantian philosophy, the basic tradition of the field of geography has persisted in his writings, i.e. "geography is a description of the whole earth, where the earth is understood to mean the surface of the earth and the dwelling place of the man". The exposition of geography as 'chorological' in nature by Alfred Hettner and Richard Hartshorne in the late nineteenth and early twentieth century had closely followed the Kantian philosophy.

After Kant, Alexander von Humboldt and Carl Ritter revived the tradition of Bernard Varenius. Both were contemporary, and they together tried to mould the subject matter of geography into scientific form, following the tradition of Varenius. The prevalent philosophies at their time were those of natural and moral; the former related to the physical world, envisaging the earth as an organic whole, inseparable and mutually interdependent; whereas, the latter were related to man's inner world and human thinking. Humboldt's basic inclination was towards natural, whereas Ritter had anthropocentric view. Ritter envisaged the earth as a whole, an organic whole, all parts of which being mutually interdependent. Like Kant, he also attempted on a division of knowledge, but the classification he contemplated fell into three groups, viz. Systematic, Historical and Earth Sciences. Although, the basis of this classification was close to that of Kant, but, Humboldt accorded all kind of subject-matter (including history and geography) the status of sciences. While dealing with the subject matter of geography, Humboldt coined a term 'cosmography', the study of the Universe; and divided it into Uranography and Geography, the former concerned with mapping the stars, galaxies or other celestial bodies; and the latter studying the terrestrial part with the interrelationships of phenomena that exist together in an area. Humboldt followed empirical, inductive, systematic and comparative methods of study. Humboldt conceived man and his activities as an integral part of the nature. This theme runs through the 'Cosmos', his best-known work. Contrary to and complementing Humdoldt, Ritter exercised profound influence on geography by adopting a different methodology. He emphasized more on human, and favoured regional or synthetic approach, working basically as an armchair geographer. Just like 'Cosmos' of Humboldt, Ritter also authored a great work entitled '*Die-Erdkunde*'. Since Ritter was deeply a religious man, he thought that divine has created the earth as a home of man. He also had a vision of an ordered and harmonious universe, but, his approach was teleological. He asserted that all phenomena are spatially distributed according to the plan of God for mankind. In spite of being an armchair scholar himself, Ritter did identify geography an empirical and descriptive science. According to him, geography deals with local conditions and embraces the attributes of a place in the form of topographical, formal and material characteristics. These three attributes constitute the basic content of his most important work, *Die-Erdkunde*, as well. '*Die Erdkunde*' is based on the regional descriptions of the earth covering largely Asia, Africa and Europe. This work was written with a distinct theme, i.e. comparative and inter-relationships. Ritter claimed that the central principle of geography is the relation of all phenomena and forms of nature to the human race, examined and organized within the framework of the unique geographical associations of land and man on the earth's surface". Through his Erdkunde, Ritter made a rhetorical claim for geography as the science of the earth. He believed the earth to be an organism made, even in its smallest details, with divine intent, to fit the needs of the man to perfection; as the body is made for the soul, so is the physical globe made for mankind. Besides, Ritter is also popular among geographers for his Regional Concept. His division of the Earth consists of four levels in hierarchical order: Earth or *Erdkunde* (First Order); Continental Units or *Erdteile* (Second Order); Division of each Continent into various relief or Orographic regions (Third Order); and Smaller Areal or terrestrial units arrived at inductively (fourth Order). His plan of study, based on the concept of regional associations of terrestrial phenomena, became the model for geographical presentations and had far-reaching influences.

This classical period of German geography terminated with the death of both Humboldt and Ritter in the year 1859. Both Humboldt and Ritter worked towards providing geography as the status of a scientific discipline and appeared, as such, inspired by Bernard Varenius. The writings of Humboldt and Ritter had a thrust on inter-relationships, spatial associations or spatial integration of phenomena. The Physical Geography was considered as the base of all geographical studies by Humboldt and Ritter. Both emphasized on the importance of empirical method of research in geography. Although both believed in the Unity of Nature, they differed in their philosophical approach to geographical studies. For Humboldt, this unity of nature was a balanced unity of the whole of nature, of which man was a part; whereas for Ritter, it was teleological, as he spoke of discovering the divine secrets. In spite of all the differences, the works of Humboldt and Ritter together, particularly 'Cosmos' and 'Erdkunde' did provide an almost complete explanation for geography, in terms of its definition, content, scope, approach and method of study. Coincidently, both of them died in the same year. The year 1859 marked not only the death of Humboldt and Ritter but also the publication of Darwin's "On the Origin of Species". *Charles Darwin* (1809-1882) was an English naturalist who established the theory of organic evolution.

From the early nineteenth century to mid twentieth century, geography was developing in several directions due to distinct schools of thought developed under the influence of Humboldt, Ritter and Darwin, and there were followers of each. The geographical studies were basically bifurcated in two directions: Study of earth as a natural body and Study of earth as a dwelling place of man. At the same time a number of dualisms also overshadowed the discipline for about a century after the emergence of German School of Thought. It is because, in the latter half of the nineteenth century, the followers of Humboldt and Ritter exaggerated certain aspects of views of each of the founders. However, in the late nineteenth century, the more expanding frontiers were those of physical and biological sciences than human under the impact of Darwin. Darwinism, dualistic tendencies and a physical bias in geographical studies brought another turning point (Crisis Phase III) in the history of the discipline. The philosophy of Darwin caused considerable intellectual ferment and had a tremendous influence on geographic thought. This influence came through his theory of 'evolution by natural selection' and the concept of 'survival of the fittest'. The cosmographic way of thinking was replaced by the causal explanations characteristic of natural sciences. Darwin was primarily concerned with the mechanism of change and he rejected all supernatural phenomena and causations. Eliminating God from science made room for strictly scientific explanations of all natural phenomena; it gave rise to positivism. It produced a powerful intellectual and spiritual revolution, the effects of which have lasted to this day. This intellectual ferment led to a great reappraisal of thinking in both the natural and social sciences. As a result, traditional natural philosophy professed by amateurs became transformed into modern science developed by professional scientists. A new term 'Social Darwinism' emerged in the second half of the nineteenth century to refer to various ways of thinking and theories that tried to apply the evolutionary concept of natural selection to human society.

The dualistic tendencies that have always been part of geographical studies since beginning became more conspicuous after the death of Humboldt and Ritter. The major controversies that emerged in geography during modern period were related to its subject matter and approach to study, viz. 'Physical *versus* Human' and 'Systematic *versus* Regional' respectively. It is particularly with the writings of German scholars like Kant in the initial stage and of that of Humboldt later that physical aspect of geographical studies got greater emphasis than human side. Humboldt conceived man as the part of nature, dependent upon nature-minus-man, which existed before man and which could exist independent of man, whereas the reverse cannot hold true. His 'physical geography' was directly comparable to the 'general geography' or the 'systematic geography' of Varenius. Carl Ritter, on the other hand, had an anthropocentric approach to geographical studies. Ritter conceived Earth as the home of man. According to him, the physical surroundings get meaning only with reference to human beings. They are required to be known only in relation to human beings. Because, it is the man who interacts with the physical world. The relation between the world of man and the non-human world is of prime concern in geography. Since the earlier times, the human and the non-human aspects have contributed to the content of geography. The dichotomy is what aspect is more important to study in geography, physical or human. Originally the controversy between man and physical (nature) was external to geography. It appears to have entered into the discipline first with the writings of German scholars, and was later supported by the rise of natural sciences and Darwinism. The important sources of problem may, therefore, be identified as: the works of German scholars; Environmentalism (Determinism and Possibilism); and Systematic versus Regional Dualism.

The terms 'General' and 'Specific', as suggested by Varenius, were caught by various geographers later and the dichotomy between the two widened to the extent that there had been periods when there existed double dualism in geography, i.e. 'Systematic and Physical' versus 'Regional and Human". One such time was that of Emmanuel Kant. Kant substituted the word 'physical' for 'general', and classified all the general or systematic studies under 'physical'. It was under his influence that the geographers in late eighteenth century concentrated on systematic or general studies at the expense of regional or specific studies. The close of nineteenth century found an increased interest on physical geography; and the systematic studies in their new guise were highly specialized. They were undertaken by scientists who tended to be geomorphologists and climatologists. This was a time when 'systematic' and 'physical' became synonymous and the study of landforms emerged as the dominant and most appealing part of geography. This attracted the researchers not only from Germany but also from outside, especially U.S.A. The development that started after 1859 culminated in the works of W.M. Davis. In comparison, Human Geography lagged behind, because some geographers wanted to make geography a preserve of pure science. Since human behaviour did not appear to be subject to general laws, they proposed to exclude man entirely from the field of geography. The early twentieth century saw a preponderance of physically-biased studies in geography, particularly brought in by the German and American Schools. Thus, after the death of

Humboldt and Ritter in 1859 till the First World War, the emphasis of geographical studies kept on shifting in the writing of various scholars. Although this shift was evident in the form of Physical and Systematic (Peschel, Gerland), to Physical and Regional (Richthofen, Penck), to Systematic and Human (Ratzel), to primarily Regional (Hettner, Schluter) and predominantly Physical (Wegener, Penck) during the late nineteenth and early twentieth century, the physical bias was very strong in most of these works.

It was the Darwinian tradition that led geography into next phase of professionalism, i.e. Paradigm Phase IV. Rise of Determinism was the major philosophical outcome of this development which happened in late nineteenth and early twentieth century. Darwin's paradigm in conjugation with that of Davis's replaced all the Old traditions. Davis sought to integrate the two elements, viz. Nature and Man. This he did after a period in which a marked dualism prevailed in the discipline, i.e. between exclusively physical geography (including plants and animals) and the geography of man. Davis found the key at this time in the Darwinian evolutionary approach that regarded the facts of nature and man as existing in a 'harmonious relationship' in a continuous chain of cause and effect. The writings of W.M. Davis played a great role in strengthening professionalization in the discipline during this period. The 'Theory of Cycle of Erosion' was his most important contribution. Representing the prevailing deterministic thinking Davis restated that the core of geography is the study of the relationship between the physical environment—the control and the human behaviour—the response. Davis sought to classify the facts of man in reference to the facts of the physical earth, calling them respectively 'Ontography' and 'Physical Geography'. Davis identified 'Ontography' as the other half of geography, defining it as geographical description of life forms on the earth. This was an extension of the evolutionary interpretation of man on the earth in so far as human characteristics show direct results of geographical (physical) influences and controls. Davis believed that such a system would form the framework of an integrated science. This viewpoint very clearly set the pattern for the pursuit of man-land relationships on a deterministic basis. Besides Davis, the contemporary thinking was also dominated by the ideas of his pupils like Ellen Churchill Semple, Ellsworth Huntington and Harlan Barrows. The environmentalist approach was taken up in their far ranging researches. Semple was an environmental determinist. Huntington was known for his studies on climatic determinist and examined the influence of climate and climatic change on men and the course of history. Harlan Barrows, on the other hand, presented the concept of geography as human ecology, stating that man's adaptation to the physical environment is the function of human choice. He proposed human ecology as the unifying theme for geographical work.

Among geographers influenced by Darwinism, it was Friedrich Ratzel, the German scholar, who explored the influences of physical environment on mankind. He corrected the prevailing tendency of overstressing the physical aspects of geography and established a more balanced viewpoint. Influence of the thinkers like Charles Darwin and the zoologist like E.H. Haekel, reflects the writings of Ratzel. In his two most important literary contributions, 'Anthropogeographie' and 'Politische Geographie', Ratzel treated human geography from Darwinian's point of view. His works served to fix interest still more firmly on systematic studies, and regional geography continued to receive scant attention. Scholars like Huntington, Semple and Davis further extended and elaborated his work into what become known as Environmental Determinism; a view that dominated geography for decades. The basic theme of the first volume of Ratzel's Anthropogeographie is that the man and his actions are not free from the influences of the surroundings. They are controlled by nature. He believed that man is the creature of his environment in much the same way as Darwin had demonstrated the adaptation and survival of the fittest in the animal world. However, in its second volume, written with more liberal views on man-nature relationships, he showed a different spirit, where the emphasis was on distribution and density of population, forms of settlements, migrations, and diffusion of cultural traits. Here, he added that the survival of man and his adaptation to the environment also depends on fact that how fit and skilful he is in moulding the natural circumstances. Man's reactions to the nature differ sharply accordingly to his culture and historical background. If belonging to primitive culture then nature has a stronghold; and if advanced technologically, then the nature's hold becomes weaker. But, even then the nature is dominant. However, according to Ratzel, whether man adapts or modifies the surroundings, in any case, is the creature of his environment. Because, even the act of moulding is a kind of adaptation prompted by the nature in an indirect manner. Ratzel saw man as the end product of evolution, in which the mainspring was the natural selection of types according to their capacity to adjust themselves to the physical environment. Through Politische Geographie, his other book, Ratzel reshaped the old topic along the lines conformable with the principles he had enunciated in his Anthropogeographie. In this work, he tried to make a comparative investigation of the relation between the political state and the earth's surface. The influence of revolutionary biology led Ratzel to adopt the organic theory of state and society. The political entities (nations or states) expand or shrink like organisms. Based on this viewpoint, he postulated the 'Organic Theory of State', the concept of state as an organism, a piece of humanity on the surface of the earth. He believed that a nation or state is not only a spatial unit, but also a human entity, and therefore, it is a living unit. From these ideas, Ratzel developed the concept of 'lebensraum' or living space, and hypothesized that the state naturally seeks to increase its size. As evidenced, Ratzel believed that space was a great political force.

Ratzel is known as the propounder of Environmental Determinism in geography. According to this viewpoint, man is the product of nature. It is the nature, which governs man and his activities, and determines his way of life, living styles and also physical appearance. Determinism emerged as one of the ways to study man and environment relationship. The view persisted up to the Second World War in one form or the other. Supporters of the determinism believed that nature is superior to man and man is a passive agent on whom the physical factors are acting and determining his attitudes and the process of his decision-making. The basic argument of the environmental determinists was that aspects of physical geography, particularly climate, influenced the psychological temperament of individuals, which in turn shaped the culture and society that those individuals formed.

However, the deterministic thinking was not new at this point in time. Its origin goes back to antiquity, when various Greek and Roman scholars made first attempts to explain the physical features and character traits of various people and their culture with reference to the influence of natural conditions. The geographical determinism continued to reflect the writings of Arab Scholars in the medieval period as well. Determinism was also strongly observed in the work of German Scholars. The philosophy of determinism was greatly influenced by the high profile of evolutionary biology of Charles Darwin later. The pre-Darwinian views on Determinism were found relatively more modest in comparison to that in post-Darwinian period, when they attained more extreme form. The maximum influence of deterministic thinking was exercised after the popularity of Darwin's Theory.

The scientific developments in geography that followed Darwin were deeply influenced by the role of nature on organisms, including man. This role of nature was deeply studied in human affairs. Gradually the term 'environmental influence' was replaced by the 'environmental control', i.e. the environment does not simply influence but determines. The observations of various scholars caused them to conclude that 'man's immediate surroundings have a decisive (deterministic) effect not only on his occupations, dwellings and the materials he uses in everyday life, but also on his personal traits, his general character, and even his religious beliefs. They found environment to be superior to man and mastering all of his actions. Such generalizations were called 'extreme' and not as the result of 'cause and effect 'relationship between man and environment. Thus, there emerged extreme form of determinism in the late nineteenth century. The scientific milieu in the later half of the nineteenth and the early twentieth centuries was dominated by Darwin's idea and the deductive approaches. Fitting well into this intellectual environment, the theme of environmental determinism was prevailing view not only in Germany but also in American geography. Ratzel supplemented 'classical' geographical determinism with the elements of 'Social Darwinism' and developed a 'theory of the state as an organism' that owed its life to the earth, which was ever striving to seize more and more territory. In spite of its vigourous criticism the school of Determinism did not die away completely, its central theme continued to survive in the form of Neo-Determinism. Where some of the early environments were apt to make unsound generalizations, the modern determinists were more careful, to seek and record a true sequence from cause to effect. Griffith Taylor used the term 'Stop-Go-Determinism' for Neo-Determinism.

Geography encountered another crisis phase in early twentieth century when there was widespread criticism to Determinism. Resultantly, there took place a paradigmatic shift to Human from Physical. However, this turned out to be relatively a very short period in transition, as the discipline soon witnessed the next Paradigm phase. By 1900, the philosophy of determinism had become particularly widespread in Germany and the thinking also percolated in U.S.A. But, from the beginning of twentieth century this doctrine came under attack. The chief fault in the assumption of the extreme determinists was found in their over-simplication of the man-environment relationships. Their generalizations were accused of being 'too-sweeping', without being backed by careful investigations and cause-and-effect relationships. The determinists were blamed for over-emphasizing the role of physical environment in human affairs. Man was seen only as a passive element of the environment. Such works of man were overlooked for which environmental forces alone give no satisfactory explanation. Although, the school of determinism, as such, had a short span of life, its central theme did not die out completely. It survived and continued to persist in the form of 'Neo-Determinism' or 'Stop-Go-Determinism'. In spite of the emergence of Neo-determinism, the move towards human studies in geography continued to grow stronger. Largely, as a reaction to the exclusive concentration on systematic studies on the part of physical geographers, the interest began to shift towards social or human geography. The French scholars owe the credit for this change, largely under the influence of Vidal de La Blache, a great human geographer and the leader of French School of Thought. He was particularly influenced by the second volume of 'Anthropogeographie' of Ratzel. Blache insisted on the study of the geographical environment and Man. He considered that the peculiar task of the geographer lies in the correlation of physical and human conditions on their spatial basis. The physical (natural) environment provides a range of possibilities which Man turns to his use according to his needs, wishes and capacities, in creating his habitat. Hence, as opposed to the "environmental determinism" that was prevalent at the time as a Darwinian heritage, he established a conceptual framework of 'Possibilism' in its place. The major thrust of his work was on the study of man-environment relationships from the point of view of man. Among other scholars, working on the lines of Blache, the contributions of Himly, Reclus, Bruhnes, Demangeon, Seigfried, Vilar and Gottaman are noteworthy. Their attempts also proved instrumental in bringing a humanistic tilt in geographical studies. Along with the French, the German and American schools also showed a gradual inclination towards the use of humanistic viewpoint in geography almost simultaneously. Two scholars are known to have brought this change in Germany and they are Alfred Hettner and Otto Schluter; and in America it was largely due to the efforts of Isaiah Bowman and Carl Sauer. The Berkeley School of Carl Sauer, for instance, is based on areal characterization, rather than on environmental relationships, which he considered too general and wider field to be enfolded within the content of geography. All these scholars in modern period have contributed to human studies with regional approach, particularly emphasizing on the historical, economic, political, colonial and electoral aspects. Their attempts have been successful in bringing a shift in geographical studies from physical to human, to the extent that a new paradigm was established, getting popularity as 'Videlian Tradition'.

The development of geography during the Second quarter of twentieth century, thus, has been marked by an increasing interest in regional geography. The climax was witnessed in the views of Richard Hartshorne and Alfred Hettner, when the concept of geography as a chorological science was advocated by full force. This led to the development of concept of geography as a science of 'areal differentiation'. The idea of region was further strengthened by its usage in different forms, e.g. 'Paysage' in France (Blache), 'Landschaft' in Germany (Otto Schluter) and 'Landscape' in America (Carl Sauer). This was a period when the 'Possibilistic' thinking was at its peak.

Paradigm Phase V was devoted to research in human and regional (chorological) studies in geography in post First World War period. During this period, the deterministic thinking was replaced by French school of the thought. As the dominating paradigm in human geography, determinism had a short life, being challenged by the possibilists. Blache and his followers stressed the idea that humans have free will to evaluate different options and participate in the development of each landscape through unique historical processes. With the fusion of regional and human approaches to geographical studies, Blache was able to put forward three interrelated doctrines, viz. 'Possibilism', '*Genre de Vie*' and 'Paysage'. These viewpoints constitute the 'Videlian Tradition'. The main theme of 'Possibilism' is that the physical surroundings provide a range of possibilities to man and man makes a choice. This choice depends on his needs, aspirations and capacities. Secondly, man establishes relationships with nature not as an individual, but as a group or society. In other words, man views this relationship through the prism of his culture. 'Culture' here includes his beliefs, institutions, traditions, attitudes, religion, language, diet, habits, customs, skills and technology etc. In this way, man creates his own habitat, the local modified environment within which the man lives, works or acts. This habitat is a distinct landscape created or developed because of man's occupancy, imprint and usage. This landscape has been termed as 'pays' at micro level and 'paysage' at or macro level in French by Blache. Thirdly, with time and the cultural advancement the level of interaction of man and environment increases. This brings changes in human life-styles as well. Thus, the possibilistic approach to the study of man-environment relationships demands a three-fold interpretation, i.e. in terms of spatial, cultural and historical (temporal) perspectives. In other words, the present ways of life of a society, its cultural, socio-economic and overall development, is the result of the history of its occupation of the physical surroundings. As a regional geographer, Blache preferred to view man within a physical milieu, or pays, and within his cultural milieu, or *Genre de Vie*. A 'pays' is a natural region with some homogeneous physical characteristics, while a '*Genre de Vie*' is a way of living or local culture. It includes the traditions, institutions, language, habits, foods etc. of a people. In other words, when ones studies geography regionally, one may identify a region with regard to both, its physical characteristics and its human characteristics. The same environment carries different meanings to people with different ways of life or culture. He drew attention to the fact that the same environment frequently leads to different responses from people with different *Genres de Vie,* since the complex of institutions, traditions, attitudes and technical skills which make up a *Genre de Vie,* largely determine which of the many possibilities offered by nature in any given area shall be selected for development. Blache opposed to the dichotomy between natural and cultural aspects of the earth's surface; instead, he considered them as the two ends of the same continuum. To him, natural and cultural aspects of the earth's surface cannot be separated from each other. It is because, he reasoned, in every inhabited part of the earth's surface, the original landscape is significantly transformed as a result of human habitation. Such changes are greater in the case of culturally advanced societies where, owing to the more developed technology, the degree of man's intervention in nature is more far-reaching. It is, therefore, impossible to study landscapes meaningfully without due reference to the interlocking roles of the nature and culture. Possiblism continued to grow and spread on the both in France and outside France. According to possibilists it is man who is the primary architect of his environment. It depends on the choices he makes among the numerous possibilities offered by the environment. In order to use these possibilities he makes two types of adjustments with the environment, i.e. in the form of 'adaptation' and 'modification'. For the possibilists, the works of man, not the earth, are the starting point and the most important is the freedom of man to choose. There are no necessities, but everywhere possibilities and man as a master of these possibilities is the judge of their use. In spite of the fact that man has numerous possibilities in a given physical setting, he cannot go against the directions laid by the physical environment. On this ground, the possibilistic approach was also criticized, and the possibilism, too, did not hold for many years.

The concepts of Landscape (American) and Landschaft (German) emerged as parallel viewpoints of the time. The Concept of Landscape of Sauer served as an important Paradigm in the Second Quarter of twentieth century. The term Landscape was accepted by many of the followers of Sauer as the basic term in geography. This landscape is an area made up of a distinct association of forms (physical and cultural). The design of a landscape includes both: the features of natural area; and the forms superimposed over it by the activities of man, converting it into cultural landscape (or *Kulturlandschaft* as identified by the Germans). Originally the Landscape is a natural landscape, and later it turns into cultural landscape or human habitat. Thus, a landscape at a later stage is the creation and workplace of human beings. The relationships of human beings with their habitat are conditioned by their group habits and cultural habits. There may be a succession of these landscapes with a succession of cultures. An alternative view built around 'the concept of landscape' was that of Landschaft by Otto Schluter, the German scholar. Landschaft also means "a piece of land (region) with certain physical and human attributes". Schluter's Kulturlandschaft (cultural landscape) appears to have a greater resemblance with the French 'Pays'—small homogeneous well-defined regions manifesting intimate and reciprocal relationships between man and his immediate surroundings. A 'pay' is, therefore, an areal expression of the *genre de vie* of a human community whose occupation of the area over centuries seems to have created the landscape. An area of the Earth's surface with a distinctive appearance, natural, man-made, or both is called a landscape. Any tract of the Earth's surface with either natural or man-made distinctive characteristics is determined by these criteria.

As part of Videlian tradition, the major concern of geographers became to study the region. For a long time, geographers continued to stress the central position of Regional Geography that especially flourished in France and other European countries. Hettner and Hartshorne, both, regarded the regional geographical synthesis as core to geography. In conformity with the Kantian philosophy, geography came to be regarded as a chorological science. In the writings of the modern geographers, geography emerged as the integrating science, completely different from other disciplines. The goal of this chorological point of view was to know the character of regions and places through comprehension of existence together and interrelation among the different realms of reality and their varied manifestations. Hartshorne used the term areal differentiation to characterize the way in which geographers dealt with the wide variety of phenomena: physical, economic and social, which exist together in the area and distinguish them from other areas. Regional geography flourished in France and got diffused in its neighbouring countries. But later on this approach also became inadequate to explain the regional personality and, therefore, a period of crisis in the discipline emerged again.

Postmodernism brought with it the questioning of previous approaches of knowing, and introduction of new methods of study. Geography witnessed a number of critical phases with revolution one after the other, resulting in faster paradigmatic shifts in comparison to earlier phases of development. The nomothetic twists, emergence of spatial (locational) perspective and anthropocentricism (behaviouralism, radicalism and humanistic trends) have been instrumental in bringing about the paradigmatic shifts in geography in this period. The crises in the discipline, on the other hand, have been brought about by Schaefer-Hartshorne Debate, Quantitative Revolution, Critical Revolution and Digital Revolution. Interestingly Post Modern period in Geography began with a Crisis Phase and the discipline is still undergoing another one.

Post-modernism in geography initiated with the criticism to chorological view and installation of more systematic and nomothetic perspective. The philosophy and methodology of geography were apparently debated most earnestly in the writings of two American scholars, namely Richard Hartshorne and Fred K. Schaefer during mid twentieth century. Geography's identity and methodology were called into question in this debate. Following Hartshorne, Regional geography was the major philosophical school during 1940s and 1950s. All of these events presented a great threat to geography's position as an academic subject and thus geographers began seeking new methods to counter critique. Schaefer advocated the use of spatial laws as a basis for geographic explanation. Hartshorne's work, on the other hand, advocated for chorological (regional) approach in geography. However, the dissatisfaction developed with chorological perspective and the new generation of geographers was increasingly drawn to the systematic studies. Main reason was that the chorological viewpoint was promoting the concept of uniqueness of areas, neglecting theory and generalizations. The paradigm had become sterile in geography and was not yielding laws and theories. Schaefer criticized the claim made for geography as a chorological science, and put forward a strong case for the discipline to adopt the philosophy and methodology of scientific positivism. The essential difference between the positions taken by Hartshorne and Schaefer was that Hartshorne's was a positive view of geography, i.e. geography is what geographers have made it. Schaefer's view, on the other hand, was a normative one, i.e. what geography should be irrespective of what it had been. Geographers in increasing numbers had, by the end of 1950s, turned to Schaefer's view of geography as a spatial (locational) science. For this they used methods of other systematic sciences and were increasingly concerned with quantification and development of theory. The net outcome of Schaefer-Hartshorne Debate was that geography, by the end of the 1950s, had come increasingly to be viewed as a science requiring the use of the 'scientific method' so that like other sciences, it could also develop laws and theories relevant to its field of study. This brought about a distinctive shift in emphasis from 'regional' to 'systematic' studies. This meant that geography thereafter began increasingly to be viewed in a nomothetic perspective, i.e. it required developing the habit of seeking the general in the particular. This also involved a shift from 'areal' to 'locational' studies; from 'absolute' to 'relative' locations; and from 'areal differentiation' to 'spatial interaction'. The mechanistic approach of environmentalism, which dominated much of the nineteenth century, was soon replaced by a more nomothetic and scientific approaches.

The dualism developed between Nomothetic and Idiographic got resolved after Schaefer-Hartshorne Debate. Nomothetic method advocated for the aspect of science that can be expressed in universals and Idiographic that which is concerned with the individual attribute or object. It was agreed upon that geography 'like other sciences' can develop 'the knowledge of principles, laws, and general truths'—and thus lay claim to the name of science. The idiographic view was very strong in the earlier part of the twentieth century, but from 1950s onwards various objections were raised against this method of study. In the light of vast geographical knowledge accumulated by the end of Age of Discovery, the idiographic method proved inefficient to handle this information. The Nomothetic method was, therefore, was proposed as the key to the problem.

The Quantitative Revolution began under the banner of the scientific method. Apparently new perspective was opened up under the impact of Revolution during mid twentieth century. It sought to develop a more rigorous and systematic methodology for the discipline that later lead to a paradigm shift. It came as a response to the inadequacy of regional geography to explain general spatial dynamics. Then, during the Second World War, technology became very influential in our society. This culminated in the Quantitative Revolution that occurred during the 1950s and 1960s, and marked a rapid change in the method behind geographical research. The revolution was geography's attempt to redefine itself as a science. The revolutionaries declared that the purpose of geography is to test general laws about the spatial arrangement of phenomena. Adopting the philosophy of positivism from natural sciences, they turned to mathematics, especially statistics, as a way of proving solutions to their problems. The mathematical and statistical methods were introduced to attain a desired level of objectivity for the search of laws and theories in geography. All these efforts were devoted to the fundamental conception of geography as a spatial science. Although, this movement is over now, it did lead to mathematization of much of our discipline, with an attendant emphasis on the construction and testing of theoretical models. The main claim for the Quantitative Revolution was that it led to a shift from a descriptive (ideographic) geography to an empirical law making (Nomothetic) geography. During this phase of geographic development, science and mathematics enjoyed unrivalled importance. Nomothetic aspects of geography heightened in importance and quantitative methods were applied to virtually all investigation and experimentation. By the mid-1960s the quantitative revolution had successfully displaced regional geography from its dominant position. The paradigm shift was evident by a myriad of publications in academic journals and textbooks of geography. However, the greatest impact of the quantitative revolution was not the revolution itself but the effects that came afterwards in the form of spread of positivist (post-positivist) thinking and counter-positivist responses. The counter-positivist response was created in the form of introduction of anthropocentricism in discipline later.

The quantitative revolution had enormous implications in shaping the discipline of geography into what it looks like today. It replaced the 'idiographic' approach based on areal-differentiation by 'nomothetic' one, which had its roots in the search for models of spatial structure and phenomenon. The quantitative revolution led the basis of geography as a spatial science that dealt with the spatial analysis of phenomena that existed on the earth surface. In simpler words, it gave geography a scientific vision through the application of methodology rooted in statistical methods. The most obvious change has occurred due to the quantitative revolution that brought changes in the methods and techniques used to explain the geographical phenomenon in a spatial framework. The new method of inquiry led to the development of generalizations about spatial aspects in a wide range of natural and cultural settings. Attention, since then, has focused on the spatial organization of economic, social, physical, political and urban processes, and on how outcomes generated by these processes are evidenced at given times and in particular places. Special attention has been placed on the spatial aspects of social processes. The theoretical approaches and quantitative methodologies evolved in these decades lead to the evolution of Spatial Science School of Geography by 1970s. The School emphasized the study of aggregate spatial patterns, including spatial behaviour, within clear theoretical frameworks, using quantitative methods to evaluate models and hypotheses. The general acceptance of quantitative techniques, the more complete mathematical training of a new generation made possible this development. The Spatial science approach dealt with aggregate patterns in space and their generating processes.

However, within two to three decades of the beginning of Quantitative Revolution, the overwhelming focus on statistical modeling itself eventually became responsible for the undoing of this revolution. Many geographers became increasingly concerned that these techniques simply put a highly sophisticated technical gloss on the approach to study that was barren of theory. Other critics argued that it removed the human dimension from a discipline that always prided itself on studying the human and natural world alike. As the 1970s dawned, the Quantitative Revolution came under direct challenge. The mood changed with a significant number of geographers beginning to question the wisdom and even the morality of this quantitative turn. Critical Revolution came as a critique on positivism introduced by the Quantitative revolution. This again proved as one of the major turning points in the history of geography.

The discipline of Geography saw another phase of transformation since early 1970s that was associated with opposition to quantification in geography. This was due to the so-called Critical Revolution. By mid 1960s, the initial euphoria about quantitative revolution had receded. Many had begun to question its relevance to real life. The backlash against the quantification trend led to Revolution under the humanistic and welfare approaches in geography. Quantitative methodologies, undoubtedly, are a powerful research technique in human geography that can provide valuable and accurate insights if used appropriately and with an understanding of the limitations. However, during the 1980s and 1990s there was a downturn in the popularity of such methodologies. The overwhelming focus on statistical modelling was, eventually, responsible for the undoing of the quantitative revolution. The counter-positivist response came as geographers began to expose the inadequacy of quantitative methods to explain and address issues regarding race, gender, class and war. The positivist assumptions upon which the approach was based were attacked as geography experienced its `anthropocentric turn' and a split into a range of modes of thought. The Critical geography of this phase, in turn, lead to three main schools of thought, viz. Behaviouralism, Radicalism and Humanism. Challenging the quantitative side of geography, Behaviouralism tried to provide a greater understanding of how people perceive space and place; Radicalism argued that quantitative research is not useful when there are no solutions nor alternatives to human problems; and the last and most accepted ideology of Humanism argues that human research should always be qualitative instead of quantitative.

During 1970s, when the need for precision in research through increased use of statistical-mathematical techniques had been commonly accepted, and the controversy regarding systematic versus regional geography resolved, with the scale tilting in favour of the former, but both being accepted as independent and equally useful components of the discipline, the euphoria about quantitative revolution had also greatly receded. Now, many workers had begun to feel disillusioned with the quantitative models being developed and used by human geographers of the positivist orientation, since they failed to relate to real-life situations. Consequently, geography again witnessed a paradigmatic shift, as many geographers began to feel that a deeper understanding of man-environment interaction could be achieved if geographers paid greater attention to the study of the mental images that the decision-makers have of the environment to which their decisions relate since it is the perceived reality that leads to action. This shift in focus inevitably led to a shift to the behavioural sciences. The basic philosophy of behaviouralism is that man and environment are dynamically interrelated. Man is viewed as a motivated social being, whose decisions and actions are mediated by his cognition of the spatial environment and creation of an environmental imagery. As an approach, it looks at man-environment relationships with a new and sharp outlook. It aimed at replacing the simplistic and mechanistic conception that previously characterized man-environment theory with new version that explicitly recognizes the complexities of human behaviour within a given set of environment; and the degree of their impact on space. The objective of behaviouralism was to replace the simplistic and mechanistic conceptions of man-environment relationships with a new theory that views human actors in space as thinking beings whose actions are mediated by cognitive processes by means of which people acquire, organize and use knowledge. The focus was on studying how people come to terms with their physical and social milieu. The behaviouralist rejected the assumption that decisions are always based upon a sequence of reasoned thoughts involving systematic evaluation of all available options. The behaviouralist viewed the human, subjects of his studies, as thinking beings whose actions are intervened by mental processes. In between the environmental influences and the man's actions there are mental processes of various kinds, which govern his behaviour in a given set of environmental elements. The behaviouralist is interested in the way in which people come to terms with their physical and social milieu. He is also interested in the factors that influence the interrelationships of thought and action of man. The Environmental Perception and Mental Maps are recognized as the basic concepts of Behavioural Geography. Environmental Perception is the total perception an individual has of his surroundings, which creates mental maps, whereas a mental map is an individual's own internal map of their known world.

Radicalism that also grew as a major criticism of quantitative geography, suggested an approach to geographic research that seeks to understand social and spatial problems, and advocate solutions. Radical geographers are interested in the issues of relevance to everyday social life: the lived experiences of members of society. Radical geographers tried to transform the scope of the discipline of geography by responding to the great issues of the time, viz. civil rights, environmental pollution and war. A quest for social justice permeates much of radical geography. The radicalists put emphasis on the need for a revolution in both theory and practice of geography. The radical approach is value based, especially the theory of labour value, as against the supposedly value-free approaches. Most of the radicalists have a strong Marxist base and take a holistic view of economics, society and polity. Radical geographers reject the spatial determinism implicit in spatial science; instead they seek to place questions of geography within broader social and political contexts. They emphasize context and relationships between places and people and broaden the geographical research agenda by considering previously neglected issues including poverty, hunger, urban decay, and social inequality.

Humanism in geography also emerged due to a deep dissatisfaction with the mechanistic models of spatial science that had developed during the quantitative revolution. It evolved as a criticism against positivism and quantitative revolution. The basic objection of humanists against quantitative revolution is that its tools and assumptions do not adequately explain human world and human issues, especially relating to social institutions, attitudes, morals, customs, traditions and aesthetics. Humanistic geography deals with the meaning, value and human significance of life events. The Humanists are not in favour of reducing space to mere geometrical concepts of surface as viewed by the positivists. From a humanistic point of view, space is inseparable from consciousness of humans who inhabit it. Humanism is a rejection of the 'geometric determinism' which views human beings as always dictated by 'universal spatial structures and abstract spatial logics'. Geographers have become increasingly active in applying their skills of spatial analysis to finding equitable solutions to the many problems that now confront humankind. The humanist epistemology sees that studying the human world is very different from studying the physical, biological or material world. The contemporary humanism in geography emphasizes the study of meanings, values, goals and purposes. Within this humanistic perspective concepts of traditional significance in geography are given new meanings. Humanistic geography gives central and active role to human awareness and human agency, human consciousness and human creativity. In the humanistic strand, the intent has been to understand and recognize the dignity and humanity of the individual. The focus of humanistic geography is on people and their condition. Humanists explain and interpret man and space relationship with historical approach. Humanism does not treat humans as machines. The followers of this approach consider geography as "the study of the earth as home of man." Humanistic geography, as such, is not an earth science in its ultimate aim. It belongs to the humanities and to the social sciences to the extent that they all share the hope of providing an accurate picture of the human world. In humanities the scholars gain insight into the human world by focusing what man does supremely well, in the arts and logical thought. Humanistic geography achieves an understanding of the human world by studying people's relation with nature, their geographical behaviour as well as their feelings and ideas in regard to space and place. Humanists reject the reduction of space and place to geometrical concepts of surface and point as perceived and preached by the positivists through the methodology of quantitative techniques.

Geography, in the twenty first century, is experiencing another revolutionary phase. With the widespread introduction of computer technology, geography has entered a new methodological revolution. In 1950s, such a change has been brought by the quantitative revolution, and now it is Digital Revolution. Computers, remote sensing, GIS and related fields are opening up new avenues for geographers. With the use of these modern techniques there may be collected and interpreted a variety of information about the earth's surface. Since geographers are directly concerned with analyzing the variable data on the earth's surface, this is going to open up vast areas for analysis and research. The Digital Revolution refers to the advancement of technology from analogue electronic and mechanical devices to the digital technology available today. The Revolution also marks the beginning of the Information Era. The Digital Revolution is sometimes also called the Third Industrial Revolution. The Revolution is the shift from mechanical and analogue electronic to digital electronics. Although it had actually begun in late 1950s, its adoption and the proliferation of digital computers and digital record keeping have become more widespread in the present times. Implicitly, the term refers to the sweeping changes brought about by digital computing and communication technology during and after the latter half of the twentieth century. Central to this revolution is the mass production and widespread use of digital logic circuits, and its derived technologies, including the computer, digital cellular phone, and the Internet. These technological innovations have transformed traditional practice of data collection, presentation and analysis. The most visible representation of the Revolution is the Internet, which integrates telecommunications and computing. The information may be more easily collected, stored, processed, accessed and communicated after the conversion of technology from analogue to digital format. Digital information is compact, transportable, and, therefore, more efficient to use. There have been huge benefits to society from the digital revolution, especially in terms of the accessibility of information. Expanded powers of communication and information sharing, increased capabilities for existing technologies, and the advent of new technology brought with it many potential opportunities for exploitation. The digital revolution helped usher in a new age of mass surveillance, generating a range of new civil and human rights issues.

The computer has drastically transformed both the world of geography as an academic discipline and the geography of the world in which we live. From being a tool for geographers to collect, analyze, map and visualize data since 1950s, the computers have increasingly become an integral part of the geographers' study by the end of twentieth century. No other technological innovation in human history has affected the practice of geography in such a profound way as the computer. The traditional geography has been radically changed by the past fifty years of the computer age, and especially by the last twenty years. Computers have fundamentally changed not only how geographers study the world, but also what we study, as spaces and places, the natural and human landscapes, are constantly being refigured by computers. The increasing use of computers has allowed geographers faster processing of larger sets of data. The increasing speed and capacity of computers allow and encourage the use of datasets that were simply unmanageable and tedious only twenty years ago. Not only in physical geography but also in human geography, far more comprehensive datasets can be used and far more complex models can be run. The computers have vastly eased the coding of qualitative data, i.e. recorded interviews, answers to open-ended survey questions, through direct computer answering of surveys, data entry and analysis, and textual search. The Computers have also stimulated the growth of new branches of geography, as quantitative geography), computer-assisted cartography, analytical cartography, remote sensing, and GIS; the process culminating in building of the Virtual Geography Departments the world over. In this virtual world, everything geographers study has become bits of information, which can be transmitted across the globe instantaneously.

The most ambitious concept to exemplify the potential of computers both as tools and as reality now reflects itself in the Concept of Digital Earth. Indeed, the entire earth is now becoming digital. The digital earth is envisioned as a multi-resolution, three-dimensional representation of the planet, into which we can embed vast quantities of georeferenced data. Digital Earth is inherently an intriguing concept for geographers that challenges our state of knowledge about the planet, not only in terms of raw data, but also in terms of data access and the ability to communicate data through visualization. Moreover, it challenges our understanding of process of model building and predicting, since the concept is not limited to static portrayal. Second, Digital Earth is interesting because of its implications for the organization of information. As a consequence, the content of geography has changed, for example, *via* the very concept of globalization space-time compression, and space-time distanciation. The rapid, cheap worldwide transfer of messages, documents, data, and images has brought about yet another shrinking of distance and abetting of globalization in all its economic and cultural forms. Digital Earth is regarded as a materialization of what has been called the Noosphere, dominated by flows of digital information. Together with geographers' focus on the atmosphere, hydrosphere, and biosphere, the study on the geography of Noosphere is a new challenge for geographers in the twenty first century.

To conclude it may be stated that geospatially literate population is the goal of geographers in the twenty first century. Such a population is better equipped to recognize, understand, and resolve those critical issues, whether local or global, that will confront us today and in the future. New technologies have always been important in advancing geographic understanding, but never have they been so thoroughly and rapidly transformative of the discipline as at this stage in geography's evolution. The revolutionary new geographic technologies developed during the past few decades extending frontiers in geographic research, education and applications. They are also creating new and resurgent roles for geography in both society and in the university. This trend is still accelerating, as the integration of geographic technologies, such as the global positioning system and geographic information systems (GPS/GIS), is creating an explosion of new "real-time, real-world" applications and research capabilities. The study is now being set for another paradigmatic shift.

Bibliography

Ackerman, E.A. (1958). *Geography as a Fundamental Research Discipline.* Univ. of Chicago, Department of Geography, Research paper no. 53,

Adhikari, Sudeepta (2015). *Fundamentals of Geographic Thought,* Orient Blackswan, pp. 161-63.

Agamben, Giorgio (2002). "What is a Paradigm", http://www.youtube.com, August 2002.

Agger, Ben (1988). *Critical Social Theories: An Introduction.* Avalon Publishing. ISBN 9780813321745.

Agnen, J.R. and J.S. Dunction, (2016). *Human Geography,* Wiley Blackweel.

Aitken, S. and Valentine, G. (Eds.) (2006, 2009). *Approaches to Human Geography.* California, London, New Delhi, Singapore: Sage.

Andrzej Diniejko, "Charles Darwin's Theory of Evolution and the Intellectual Ferment of the Mid- and Late Victorian Periods" (Dr. Andrzej Diniejko, Senior Lecturer in English Literature and Culture, Warsaw University; Contributing Editor, Poland), *[Victorian Web Home — Science — Biology — Evolutionary Theory], Internet*

Armstrong, M. (2000). "Geography and Computational Science", *Annals of the Association ofAmerican Geographers* 90: 146-56.

Balchin, W. and Coleman, A. (1967). "Cartography and Computers", *The Cartographer* 4: 120-27.

Bansal, Taruna, "Quantitative Revolution in Geography", *E-text.*

Barnes, B., 1982. *T.S. Kuhn and Social Science,* London: Macmillan.

Barnes, T.J. (2003). "The Place of Locational Analysis: A Selective and Interpretive History", *Progress in Human Geography* 27: 69–95.

Bennet, R., (1985). "Quantification and Relevance". In: Johnston, R. (ed) *The Future of Geography.* London: Methuen.

Bertrand Russell, (1967). *A History of Western Philosophy,* New York: Simon and Schuster, ISBN 9780671201586.

Best, U. (2009), "Critical Geography", In: Kitchin, R. and Thrift, N. (eds.) *International Encyclopedia of Human Geography,*Volume 2, pp. 345–357. Oxford: Elsevier. ISBN: 978-0-08-044911-1

Blaikie, N. (1993). *Approaches to Social Enquiry,* Cambridge: Blackwell.

Blaikie, N. (2000). *Designing Social Research,* Cambridge: Blackwell.

Blomley, Nicholas (2006). "Uncritical critical geography?", *Progress in Human Geography,* **30** (1): 87–94.

Boulding, K.E. (1956). *The Image: Knowledge in Life and Society,* University of Michigan Press.

Bourbaki, Nicolas (1998). *Elements of the History of Mathematics,* Berlin, Heidelberg, & New York: Springer-Verlag, ISBN 3-540-64767-8.

Bowman, John S. (2000). *Columbia Chronologies of Asian History and Culture,* New York: Columbia University Press.

Boyer, Carl Benjamin (1949). *The History of the Calculus and its Conceptual Development,* Hafner. Dover edition 1959, ISBN 0-486-60509-4.

Braithwaite Richard Bevan, (1953). *Scientific Explanation: A Study of the Function of Theory, Probability and Law in Science,* The Cambridge University Press, London and New York.

Breasted, J.H. (1916). *Ancient Times: A History of the Early World: An Introduction to the Study of Ancient History and the Career of Early Man.* Boston: Gin and Company.

Brunn, S.D. (1998). "The Internet as the New World of and for Geography: Space, Structures, Volumes, Humility and Civility", *GeoJournal* 45: 5-15.

Brunn, S.D. and Dodge, M. (2001). "Mapping the Worlds of the World Wide Web: (Re) Structuring Global Commerce Through Hyperlinks", *American Behavioural Scientist*, 44: 1716-39.

Brunn, S.D. and Leinbach, T.R. (1991). *Collapsing Space and Time: Geographic Aspects of Communication and Information*. London: Harper Collins Academic.

Bryman, A. (2004). *Social Research Methods*, Oxford: Oxford University Press.

Bunge, W. (1962). *Theoretical Geography*, Lund Studies in Geography. Lund, C.W.K. Gleerup.

Burton, I. (1963). "The Quantitative Revolution and Theoretical Geography", *Canadian Geographer*, 7, 4, pp. 151-182.

Buttimer, Anne (1976). "Grasping the Dyanism of Life World", *Annals of the Association of American Geographers*, **66**. pp. 277–292.

Carroll, Joseph, (2004). *Literary Darwinism: Evolution, Human Nature, and Literature*, Routledge, New York.

Cartledge, Paul, (1997). *The Cambridge Illustrated History of Ancient Greece*, Cambridge University Press.

Chadwick, Henry (1981). *Boethius: The Consolations of Music, Logic, Theology and Philosophy*, Oxford: Clarendon Press.

Chan, Alan Kam-leung and Gregory K. Clancey, Hui-Chieh Loy (2002). *Historical Perspectives on East Asian Science, Technology and Medicine*, Singapore: Singapore University Press. ISBN 9971-69-259-7.

Chemla, Karine; Cohen, Robert S.; Renn, Jürgen *et al.* (2005), *History of Science, History of Text (Boston Series in the Philosophy of Science)*, Dordrecht: Springer Netherlands

Chen, C. (2002). *Mapping Scientific Frontiers: The Quest for Knowledge Visualization*, London: Springer.

Chisholm, Hugh, (ed.) (1911). "Varenius, Bernhardus". *Encyclopedia Britannica*, **27** (11th ed.). Cambridge University Press.

Chopra, Pran Nath (1982). *Religions and Communities of India*, Vision Books, ISBN 9780856920813.

Chorley, R.J. (1957). "Climate and Morphometry", *Journal of Geology*, 65, 628-638.

Chorley, R.J. and Haggett, P. (eds.) (1965). *Frontiers in Geographical Teaching*. London: Methuen.

Chorley, R.J. and Haggett, P. (eds.) (1965). *Models in Geography*, London: Methuen.

Chorley, R.J. and P. Hagget (1967). *Integrated Models in Geography*, Part IV, Methuen & Co. Ltd., London.

Christensen. K, (1982). "Geography as a Human Science: a Philosophic Critique of the Positivist-humanist split", in Gould, P., and Olsson, G. *A Search for Common Ground*. London: Pion. pp. 37-57.

Christopher, R.W. (1994). *Geosystems: An Introduction of Physical Geography*, Macmillan College Pub. Co., New Jersey: Englewood Cliffs.

Cliff, A. (1970). "Computing the Spatial Correspondence Between Geographic Patterns", *Transactions of the IBG* 50: 143-54.

Cloke, P., *et al.*, (2014). *Introduction to Human Geogrphy*, Routledge Library Edition.

Cloke, P.; Philo, C. and Sadler, D. (1991). *Approaching Human Geography*. London: Chapman.

Copleston, Frederick Copleston (1950). *A History of Philosophy*. Vol. II: *Medieval Philosophy: From Augustine to Duns Scotus*. Westminster, Md.: The Newman Press.

Copleston, Frederick Copleston (1953). *A History of Philosophy*. Vol. III: *Late Medieval and Renaissance Philosophy*. Westminster, Md.: The Newman Press.

Crawford, O.G.S. (1927). *Antiquity*, England: Glou Cester.

Crespigny, Rafe de. (2007). *A Biographical Dictionary of Later Han to the Three Kingdoms (23-220 CE)*. Leiden: Koninklijke Brill. ISBN 90-04-15605-4.

Crook, Paul (1999). "Social Darwinism in European and American Thought, 1860–1945". *The Australian Journal of Politics and History*. 45.

Curry, M.R. (1998). *Digital Places: Living with Geographic Information Technologies*, London and New York: Routledge.

Daniel Sui, D. and Richard Morrill (2004). "Computers and Geography: From Automated Geography to Digital Earth", in S.D. Brunn, S.L. Cutter, and J.W. Harrington (eds.), *Geography and Technology*, pp. 81-108, Wikipedia, the free encyclopedia.

Darwin, Charles (1859). *On the Origin of Species by Means of Natural Selection,* John Murray, London.

David Knowles, (1964). *The Evolution of Medieval Thought,* London: Longmans.

Davies, W.K.D. (1972). *Conceptual Revolution in Geography,* New Jersey: Rowman & Littlefield.

De Blij, H.J. and P.O. Muller (1971). *Geography: Regions and Concepts,* New York: John Wiley & Sons Inc., pp. 1-18.

Dickinson, R.E. (1969). *The Makers of Modern Geography,* New York: Praeger Publishers.

Dikshit, R.D., (1997). *Geographical Thought: A Contextual History of Ideas,* Prentice Hall of India.

Dobson, J.E. (1983). "Automated Geography", *The Professional Geographer* 35: 135-143.

Dobson, J.E. (1993). "The Geographic Revolution: A Retrospective on the Age of Automated Geography", *The Professional Geographer* 45: 431-39.

Dodge, M. and Kitchin, R. (2000). *Mapping Cyberspace.* London: Routledge.

Durant, Will (1926). *The Story of Philosophy.* New York: Simon & Schuster, ISBN 0-671-69500-2.

Dury, (1954). "Contribution to a General Theory of Meandering Valleys", *American Journal of Science,* 123, pp. 167-176.

Ebrey, Patricia Buckley, Anne Walthall, and James B. Palais (2006). *East Asia: A Cultural, Social and Political History.* Boston: Houghton Mifflin Company. ISBN 0-618-13384-4.

Edward Gibbon (1776). *The History of the Decline and Fall of the Roman Empire.* I.W. Strahan and T. Cadell.

Encyclopedia Britannica.

Encyclopedia of Philosophy, Volume 3, (1967), Macmillan, Inc.

Entrikin, J. Nicholas (1976). "Contemporary Humanism in Geography", *Annals of the Association of American Geographers,* Vol. 66, No. 4, (December, 1976), pp. 615-632.

Ernst Mayr, (2000). "Darwin's Influence on Modern Thought', *Scientific American,* July Issue.

Eves, Howard (1969). *An Introduction to the History of Mathematics, holt,* Rinehart and Winston.

Eyre, S.R. (1964). "Determinism and Ecological Approach to Geography", *Geography,* Vol. I (49) : 369-376.

Fay, Brian (1987). *Critical Social Science: Liberation and its Limits,* Cornell University Press. ISBN 9780801494581.

Feldman, M.P. (2000), "Location and innovation: the new economic geography of innovation, spillovers, and agglomeration" in G. Clark, M. Feldman and M. Gertler, editors. *Oxford Handbook of Economic Geography,* Oxford: Oxford University Press.

Feldman, Maryann P. (2009), "The Internet Revolution and the Geography of Innovation", *Internet.*

Fielding, G.J (1974), *Geography as a Social Science,* Oxford University Press, New York: Harper & Row Publishers.

Finkel, Irving L. (1995), "A join to the Map of the World: a notable discovery", *British Museum Magazine: the Journal of the British Museum Friends* (Winter 1995), pp. 26-27;

Forer, P. (1978), "A Place for Plastic Space", *Progress in Human Geography,* Vol. 2, pp: 230-67.

Fortheringham, A.S., Brunsdon, C., and Charlton, M. (2000), *Quantitative Geography: Perspective on Spatial Data Analysis,* London: Sage.

Fortheringham, A.S., Brunsdon, C., and Charlton, M. (2000). *Quantitative Geography: Perspective on Spatial Data Analysis,* London: Sage.

Foster, J. (2003). "Between economics and ecology: Some historical and philosophical considerations for modelers of natural capital". *Environmental Monitoring and Assessment,* **86.**

Fotheringham, A.S. (2006). "Quantification, evidence and positivism", In S. Aitken and G. Valentine, editors, *Approaches to Human Geography.* London: Sage, 237-250.

Fuller, Duncan; Kitchin, Rob (2004). "Radical Theory/critical Praxis: Making a Difference Beyond the Academy?" *Praxis (e) Press.* ISBN 9780973456103.

Gaits, G. (1969). "Thematic Mapping by Computer", *The Cartographic Journal* 6: 50-68.

Garrison, W.L. 1959: "Spatial structure of the economy II", *Annals of the Association of American Geographers* 49: 471-482.

Getis, A. and Ord, J.K. (1992). "The Analysis of Spatial Association by Use of Distance Statistics", *Geographical Analysis* 24: 189-206.

Gibson, Margaret, (ed.) (1981). *Boethius: His Life, Thought and Influence.* Oxford: Basil Blackwell.

Gilbert, D. (1995). "Between Two Cultures: Geography, Computing, and the Humanities", *Ecumene,* 2: 1-13.

Gilson, E. (1955). *History of Christian Philosophy in Middle Ages,* New York.

Glenn R. (1997), "The population of ancient Rome", *Antiquity,* Vol. 71, Issue 274, December, pp: 966-978.

Gold, J.R. (1980). *An Introduction to Behavioural Geography,* Oxford: Oxford University Press.

Golledge, G., and Timmermans, H. (1990). "Applications of behavioural research on spatial problems". *Human Geography,* Vol. 14. 1990.

Goodchild, M.F. (2000). "Communicating Geographic Information in a Digital Age", *Annals of the Association of American Geographers,* 90: 344-55.

Gould, P. (1970). "Computers and Spatial Analysis: Extensions of Geographic Research", *Geoforum,* 1: 53-69.

Gould, P. (1985). *The Geographer at Work,* Boston: Routledge and Kegan Paul.

Gould, P. and White, R. (1974). *Mental Maps,* Harmondsworth: Penguin Books.

Gould, P.R. (1966). *On Mental Maps,* Michigan Inter-University Community of Mathematical Geographers, Discussion Paper 9, reprinted in Downs, R.M. and D. Stea (Eds.) (1973). *Image and Environment,* London: Edward Arnold.

Gracia, Jorge J. E. and Timothy Noone [2003]. *A Companion to Philosophy in the Middle Ages,* Blackwell Companions to Philosophy. Oxford: Blackwell Publications.

Grayling A.C. (1998), *Philosophy: A Guide through the Subject,* Oxford University Press.

Grayling, A.C. (1998). *Philosophy: A Guide through the Subject,* Oxford: Oxford University Press.

Gregory, Derek (1994), *Geographical Imaginations,* Blackwell ISBN 0631183310

Gregory, Derek; Johnston, Ron; Pratt, Geraldine; Watts, Michael J.; Whatmore, Sarah (2009). *The Dictionary of Human Geohraphy* (5th ed.). USA & UK: Willey-Blackwell. p. 612.

Gregory, S. (1963). *Statistical Methods and the Geographers,* London: Longman.

Guruprasad, P.R. (2011), "Scientific Reasoning in Ancient India", *Internet.*

Hagerstrand, T. (1967). "The Computer and the Geographer", *Transactions of the Institute of British Geographers,* 42: 1-20.

Hagget, P. (1979). *Geography: A Modern Synthesis,* New York.

Hagget, P. and R.J. Chorley (1965). *Frontiers in Geographical Teaching,* Mathuen.

Haggett, P. (1966). *Locational Analysis in Human Geography,* Edward Arnold.

Haggett, P. (1969). On Geographical Research in a Computer Environment", *Geographical Journal,* 135: 497-505.

Haggett, P. (1990). *The Geographer's Art,* Oxford: Baickwell.

Haggett, P., Cliff, A.D. and Frey, A.E. (1977). *Locational Analysis in Human Geography* (second edition). London: Edward Arnold.

Haggett, P., Richard J., Chorley (1967). *Models in Geography* (2nd ed.). Michigan: University of Michigan Press.

Hall, S.S. (1993). *Mapping the Next Millennium: How Computer-Driven Cartography is Revolutionizing the Face of Science,* New York: Vintage Books.

Harald, B., Engel-Di Mauro (2008). *Critical Geographies: A Collection of Readings,* Praxis(e) Press, Critical Topographies Series.

Hartshorne, R. (1939). *The Nature of Geography: A Critical Survey of Current Thought in the Light of the Past,* Association of American Geographers, Lancaster.

Hartshorne, R. (1955). "'Exceptionalism in Geography' Re-examined", *Annals of the Association of American Geographers,* Vol. 45, pp. 205-44.

Hartshorne, R. (1958). "The concept of geography as science of space from Kant and Humboldt to Hettner", *Annals of the Association of American Geographer,* Vol. 32 : 97-108.

Hartshorne, R., (1959). *Perspective on the Nature of Geography,* Chicago: Rand Mcnally & Co.

Hartshorne, Richard (1954). "Comment on 'Exceptionalism in geography'". *Annals of the Association of American Geographers,* Vol. 38 : 108–9.

Harvey, D. (1969). *Explanation in Geography.* London: Edward Arnold.

Harvey, D. (1973). *Social Justice and the City,* London: Arnold.

Harvey, D. (1984). "On the history and present condition of geography: an historical manifesto". *The Professional Geographer,* 36 (1), 1-11.

Harvey, David (2000). *Spaces of Hope.* University of California Press. ISBN 9780520225787.

Harvey, F. (1997). "From Geographic Holism to Geographic Information System", *The Professional Geographer* 49: 77-85.

Harvey, M.E. and Holly, B.P. (Eds.) (1981). *Themes in Geographic Thought,* Croom Helm, Beckenhem.

Hayashi, Takao (2003). "Indian Mathematics", in Grattan-Guinness, Ivor, *Companion Encyclopedia of the History and Philosophy of the Mathematical Sciences,* 1, pp. 118-130, Baltimore, MD: The Johns Hopkins University Press.

Hayashi, Takao (2005). "Indian Mathematics", in Flood, Gavin, *The Blackwell Companion to Hinduism,* Oxford: Basil Blackwell.

Henry, Desmond Paul (1967). "Medieval Philosophy", *The Encyclopedia of Philosophy.* New York: Macmillan. Vol. 5, pp. 253–58.

Herbertson, A.J. (1905). "The Major Natural Regions: An Essay in Systematic Geography", *Geography,* pp. 300-310.

Himmelfarb, Gertrude (1959). *Darwin and the Darwinian Revolution,* Garden City, NY: Doubleday.

'History of China', Wikipedia, the Free Eencyclopedia.

Holt-Jenson, A. (1999). *Geography History and Concepts.* London: Sage.

Holt-Jenson, Arild (2009). *Geography: History and Concepts: A Systematic Guide,* 4th Edition, Sage.

Howard Eves (1990). *An Introduction to the History of Mathematics* (6 ed.). New York: Saunders College Publishing House.

Hsu, Mei-ling (1993). "The Qin Maps: A Clue to Later Chinese Cartographic Development", *Imago Mundi,* Volume 45, pp. 90-100.

Huffman, Carl A. (1993) *Philolaus of Croton: Pythagorean and Presocratic,* Cambridge: Cambridge University Press.

Hugh, Thurston (1996). *Early Astronomy,* New York: Springer-Verlag.

Husain, M. (1988). *Evolution of Geographical Thought,* Second Revised Edition, Jaipur: Rawat Publications.

Hussain, M. (1995). "Paradigm in Geography", *Evolution of Geographic Thought,* New Delhi: Rawat Publications, pp. 300-340.

Ifrah, G. (2000). *The Universal History of Numbers.* New York: John Wiley & Sons.

Irena, Bojanova (2014). "The Digital Revolution: What's on the Horizon?". *IT Professional* (January-February 2014). **16** (1): pp. 8–12.

Irving L. Finkel, (1995). "A Join to the Map of the World: A Notable Discovery", *British Museum Magazine*: The Journal of the British Museum Friends, pp. 26-27.

James, P.E. and C.F. Jones (1954). "The Regional Concept and the Regional Method", *American Geography: Inventory and Prospects,* Syracuse: Syracuse University Press, pp. 19-68.

James, P.E. and Martin, G. (1981). *All Possible Worlds: A History of Geographical Ideas,* New York: John Wiley & Sons.

Janelle, D. and Hodge, D. (1999). *Information, Place, and Cyberspace: Issues in Accessibility,* Berlin: Springer-Verlag.

Jewish Encyclopedia.

John Marenbon (1983 & 1988). *Early Medieval Philosophy (480-1150): An Introduction,* London: Routledge.

Johnston, R., "Spatial Science School of Geographical Sciences", University of Bristol, Internet.

Johnston, R.J. (1978). "Paradigm and revolution: observations on human geography since the Second World War". *Progress in Human Geography,* 2, pp. 189-206.

Johnston, R.J. (1978). *Multivariate Statistical Analysis in Geography: A Primer on the General Linear Model,* London: Longman.

Johnston, R.J. (1979, 1997, 2004). *Geography and Geographers: Anglo-American Human Geography Since 1945,* Edward Arnold, London.

Johnston, R.J. (1991). "The nature of an academic discipline", *Geography and Geographer,* 5-27.

Johnston, R.J. (Ed.) (2014). *The Future of Geography,* Vol. 6, Routledge Library Edition: Social and Cultural Grography.

Johnston, Ron and Sideway James (2004, 2016). *Geography and Geographers: Anglo-American Human Geography since 1945*. New York: Routledge.

Jones, E. 1956. "Cause and effect in human geography", *Annals of the Association of American Geographers* 46: 370-377.

Joseph, Brian (Ed.); Janda, Richard (Ed.) (2008). *The Handbook of Historical Linguistics*. Blackwell Publishing (published 30 December 2004). p. 163. ISBN 978-14051274.

Joseph, G.G. (2000). *The Crest of the Peacock: The Non-European Roots of Mathematics*, Princeton, NJ: Princeton University Press.

Katz, Victor J. (1995). "Ideas of Calculus in Islam and India", *Mathematics Magazine (Math. Assoc. Amer.)* 68 (3): 163–174.

Katz, Victor J. (2007). *The Mathematics of Egypt, Mesopotamia, China, India, and Islam: A Sourcebook*. Princeton: Princeton University Press. ISBN 0-691-11485-4.

Kellerman, A. (2002a). *The Internet on Earth: A Geography of Information*, New York: John Wiley.

Kennedy, Hugh (1995). "The Muslims in Europe". In McKitterick, Rosamund, *The New Cambridge Medieval History: c.500-c.700*, Cambridge University Press.

Kerski, Joseph (2015). "Teaching Geography in the Twenty-First Century", *Internet*, January 26, 2015.

Kimble, G.H.T. (1951). *The Inadequacy of the Regional Concept*, London.

Kirk, W. (1951). "Historical geography and the concept of behavioural environment," pp. 152-160 in Silver Jubilee Souvenir and N. Subrahmanyam Memorial Volume. Madras: Indian Geographical Society.

Knowles, David (1964). *The Evolution of Medieval Thought*. New York: Vintage Books.

Kotkin, J. (2001). *The New Geography: How the Digital Revolution is Reshaping the American Landscape*, New York: Random House.

Kuhn, T. (1962). *The Structure of Scientific Revolutions*, Chicago: University of Chicago Press.

Kuhn, T. (1970). *The Structure of Scientific Revolutions* (2nd ed.). Chicago: University of Chicago Press.

Kuhn, T. (1980). "The halt and the blind: philosophy and history of science", *British Journal for the Philosophy of Science*, 31, 181-192.

Kurt, A. Raaflaub and Richard, J.A. Talbert (2009), *Geography and Ethnography: Perceptions of the World in Pre-Modern Societies*, John Wiley & Sons, ISBN 1405191465.

Lange, L. (1999). "The Internet." *IEEE Spectrum*, 36 (1), January: 35-40.

Latour, B. (1993). *We Have Never Been Modern*, Trans. C. Porter, Cambridge: Harvard University Press.

Litan, Robert, E., and Alice M. Rivlin, (2001). "Projecting the Economic Impact of the Internet." *American Economic Review*, 91 (2): 313-317.

Livingstone, D.N. (1992). "Statistics don't bleed: quantification and its detractors", *The Geographical Tradition*, Ch. 9.

Livingstone, David N., (1993). *The Geographical Tradition: Episodes in the History of a Contested Enterprise*, Blackwell Publishers ISBN 0-631-18586-0.

Loewe, Michael. "The Oracles of the Clouds and the Winds", *Bulletin of the School of Oriental and African Studies*, University of London (Volume 51, Number 3, 1988): 500–520.

Longley, P., Brooks, S.M., McDonnell, R., and Macmillan, B. (1998). *Geocomputation: A Primer*, Chichester, John Wiley and Sons.

Luscombe, D.E. (1997). *Medieval Thought*, History of Western Philosophy, Vol. 2. Oxford: Oxford University Press.

MacDonald, Scott, and Kretzmann, Norman [1998]. "Medieval Philosophy." *Routledge Encyclopedia of Philosophy*, London: Routledge. Vol. 6, pp. 269–77.

MacDougal, E.B. (1976). *Computer Programming for Spatial Problems*, London: Edward Arnold.

Macionis, J. and Plummer, K. (2008). *Sociology: A Global Introduction*. 4th edition, Pearson Education Ltd.

Mackay, (1963). *The Mackenzie Delta Area*, N.W.T., Ottawa: Queen's Printers.

Malecki, Edward J. (2009). *The Economic Geography of the Internet's Infrastructure*, Wiley.

Malik, Anand (2014). "Thomas Kuhn and Changing Paradigm in Geography: Critical Review", *Asian Journal of Multidisciplinary Studies*, Volume 2, Issue 4, April 2014.

Marcia L. Colish, (1997). *Medieval Foundations of the Western Intellectual Tradition: 400-1400.* New Haven, CT: Yale University Press.

Marshall, Alan, "A critique of the development of quantitative methodologies in human geography", *E-Book, Internet.*

Martin, G. (1989). The nature of Geography and Schaefer-Hartshorne Debate, reflection's on Richard Hartshorne's *The Nature of Geography.* Annals of the Association of American geographers. 69-88.

Martin, G.J. (1985). "Paradigm Change: A History of Geography in United States", *National Geographic Research,* Spring, pp. 217-35.

Masterman, M. (1970). *"The nature of a paradigm"*, In: *Proceedings of the 1965 International Colloquium in the Philosophy of Science* 4 (3ed.), Cambridge: Cambridge University Press.

Masterman, M. (1970). "The nature of a paradigm", In: *Proceedings of the 1965 International Colloquium in the Philosophy of Science* 4 (3ed.), Cambridge: Cambridge University Press.

Maxwella, E. (1961). *Analyzing Qualitative Data.* London: Methuen and Co.

Miller, H.J. and Han, J.W. (2001). *Geographic Data Mining and Knowledge Discovery,* London: Taylor and Francis.

Minshull, R. (1970). *The Changing pattern of Geography,* London.

Mohanty, J.N. (2000). *Classical Indian Philosophy of Induction,* Rowman & Littlefield Publishers.

Müller, Max (1899). *Six Systems of Indian Philosophy; Samkhya and Yoga, Naya and Vaiseshika.* Calcutta: Susil Gupta (India) Ltd. ISBN 0-7661-4296-5.

Needham, J. (1986). *Science and Civilization in China: Volume 3, Mathematics and the Sciences of the Heavens and the Earth,* Taipei: Caves Books Ltd.

Needham, Joseph (1986). *Science and Civilization in China: Volume 1, Introductory Orientations.* Taipei: Caves Books Ltd.

Needham, Joseph (1986). *Science and Civilization in China: Volume 4, Physics and Physical Technology, Part 1: Physics.* Taipei: Caves Books Ltd.

Needham, Joseph (1986). *Science and Civilization in China: Volume 4, Physics and Physical Technology, Part 2: Mechanical Engineering.* Taipei: Caves Books Ltd.

Needham, Joseph (1986). *Science and Civilization in China: Volume 4, Physics and Physical Technology, Part 3: Civil Engineering and Nautics.* Taipei: Caves Books Ltd.

Needham, Joseph (1986). *Science and Civilization in China: Volume 5, Chemistry and Chemical Technology, Part 1: Paper and Printing.* Taipei: Caves Books Ltd.

Needham, Joseph (1986). *Science and Civilization in China: Volume 5, Chemistry and Chemical Technology, Part 7: Military Technology: The Gunpowder Epic.* Taipei: Caves Books, Ltd.

Needham, Joseph (1986). *Science and Civilization in China: Volume 6, Biology and Biological Technology, Part 1: Botany.* Taipei: Caves Books Ltd.

Nokolaos, Bakalis (2005). *Handbook of Greek Philosophy: From Thales to the Stoics Analysis and Fragments,* Tafford Publishing, ISBN 1-4120-4823-5.

Norman Cantor, (1994). *The Civilization of the Middle Ages,* New York: Harper Collins.

North, D.C. (1961). "Quantitative Research in American Economic History". *American Economic Review.,* 53, pp. 128-30.

Norton, W. (1997). Human geography and behaviour analysis: An application of behaviour analysis to the evolution of human landscapes. *The Psychological Record,* 47, 439–460.

Olson, E. and Sayer, A. (2009). *Radical Geography and its Critical Standpoints: Embracing the Normative.* Antipode, 41: 180–198.

Openshaw, S. (1994). "Computational Human Geography: Towards a Research Agenda", *Environment and Planning,* A 26: 499-505.

Parker, I. (1999). "Qualitative data and the Subjectivity of Objective Facts", in Simpson, L. and Dorling, D., *Statistics in Society.* London: Hodder.

Pearcy, E. (1957). *World Political Geography,* Crowell Company, New York.

Peet, J. Richard, (Ed.) (1977). *Radical Geography: Alternative Viewpoints on Contemporary Social Issues.* Maaroufa Press.

Peet, R. (1998). *Modern Geographical Thought.* Oxford: Blackwell.

Peet, Richard (2000). "Celebrating Thirty Years of Radical Geography", *Environment and Planning,* A. **32** (6): 951–953.

Perez, C. (1983). "Structural change and assimilation of new technologies in the economic and social systems", *Futures*, Vol. 15 : 357–375.

Perez, C. (2004). "Technological Revolutions, Paradigm Shifts and Socio-Institutional Change" in E. Reinert, Globalization, *Economic Development and Inequality: An alternative Perspective*, Cheltenham: Edward Elgar, pp. 217–242.

Pickerill, Jenny (2017). 'Radical Geography', Published Online: 6 March, 2017, DOI: 10.1002/9781118786352.wbieg0506.

Pingree, David (1992). "Hellenophilia versus the History of Science", *ISIS* 83 (4).

Pitts, F. (1967). "Chorology Revisited", *The Professional Geographer*, 14: 8-11.

Plofker, Kim (2009). *Mathematics in India: 500 BCE–1800 CE*, Princeton, NJ: Princeton University Press. ISBN 0691120676.

Poon, J., (2004). "Quantitative methods: past and present", *Progress in Human Geography*, Vol. 28 : 807-814.

Pounds, N.J.G. (1972) *Political Geography*, Ed. 2, McGraw Hill, New York.

Quereshi, M.H. Ed. (2013). *Paradigm Shift in Geography*, New Delhi: Manak Publications Pvt. Ltd.

Quereshi, M.U. (2005). *Encyclopedia of Geographical Thought*, Vol. I, New Delhi: Anmol Publications Pvt. Ltd.

Rana, L. (2008). *Geographical Thought: A Systematic Record of Evolution*, New Delhi: Concept Publishing Company.

Rana, L. (2014). *Ancient Tradition in Geography: The Western and Oriental Perspective*, New Delhi: Classical Publishing Company, Karampura.

Rana, L. (2014). *Geographical Thought: Classical to Contemporary*, New Delhi: Concept Publishing Company.

Relph, E. (1981). *Rational Landscapes and Humanistic Geography*. New York: Barnes and Noble.

Richard Southern (1953). *The Making of the Middle Ages*, Yale University Press.

Robinson, A. (1954). "Geographical Cartography", in James, P.E. and Jones, C.F. (Eds.) *American Geography:* Inventory *and Prospect*, 553-77. Syracuse, NY: Syracuse University Press.

Robinson, C.A. (1951). *Ancient History from Prehistoric Times to the Death of Justinian*, New York: Macmillan.

Robinson, Fred. C. (1984). *"Medieval: the Middle Ages." Speculum* 59, pp. 745–56. (Presidential address to the annual meeting of the Medieval Academy of America, 1984).

Robinson, G. (1998). *Methods and Techniques in Human Geography*. London: Hodder.

Rose, G. (1997). "Situating Knowledge: positionality, reflexivities and other tactics", *Progress in Human Geography*, Vol. 21, pp. 305-320.

Saraswati Raju, Alex Jeffrey (2017). "Critical Geography", *The International Encypedia of Geography*, Published Online: 6 March, 2017.

Sauer, C. O. (1925), *The Morphology of Landscape*, University of California Publications in Geography.

Sayer, A. (1985) "Realism and Geography", in Johnston, R. (ed.) *The Future of Geography*, London: Methuen.

Schaefer, Fred K. (1953). "Exceptionalism in Geography: A Methodological Examination", *Annals of the Association of American Geographers*, Vol. 43 (3) : 226-249.

Schoedinger, Andrew B., ed. (1996). *Readings in Medieval Philosophy*. New York: Oxford University Press.

Seidenberg (1978). *The Origin of Mathematics:. Archive for the History of Exact Sciences*, Vol 18.

Shapere, D. (1964). "The Structure of Scientific Revolutions", *Philosophical Review*, 73, 383-394.

Shapere, D. (1971). "The Paradigm Concept", *Science*, 172, 706-709.

Shu-Chiu LIU (2006). "Historical models and science instruction: A cross-cultural analysis based on students' views", *Asia-Pacific Forum on Science Learning and Teaching*, Volume 7, Issue 2, Article 1 (December, 2006).

Simpson, L. and Dorling, D. (2004). "Statistics of Racial Segregation: Measures, Evidence and Policy", *Urban Studies*, vol. 41, pp. 66, 1-681.

Sivin, Nathan (1995). *Science in Ancient China: Researches and Reflections*. Brookfield, Vermont: VARIORUM, Ashgate Publishing.

Slocum, T.A. (1990). "The Use of Quantitative Methods in Major Geographical Journals, 1956-1986", *The Professional Geographer* 42: 84-94.

Smith, M. (1998), *Social Science in Question*. London: Sage.

Spade, Paul Vincent (1994). "Medieval Philosophy." In Anthony Kenny, (ed). *The Oxford Illustrated History of Philosophy*. Oxford: Oxford University Press. Chap. 2, pp. 55–105.

Spate, O.H.K. (1956). "Quantity and Quality in Geography". *Annals of Association of American Geographers*, 46, pp. 369-77.

Stamp, D. (1966). "Ten Years On", *Transactions, of the Institute of British Geographers*, Vol. 40 : 11-20.

Stamp, L. Dudley (1957). "Geographical Agenda: A Review of some Tasks Awaiting Geographical Attention". Pres. address, *Trans. and Papers, Inst. Brit. Geog.*, 23, 1-17.

Stanford Encycopedia of Philosophy (Internet).

Stoddart D.R. (1972) "Darwin's Impact on Geography" and "Geography and the Ecological Approach" in W.K.D. Davies, *op. cit.*, pp. 52-68 and 301-311.

Stoddart D.R. (1981). "The Paradigm Concept and History of Geography" in *Geography, Ideology and Social Concern*, Blackwell, Oxford.

Stoddart, D.R. (1966). 'Darwin's Impact on Geography', *Annals of the Association of American Geographers* , Volume 56https://www.tandfonline.com/author/STODDART%2C+D+R, Issue 4, Pages 683-698.

Stoddart, D.R. (1981). "Ideas and interpretation in the history of geography" in D.R. Stoddart (ed.), *Geography, Ideology and Social Concern*. Oxford: Blackwell.

Stoddart, D.R. (1986). *On geography and its history*. New York: Blackwell.

Taaffe, E.J. (1974). *Geography*. Englewood Cliffs N.J: Prentice Hall.

Tansley, A.G. (1946). *Introduction to Plant Ecology*, Allen & Unwin, London.

Tarrant, J.R. (1968). "Computers in Geography", *IBG Newsletter* 6: 11-25.

Tatham, G. (1951) "Environmentalism and Possibilism", in Taylor, G. (Ed.) *Geography in the Twentieth Century*, London: Methuen, pp. 128-162.

Taylor, G. (1951), *Geography in the Twentieth Century*, Butlar & Tanner Ltd., London.

Taylor, P.J. (1976). "An Interpretation of The Quantification: debate in British Geography", *Transactions Institute of British Geographers*, New Sr. I, pp. 129-142.

Techopedia: "Digital Revolution".

Thompson, Bard (1996). *Humanists and Reformers: A History of theRenaissance and Reformation*, Grand Rapids, MI: Erdmans, pa: 13.

Tim Unwin (2013), *The Place of Geography*, Routledge.

Tiwari, P., "Principles of Humanism and Themes in Humanistic Geography", *Wikipedia*.

Toynbee, A. (1972). *A Study of History* (updated edition). New York: Weathervane Books.

Tuan, Yi-Fu (1976). "Humanistic Geography", *Annals of the Association of American Geographers*, **66.** pp. 266–276.

Ullman, E.J. (1953). "Human geography and area research", *Annals Association of American geographers*, 43, 54-66.

Unstead, J.F. *et al.*, (1933), "Classification of Regions of the World", *Geography*, Vol. 22, pp. 253-282.

Weinberg, Julius R. (1964). *A Short History of Medieval Philosophy*. Princton University Press, N.J.

Whittlesey, D. (1954). "The regional concept and the regional method", in P.E. James & C.F.Jones (Eds.), *op. cit.*, pp. 19-69.

Wikipedia, the Free Encyclopedia.

Wilsdon, J.P. (2001). *Digital Futures: Living in a Dot-Com World. London*; Sterling, VA: Earthscan Publications.

Wilson, A.G. (1970). *Entropy in Urban and Regional Modelling*. London: Pion.

Wilson, A.G. (2000). *Complex Spatial Systems: The Modelling Foundations of Urban and Regional Analysis*, Chichester: John Wiley.

Wilson, A.G. and Bennett, R.J. (1985). *Mathematical Methods in Human Geography and Planning*. Chichester: John Wiley.

Wilson, M.I. (2001). "Location, Location, Location: The Geography of the Dot Com Problem", *Environment and Planning* B: *Planning and Design* 28: 59-71.

Wolman, M.G. (1955). "The Natural Channel of Brandywine Creek, Pennsylvania", *Geological Survey Professional Paper*, 271.

Wright, J.K. (1947). "Terrae Incognitae: The Place of the Imagination in Geography", *Annals of the Association of American Geographers* 37: 1-15.

Wu, Kuang Ch'ing. "Ming Printing and Printers," *Harvard Journal of Asiatic Studies* (February 1943): 203-260.

Yan, Hong-sen. (2007). *Reconstruction Designs of Lost Ancient Chinese Machinery*. Dordrecht: Springer. ISBN 1402064594.

Yeung, M.W. (1997), "Critical Realism and realistic research in human geography: a method or a philosophy in search of a method?" *Progress in human Geography,* Vol. 21 : 57-74.

Index